A PHARMACOLOGY PRIMER
THEORY, APPLICATION, AND METHODS
Second Edition

A PHARMACOLOGY PRIMER
THEORY, APPLICATION, AND METHODS
Second Edition

Terry Kenakin, Ph.D.

Principal Research Investigator, Molecular Discovery
GlaxoSmithKline Research and Development
Research Triangle Park, NC 27709

AMSTERDAM • BOSTON • HEIDELBERG • LONDON
NEW YORK • OXFORD • PARIS • SAN DIEGO
SAN FRANCISCO • SINGAPORE • SYDNEY • TOKYO

Academic Press is an imprint of Elsevier

ACADEMIC
PRESS

Academic Press is an imprint of Elsevier
30 Corporate Drive, Suite 400, Burlington, MA 01803, USA
525 B Street, Suite 1900, San Diego, California 92101-4495, USA
84 Theobald's Road, London WC1X 8RR, UK

This book is printed on acid-free paper. ⊗

Permissions may be sought directly from Elsevier's Science & Technology
Rights Department in Oxford, UK: phone: (+44) 1865 843830, fax: (+44) 1865
853333, E-mail: permissions@elsevier.com. You may also complete your
request on-line via the Elsevier homepage (http://elsevier.com), by selecting
"Support & Contact" then "Copyright and Permission" and then "Obtaining
Permissions."

Library of Congress Cataloging-in-Publication Data
Application Submitted.

British Library Cataloguing-in-Publication Data
A catalogue record for this book is available from the British Library.

ISBN 13: 978-0-12-370599-0
ISBN 10: 0-12-370599-1

For information on all Academic Press publications visit our Web site at
www.books.elsevier.com

Printed in China
06 07 08 09 10   9 8 7 6 5 4 3 2 1

As always — for Debbie

... more ceterum censeo is perhaps necessary in order to rouse pharmacology from its sleep. The sleep is not a natural one since pharmacology, as judged by its past accomplishments, has no reason for being tired ...

— Rudolph Bucheim (1820–1879)

I am indebted to GlaxoSmithKline Research and Development for support during the preparation of this book and for the means and scientific environment to make the science possible.

T.P.K., *Research Triangle Park, NC*

Foreword to Second Edition

With publication of the human genome has come an experiment in reductionism for drug discovery. With the evaluation of the number and quality of new drug treatments from this approach has come a reevaluation of target-based versus systems-based strategies. Pharmacology, historically rooted in systems-based approaches and designed to give systems-independent measures of drug activity, is suitably poised to be a major (if not the major) tool in this new environment of drug discovery.

Compared to the first edition, this book now expands discussion of tools and ideas revolving around allosteric drug action. This is an increasingly therapeutically relevant subject in pharmacology as new drug screening utilizes cell function for discovery of new drug entities. In addition, discussion of system-based approaches, drug development (pharmacokinetics, therapeutics), sources of chemicals for new drugs, and elements of translational medicine have been added. As with the first edition, the emphasis of this volume is the gaining of understanding of pharmacology by the nonpharmacologist.

Terry P. Kenakin, Ph.D.
Research Triangle Park, 2006

Foreword to First Edition

If scientific disciplines can be said to go in and out of vogue, pharmacology is exemplary in this regard. The flourishing of receptor theory in the 1950s, the growth of biochemical binding technology in the 1970s, and the present resurgence of interest in defining cellular phenotypic sensitivity to drugs has been interspersed with troughs such as that brought on by the promise of the human genome and a belief that this genetic roadmap may make classical pharmacology redundant. The fallacy in this belief has been found in experimental data showing the importance of phenotype over genotype which underscores a common finding with roadmaps: they are not as good as a guide who knows the way. Pharmacology is now more relevant to the drug discovery process than ever as the genome furnishes a wealth of new targets to unravel. Biological science often advances at a rate defined by the technology of its tools (i.e., scientists cannot see new things in old systems without new eyes). A veritable explosion in technology coupled with the great gift of molecular biology have definitely given pharmacologists new eyes to see.

This book initially began as a series of lectures at GlaxoSmithKline Research and Development on receptor pharmacology aimed at increasing the communication between pharmacologists and chemists. As these lectures developed it became evident that the concepts were useful to biologists, not specifically trained in pharmacology. In return, the exchange between the chemists and biologists furnished new starting points from which to view the pharmacological concepts. It is hoped that this book will somewhat fill what could be a gap in present biological sciences; namely, the study of dose-response relationships and how cells react to molecules.

Terry P. Kenakin, Ph.D.
Research Triangle Park, 2003

Contents

1

What Is Pharmacology?

I would in particular draw the attention to physiologists to this type of physiological analysis of organic systems which can be done with the aid of toxic agents...

—CLAUDE BERNARD (1813–1878)

1.1 About This Book

Essentially this is a book about the methods and tools used in pharmacology to quantify drug activity. Receptor pharmacology is based on the comparison of experimental data to simple mathematical models with a resulting inference of drug behavior to the molecular properties of drugs. From this standpoint, a certain understanding of the mathematics involved in the models is useful but it is not imperative. This book is structured such that each chapter begins with the basic concepts, then moves on to the techniques used to estimate drug parameters, and, finally, for those so inclined, the mathematical derivations of the models used. Understanding the derivation is not a prerequisite to understanding the application of the methods or the resulting conclusion; these are included for completeness and are for readers who wish to pursue exploration of the models. In general, facility with mathematical equations is definitely not required for pharmacology; the derivations can be ignored to no detriment to the use of this book.

Second, the symbols used in the models and derivations, on occasion, duplicate each other (i.e., α is an extremely popular symbol). However, the use of these multiple symbols has been retained since this preserves the context of where these models were first described and utilized. Also, changing these to make them unique would cause confusion if these methods are to be used beyond the framework of this book. Therefore, care should be taken to consider the actual nomenclature of each chapter.

Third, an effort has been made to minimize the need to cross reference different parts of the book (i.e., when a particular model is described the basics are reiterated somewhat to minimize the need to read the relevant but different part of the book where the model is initially described). While this leads to a small amount of repeated description, it is felt that this will allow for a more uninterrupted flow of reading and use of the book.

1.2 What Is Pharmacology?

Pharmacology (an amalgam of the Greek *Pharmakos*, medicine or drug, and *logos*, study) is a broad discipline describing the use of chemicals to treat and cure disease. The Latin term *pharmacologia* was used in the late 1600s but the term *pharmacum* was used as early as the fourth century to denote the term *drug* or *medicine*. There are subdisciplines within pharmacology representing specialty areas. *Pharmacokinetics* deals with the disposition of drugs in the human body. To be useful, drugs must be absorbed and transported to their site of therapeutic action. Drugs will be ineffective in therapy if they do not reach the organs(s) to exert their activity; this will be discussed specifically in Chapter 8 of this book. *Pharmaceutics* is the study of the chemical formulation of drugs to optimize absorption and distribution within the body. *Pharmacognosy* is the study of plant natural products and their use in the treatment of disease. A very important discipline in the drug discovery process is *medicinal chemistry*, the study of the production of molecules for therapeutic use. This couples synthetic organic chemistry with an understanding of how biological information can be quantified and used to guide the synthetic chemistry to enhance therapeutic activity. *Pharmacodynamics* is the study of the interaction of the drug molecule with the biological target (referred to generically as the "receptor," *vide infra*). This discipline lays the foundation of pharmacology since all therapeutic application of drugs has a common root in pharmacodynamics (i.e., as a prerequisite to exerting an effect, all drug molecules must bind to and interact with receptors).

Pharmacology as a separate science is approximately 120 to 140 years old. The relationship between chemical structure and biological activity began to be studied systematically in the 1860s [1]. It began when physiologists, using chemicals to probe physiological systems, became more interested in the chemical probes than the systems they were probing. By the early 1800s, physiologists were performing physiological studies with chemicals that became pharmacological studies more aimed at the definition of the biological activity of chemicals. The first formalized chair of pharmacology, indicating a formal university department, was founded in Estonia by Rudolf Bucheim in 1847. In North America, the first chair was founded by John Jacob Abel at Johns Hopkins University in 1890. A differentiation of physiology and pharmacology was given by the pharmacologist Sir William Paton [2]:

> If physiology is concerned with the function, anatomy with the structure, and biochemistry with the chemistry of the living body, then pharmacology is concerned with the changes in function, structure, and chemical properties of the body brought about by chemical substances.
>
> —W. D. M. Paton (1986)

Many works about pharmacology essentially deal in therapeutics associated with different organ systems in the body. Thus, in many pharmacology texts, chapters are entitled drugs in the cardiovascular system, the effect of drugs on the gastrointestinal system, CNS, and so on. However, the underlying principles for all of these is the same; namely the pharmacodynamic interaction between the drug and the biological recognition system for that drug. Therefore, a prerequisite to all of pharmacology is an understanding of the basic concepts of dose response and how living cells process pharmacological information. This generally is given the term *pharmacodynamics* or *receptor pharmacology*, where *receptor* is a term referring to any biological recognition unit for drugs (membrane receptors, enzymes, DNA, and so on). With such knowledge in hand, readers will be able to apply these principles to any branch of therapeutics effectively. This book treats dose-response data generically and demonstrates methods by which drug activity can be quantified across all biological systems irrespective of the nature of the biological target.

The human *genome* is now widely available for drug discovery research. Far from being a simple blueprint of how drugs should be targeted, it has shown biologists that receptor *genotypes* (i.e., properties of proteins resulting from genetic transcription to their amino acid sequence) are secondary to receptor *phenotypes* (how the protein interacts with the myriad of cellular components and how cells tailor the makeup and functions of these proteins to their individual needs). Since the arrival of the human genome, receptor pharmacology as a science is more relevant than ever in drug discovery. Current drug therapy is based on less than 500 molecular targets yet estimates utilizing the number of genes involved in multifactorial diseases suggest that the number of potential drug targets ranges from 5,000 to 10,000 [3]. Thus, current therapy is using only 5 to 10% of the potential trove of targets available in the human genome.

A meaningful dialogue between chemists and pharmacologists is the single most important element of the drug discovery process. The necessary link between medicinal chemistry and pharmacology has been elucidated by Paton [2]:

> For pharmacology there results a particularly close relationship with chemistry, and the work may lead quite naturally, with no special stress on practicality, to therapeutic application, or (in the case of adverse reactions) to toxicology.
>
> —W. D. M. Paton (1986)

Chemists and biologists reside in different worlds from the standpoint of the type of data they deal with. Chemistry is an exact science with physical scales that are not subject to system variance. Thus, the scales of measurement are transferrable. Biology deals with the vagaries of complex systems that are not completely understood. Within this scenario, scales of measurement are much less constant and much more subject to system conditions. Given this, a gap can exist between chemists and biologists in terms of understanding and also in terms of the best method to progress forward. In the worst circumstance, it is a gap of credibility emanating from a failure of the biologist to make the chemist understand the limits of the data. Usually, however, credibility is not the issue and the gap exists due to a lack of common experience. This book was written in an attempt to limit or, hopefully, eliminate this gap.

1.3 The Receptor Concept

One of the most important concepts emerging from early pharmacological studies is the concept of the *receptor*. Pharmacologists knew that minute amounts of certain chemicals had profound effects on physiological systems. They also knew that very small changes in the chemical composition of these substances could lead to huge differences in activity. This led to the notion that something on or in the cell must specifically read the chemical information contained in these substances and translate it into physiological effect. This something was conceptually referred to as the "receptor" for that substance. Pioneers such as Paul Ehrlich (1854–1915, Figure 1.1a) proposed the existence of "chemoreceptors" (actually he proposed a collection of amboreceptors, triceptors, and polyceptors) on cells for dyes. He also postulated that the chemoreceptors on parasites, cancer cells, and microorganisms were different from healthy host and thus could be exploited therapeutically. The physiologist turned pharmacologist John Newport Langley (1852–1926, Figure 1.1b), during his studies with the drugs jaborandi (which contains the alkaloid pilocarpine) and atropine, introduced the concept

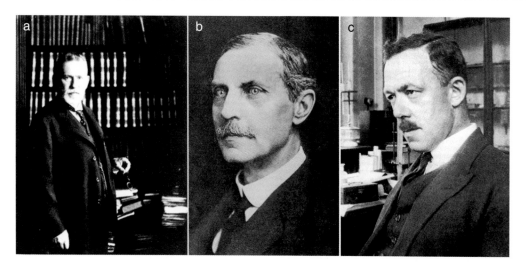

FIGURE 1.1 Pioneers of pharmacology. (a) Paul Ehrlich (1854–1915). Born in Silesia, Ehrlich graduated from Leipzig University to go on to a distinguished career as head of Institutes in Berlin and Frankfurt. His studies with dyes and bacteria formed the basis of early ideas regarding recognition of biological substances by chemicals. (b) John Newport Langley (1852–1926). Though he began reading mathematics and history in Cambridge in 1871, Langley soon took to physiology. He succeeded the great physiologist M. Foster to the Chair of Physiology in Cambridge in 1903 and branched out into pharmacological studies of the autonomic nervous system. These pursuits led to germinal theories of receptors. (c) Alfred. J. Clark (1885–1941). Beginning as a demonstrator in pharmacology in King's College (London), Clark went on to become Professor of Pharmacology at University College London. From there he took the Chair of Pharmacology in Edinburgh. Known as the originator of modern receptor theory, Clark applied chemical laws to biological phenomena. His books on receptor theory formed the basis of modern pharmacology.

that receptors were switches that received and generated signals and that these switches could be activated or blocked by specific molecules. The originator of quantitative receptor theory, the Edinburgh pharmacologist Alfred Joseph Clark (1885–1941, Figure 1.1c), was the first to suggest that the data, compiled from his studies of the interactions of acetylcholine and atropine, resulted from the unimolecular interaction of the drug and a substance on the cell surface. He articulated these ideas in the classic work *The Mode of Action of Drugs on Cells* [4], later revised as the *Handbook of Experimental Pharmacology* [5]. As put by Clark:

> It appears to the writer that the most important fact shown by a study of drug antagonisms is that it is impossible to explain the remarkable effects observed except by assuming that drugs unite with receptors of a highly specific pattern.... No other explanation will, however, explain a tithe of the facts observed.
>
> — A. J. Clark (1937)

Clark's next step formed the basis of receptor theory by applying chemical laws to systems of "infinitely greater complexity" [4]. It is interesting to note the scientific atmosphere in which Clark published these ideas. The dominant ideas between 1895 and 1930 were based on theories such as the law of phasic variation essentially stating that "certain phenomena occur frequently." Homeopathic theories like the Arndt-Schulz law and

Weber-Fechner law were based on loose ideas around surface tension of the cell membrane but there was little physico-chemical basis to these ideas [6]. In this vein, prominent pharmacologists of the day such as Walter Straub (1874–1944) suggested that a general theory of chemical binding between drugs and cells utilizing receptors was "... going too far...and...not admissable" [6]. The impact of Clark's thinking against these concepts cannot be overemphasized to modern pharmacology.

Drug receptors can exist in many forms from cell surface proteins, enzymes, ion channels, membrane transporters, DNA, and cytosolic proteins (see Figure 1.2). There are examples of important drugs for all of these. This book deals with general concepts that can be applied to a range of receptor types but most of the principles are illustrated with the most tractable receptor class known in the human genome; namely *seven transmembrane (7TM) receptors*. These receptors are named for their characteristic structure, which consists of a single protein chain that traverses the cell membrane seven times to produce extracellular and intracellular loops. These receptors activate G-proteins to elicit response thus they are also commonly referred to as *G-protein-coupled receptors (GPCRs)*. There are between 800 and 1,000 [7] of these in the genome (the genome sequence predicts 650 GPCR genes, of which approximately 190 [on the order of 1% of the genome of superior organisms] are categorized as known GPCRs [8] activated

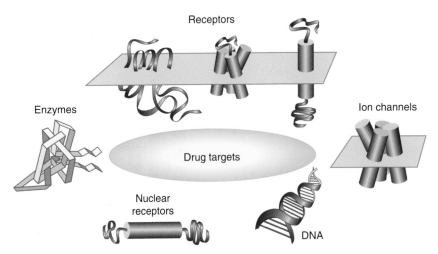

FIGURE 1.2 Schematic diagram of potential drug targets. Molecules can affect the function of numerous cellular components both in the cytosol and on the membrane surface. There are many families of receptors that traverse the cellular membrane and allow chemicals to communicate with the interior of the cell.

by some 70 ligands). In the United States in the year 2000, nearly half of all prescription drugs were targeted toward 7TM receptors [3]. These receptors, comprising of between 1 and 5% of the total cell protein, control a myriad of physiological activities. They are tractable for drug discovery because they are on the cell surface and therefore drugs do not need to penetrate the cell to produce effect. In the study of biological targets such as GPCRs and other receptors, a "system" must be employed that accepts chemical input and returns biological output. It is worth discussing such receptor systems in general terms before their specific uses are considered.

1.4 Pharmacological Test Systems

Molecular biology has transformed pharmacology and the drug discovery process. As little as ten years ago, screening for new drug entities was carried out in surrogate animal tissues. This necessitated a rather large extrapolation spanning differences in genotype and phenotype. The belief that the gap could be bridged came from the notion that the chemicals recognized by these receptors in both humans and animals were the same (*vide infra*). Receptors are unique proteins with characteristic amino acid sequences. While *polymorphisms* (spontaneous alterations in amino acid sequence, *vide infra*) of receptors exist in the same species, in general the amino acid sequence of a natural ligand binding domain for a given receptor type largely may be conserved. There are obvious pitfalls of using surrogate species receptors for prediction of human drug activity and it never can be known for certain whether agreement for estimates of activity for a given set of drugs ensures accurate prediction for all drugs. The agreement is very much drug and receptor dependent. For example, the human and mouse α_2-adrenoceptor are 89% homologous and thus considered very similar from the standpoint of amino acid sequence.

Furthermore, the affinities of the α_2-adrenoceptor antagonists atipamezole and yohimbine are nearly indistinguishable (atipamezole human α_2-C10 $K_i = 2.9 \pm 0.4$ nM, mouse α_2-4H $K_i = 1.6 \pm 0.2$ nM; yohimbine human α_2-C10 $K_i = 3.4 \pm 0.1$ nM, mouse α_2-4H $K_i = 3.8 \pm 0.8$ nM). However, there is a 20.9-fold difference for the antagonist prazosin (human α_2-C10 $K_i = 2034 \pm 350$ nM, mouse α_2-4H $K_i = 97.3 \pm 0.7$ nM) [9]. Such data highlight a general theme in pharmacological research; namely, that a hypothesis, such as one proposing two receptors that are identical with respect to their sensitivity to drugs are the same, cannot be proven, only disproven. While a considerable number of drugs could be tested on the two receptors (thus supporting the hypothesis that their sensitivity to all drugs is the same), this hypothesis is immediately disproven by the first drug that shows differential potency on the two receptors. The fact that a series of drugs tested show identical potencies may only mean that the wrong sample of drugs has been chosen to unveil the difference. Thus, no general statements can be made that any one surrogate system is completely predictive of activity on the target human receptor. This will always be a drug-specific phenomenom.

The link between animal and human receptors is the fact that both proteins recognize the endogenous transmitter (e.g., acetylcholine, norepinephrine), and therefore the hope is that this link will carry over into other drugs that recognize the animal receptor. This imperfect system formed the basis of drug discovery until human *cDNA* for human receptors could be used to make cells express human receptors. These engineered (recombinant) systems now are used as surrogate human receptor systems and the leap of faith from animal receptor sequences to human receptor sequences is not required (i.e., the problem of differences in genotype has been overcome). However, cellular signaling is an extremely complex process and cells tailor their receipt of chemical signals in numerous ways.

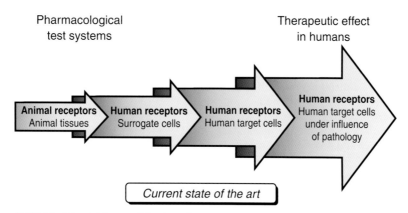

FIGURE 1.3 A history of the drug discovery process. Originally, the only biological material available for drug research was animal tissue. With the advent of molecular biological techniques to clone and express human receptors in cells, recombinant systems supplanted animal isolated tissue work. It should be noted that these recombinant systems still fall short of yielding drug response in the target human tissue under the influence of pathological processes.

Therefore, the way a given receptor gene behaves in a particular cell can differ in response to the surroundings in which that receptor finds itself. These differences in phenotype (i.e., properties of a receptor produced by interaction with its environment) can result in differences in both the quantity and quality of a signal produced by a concentration of a given drug in different cells. Therefore, there is still a certain, although somewhat lesser, leap of faith taken in predicting therapeutic effects in human tissues under pathological control from surrogate recombinant or even surrogate natural human receptor systems. For this reason it is a primary requisite of pharmacology to derive system independent estimates of drug activity that can be used to predict therapeutic effect in other systems.

A schematic diagram of the various systems used in drug discovery, in order of how appropriate they are to therapeutic drug treatment, is shown in Figure 1.3. As discussed previously, early functional experiments in animal tissue have now largely given way to testing in recombinant cell systems engineered with human receptor material. This huge technological step greatly improved the predictability of drug activity in humans but it should be noted that there still are many factors that intervene between the genetically engineered drug testing system and the pathology of human disease.

A frequently used strategy in drug discovery is to express human receptors (through *transfection* with human cDNA) in convenient surrogate host cells (referred to as "target-based" drug discovery; see Chapter 8 for further discussion). These host cells are chosen mainly for their technical properties (i.e., robustness, growth rate, stability) and not with any knowledge of verisimilitude to the therapeutically targeted human cell type. There are various factors relevant to the choice of surrogate host cell such as a very low background activity (i.e., a cell cannot be used that already contains a related animal receptor for fear of

cross-reactivity to molecules targeted for the human receptor). Human receptors often are expressed in animal surrogate cells. The main idea here is that the cell is a receptacle for the receptor, allowing it to produce physiological responses, and that activity can be monitored in pharmacological experiments. In this sense, human receptors expressed in animal cells are still a theoretical step distanced from the human receptor in a human cell type. However, even if a human surrogate is used (and there are such cells available) there is no definitive evidence that a surrogate human cell is any more predictive of a natural receptor activity than an animal cell when compared to the complex receptor behavior in its natural host cell type expressed under pathological conditions. Receptor phenotype dominates in the end organ and the exact differences between the genotypic behavior of the receptor (resulting from the genetic makeup of the receptor) and the phenotypic behavior of the receptor (due to the interaction of the genetic product with the rest of the cell) may be cell specific. Therefore, there is still a possible gap between the surrogate systems used in the drug discovery process and the therapeutic application. Moreover, most drug discovery systems utilize receptors as switching mechanisms and quantify whether drugs turn on or turn off the switch. The pathological processes that we strive to modify may be more subtle. As put by pharmacologist Sir James Black [10]:

> ... angiogenesis, apoptosis, inflammation, commitment of marrow stem cells, and immune responses. The cellular reactions subsumed in these processes are switch like in their behavior ... biochemically we are learning that in all these processes many chemical regulators seem to be involved. From the literature on synergistic interactions, a control model can be built in which no single agent is effective. If a number of chemical messengers each bring information from a different source and each deliver only a subthreshold

stimulus but together mutually potentiate each other, then the desired information-rich switching can be achieved with minimum risk of miscuing.

—J. W. Black (1986)

Such complex end points are difficult to predict from any one of the component processes leading to yet another leap of faith in the drug discovery process. For these reasons, an emerging strategy for drug discovery is the use of natural cellular systems. This approach is discussed in some detail in Chapter 8.

Even when an active drug molecule is found and activity is verified in the therapeutic arena, there are factors that can lead to gaps in its therapeutic profile. When drugs are exposed to huge populations, genetic variations in this population can lead to discovery of *alleles* that code for mutations of the target (isogenes) and these can lead to variation in drug response. Such polymorphisms can lead to resistant populations (i.e., resistance of some asthmatics to the β-adrenoceptor bronchodilators [11]). In the absence of genetic knowledge, these therapeutic failures for a drug could not easily be averted since they in essence occurred because of the presence of new biological targets not originally considered in the drug discovery process. However, with new epidemiological information becoming available these polymorphisms can now be incorporated into the drug discovery process.

There are two theoretical and practical scales that can be used to make system independent measures of drug activity on biological systems. The first is a measure of the attraction of a drug for a biological target; namely, its *affinity* for receptors. Drugs must interact with receptors to produce an effect and the affinity is a chemical term used to quantify the strength of that interaction. The second is much less straightforward and is used to quantify the degree of effect imparted to the biological system after the drug binds to the receptor. This is termed *efficacy*. This property was named by R. P. Stephenson [12] within classical receptor theory as a proportionality factor for tissue response produced by a drug. There is no absolute scale for efficacy but rather it is dealt with in relative terms (i.e., the ratio of the efficacy of two different drugs on a particular biological system can be estimated and, under ideal circumstances, will transcend the system and be applicable to other systems as well). It is the foremost task of pharmacology to use the translations of drug effect obtained from cells to provide system independent estimates of affinity and efficacy. Before specific discussion of affinity and efficacy it is worth considering the molecular nature of biological targets.

1.5 The Nature of Drug Receptors

While some biological targets such as DNA are not protein in nature, most receptors are. It is useful to consider the properties of receptor proteins to provide a context for the interaction of small molecule drugs with them. An important property of receptors is that they have a 3D structure. Proteins usually are comprised of one or more peptide chains; the composition of these chains make up the primary and secondary structure of the protein. Proteins also are described in terms of a tertiary structure which defines their shape in 3D space and a quarternary structure which defines the molecular interactions between the various components of the protein chains (Figure 1.4). It is this 3D structure that allows the protein to function as a recognition site and effector for drugs and other components of the cell, in essence, the ability of the protein to function as a messenger shuttling information from the outside world to the cytosol of the cell. For GPCRs, the 3D nature of the receptor forms binding domains for other proteins such as *G-proteins* (these are activated by the receptor and then go on to activate enzymes and ion channels within the cell; see Chapter 2) and endogenous chemicals such as neurotransmitters, hormones, and autacoids that carry physiological messages. For other receptors, such as ion channels and single transmembrane enzyme receptors, the conformational change per se leads to response either through an opening of a channel to allow the flow of ionic current or the initiation of enzymatic activity. Therapeutic advantage can be taken by designing small molecules to utilize these binding domains or other 3D binding domains on the receptor protein in order to modify physiological and pathological processes.

1.6 Pharmacological Intervention and the Therapeutic Landscape

It is useful to consider the therapeutic landscape with respect to the aims of pharmacology. As stated by Sir William Ossler (1849–1919), "... the prime distinction between man and other creatures is man's yearning to take medicine." The notion that drugs can be used to cure disease is as old as history. One of the first written records of actual "prescriptions" can be found in the Ebers Papyrus (circa 1550 B.C.): "... for night blindness in the eyes ... liver of ox, roasted and crushed out ... really excellent!" Now it is known that liver is an excellent source of vitamin A, a prime treatment for night blindness, but that chemical detail was not known to the ancient Egyptians. Disease can be considered under two broad categories: those caused by invaders such as pathogens and those caused by intrinsic breakdown of normal physiological function. The first generally is approached through the invader (i.e., the pathogen is destroyed, neutralized, or removed from the body). The one exception of where the host is treated when an invader is present is the treatment of HIV-1 infection leading to AIDS. In this case, while there are treatments to neutralize the pathogen, such as anti-retrovirals to block viral replication, a major new approach is the blockade of the interaction of the virus with the protein that mediates viral entry into healthy cells, the chemokine receptor CCR5. In this case, CCR5 antagonists are used to prevent HIV fusion and subsequent infection. The second approach to disease requires understanding of the pathological

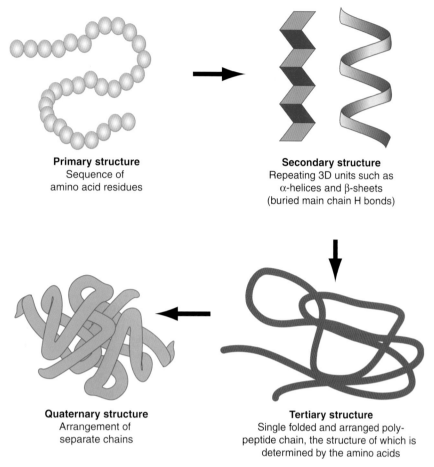

Primary structure
Sequence of
amino acid residues

Secondary structure
Repeating 3D units such as
α-helices and β-sheets
(buried main chain H bonds)

Quaternary structure
Arrangement of
separate chains

Tertiary structure
Single folded and arranged poly-
peptide chain, the structure of which is
determined by the amino acids

FIGURE 1.4 Increasing levels of protein structure. A protein has a given amino acid sequence to make peptide chains. These adopt a 3D structure according to the free energy of the system. Receptor function can change with changes in tertiary or quaternary structure.

process and repair of the damage to return to normal function.

The therapeutic landscape onto which drug discovery and pharmacology in general combats disease can generally be described in terms of the major organ systems of the body and how they may go awry. A healthy cardiovascular system consists of a heart able to pump deoxygenated blood through the lungs and to pump oxygenated blood throughout a circulatory system that does not unduly resist blood flow. Since the heart requires a high degree of oxygen itself to function, myocardial ischemia can be devastating to its function. Similarly, an inability to maintain rhythm (arrhythmia) or loss in strength with concomitant inability to empty (congestive heart failure) can be fatal. The latter disease is exacerbated by elevated arterial resistance (hypertension). A wide range of drugs are used to treat the cardiovascular system including coronary vasodilators (nitrates), diuretics, renin-angiotensin inhibitors, vasodilators, cardiac glycosides, calcium antagonists, beta and alpha blockers, antiarrhythmics, and drugs for dyslipidemia. The lungs

must extract oxygen from the air, deliver it to the blood, and release carbon dioxide from the blood into exhaled air. Asthma, chronic obstructive pulmonary disease (COPD), and emphysema are serious disorders of the lungs and airways. Bronchodilators (beta agonists), anti-inflammatory drugs, inhaled glucocorticoids, anticholinergics, and theophylline analogues are used for treatment of these diseases. The central nervous system controls all conscious thought and many unconscious body functions. Numerous diseases of the brain can occur, including depression, anxiety, epilepsy, mania, degeneration, obsessive disorders, and schizophrenia. Brain functions such as those controlling sedation and pain also may require treatment. A wide range of drugs are used for CNS disorders, including serotonin partial agonists and uptake inhibitors, dopamine agonists, benzodiazepines, barbiturates, opioids, tricyclics, neuroleptics, and hydantoins. The gastrointestinal tract receives and processes food to extract nutrients and removes waste form the body. Diseases such as stomach ulcers, colitis, diarrhea, nausea, and irritable bowel syndrome can affect this system. Histamine antagonists, proton pump blockers, opioid agonists, antacids, and serotonin uptake blockers are used to treat diseases of the GI tract.

The inflammatory system is designed to recognize self from non-self and destroy non-self to protect the body. In diseases of the inflammatory system, the self-recognition can break down leading to conditions where the body destroys healthy tissue in a misguided attempt at protection. This can lead to rheumatoid arthritis, allergies, pain, COPD, asthma, fever, gout, graft rejection, and problems with chemotherapy. Non-steroidal anti-inflammatory drugs (NSAIDs), aspirin and salicylates, leukotriene antagonists, and histamine receptor antagonists are used to treat inflammatory disorders. The endocrine system produces and secretes crucial hormones to the body for growth and function. Diseases of this class of organs can lead to growth and pituitary defects; diabetes; abnormality in thyroid, pituitary, adrenal cortex, and androgen function; osteoporosis; and alterations in estrogen/progesterone balance. The general approach to treatment is through replacement or augmentation of secretion. Drugs used are replacement hormones, insulin, sulfonylureas, adrenocortical steroids, and oxytocin. In addition to the major organ and physiological systems, diseases involving neurotransmission and neuromuscular function, ophthalmology, hemopoiesis and hematology, dermatology, immunosuppression, and drug addiction and abuse are amenable to pharmacological intervention.

Cancer is a serious malfunction of normal cell growth. In the years from 1950 through 1970, the major approach to treating this disease had been to target DNA and DNA precursors according to the hypothesis that rapidly dividing cells (cancer cells) are more susceptible to DNA toxicity than normal cells. Since that time, a wide range of new therapies based on manipulation of the immune system, induction of differentiation, inhibition of angiogenesis, and increased killer T-lymphocytes to decrease cell proliferation has greatly augmented the armamentarium against neoplastic disease. Previously lethal malignancies such as testicular cancer, some lymphomas, and leukemia are now curable.

Three general treatments of disease are surgery, genetic engineering (still an emerging discipline), and pharmacological intervention. While early medicine was subject to the theories of Hippocrates (460–357 BCE), who saw health and disease as a balance of four humors (i.e., black and yellow bile, phlegm and blood), by the sixteenth century pharmacological concepts were being formulated. These could be stated concisely as [13]:

- Every disease has a cause for which there is a specific remedy.
- Each remedy has a unique essence that can be obtained from nature by extraction ("doctrine of signatures").
- The administration of the remedy is subject to a dose-response relationship.

The basis for believing the pharmacological intervention can be a major approach to the treatment of disease is the fact that the body generally functions in response to chemicals. Table 1.1 shows partial lists of hormones and neurotransmitters in the body. Many more endogenous chemicals are involved in normal physiological function. The fact that so many physiological processes are controlled by chemicals provides

TABLE 1.1

Some endogenous chemicals controlling normal physiological function.

Neurotransmitters		
Acetylcholine	2-Arachidonylglycerol	Anandamide
ATP	Corticotropin-releasing hormone	Dopamine
Epinephrine	Aspartate	Gamma-aminobutyric acid
Galanin	Glutamate	Glycine
Histamine	Norepinephrine	Serotonin

Hormones		
Thyroid stim. hormone	Follicle-stim. hormone	Luteinizing hormone
Prolactin	Adrenocorticotropin	Antidiuretic hormone
Thyrotropin-releasing hromone	Oxytocin	Gonadotropin-releasing hormone
Growth-horm-rel. hormone	Corticotropin-releasing hormone	Somatostatin
Melatonin	Thyroxin	Calcitonin
Parathyroid hormone	Glucocorticoid(s)	Mineralocorticoid(s)
Estrogen(s)	Progesterone	Chorionic gonadotropin
Androgens	Insulin	Glucagon
Amylin	Erythropoietin	Calcitriol
Calciferol	Atrial-nartiuretic peptide	Gastrin
Secretin	Cholecystokinin	Neuropeptide Y
Insulin-like growth factor	Angiotensinogen	Ghrelin
	Leptin	

the opportunity for chemical intervention. Thus, physiological signals mediated by chemicals can be initiated, negated, augmented, or modulated. The nature of this modification can take the form of changes in the type, strength, duration, or location of signal.

1.7 System-independent Drug Parameters: Affinity and Efficacy

The process of drug discovery relies on the testing of molecules in systems to yield estimates of biological activity in an iterative process of changing the structure of the molecule until optimal activity is achieved. It will be seen in this book that there are numerous systems available to do this and that each system may interpret the activity of molecules in different ways. Some of these interpretations can appear to be in conflict with each other leading to apparent capricious patterns. For this reason, the way forward in the drug development process is to use only system-independent information. Ideally, scales of biological activity should be used that transcend the actual biological system in which the drug is tested. This is essential to avoid confusion and also because it is quite rare to have access to the exact human system under the control of the appropriate pathology available for in vitro testing. Therefore, the drug discovery process necessarily relies on the testing of molecules in surrogate systems and the extrapolation of the observed activity to all systems. The only means to do this is to obtain system-independent measures of drug activity; namely, affinity and efficacy.

If a molecule in solution associates closely with a receptor protein it has affinity for that protein. The area where it is bound is the binding *domain* or *locus*. If the same molecule interferes with the binding of a physiologically active molecule such as a hormone or a neurotransmitter (i.e., if the binding of the molecule precludes activity of the physiologically active hormone or neurotransmitter), the molecule is referred to as an *antagonist*. Therefore, a pharmacologically active molecule that blocks physiological effect is an antagonist. Similarly, if a molecule binds to a receptor and produces its own effect it is termed an *agonist*. It also is assumed to have the property of efficacy. Efficacy is detected by observation of pharmacological response. Therefore, agonists have both affinity and efficacy.

Classically, agonist response is described in two stages, the first being the initial signal imparted to the immediate biological target; namely, the receptor. This first stage is comprised of the formation, either through interaction with an agonist or spontaneously, of an active state receptor conformation. This initial signal is termed the *stimulus* (Figure 1.5). This stimulus is perceived by the cell and processed in various ways through successions of biochemical reactions to the end point; namely, the *response*. The sum total of the subsequent reactions is

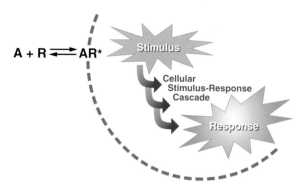

FIGURE 1.5 Schematic diagram of response production by an agonist. An initial stimulus is produced at the receptor as a result of agonist-receptor interaction. This stimulus is processed by the stimulus-response apparatus of the cell into observable cellular response.

referred to as the *stimulus-response mechanism* or *cascade* (see Figure 1.5).

Efficacy is a molecule-related property (i.e., different molecules have different capabilities to induce physiological response). The actual term for the molecular aspect of response-inducing capacity of a molecule is *intrinsic efficacy* (see Chapter 3 for how this term evolved). Thus, every molecule has a unique value for its intrinsic efficacy (in cases of antagonists this could be zero). The different abilities of molecules to induce response are illustrated in Figure 1.6. This figure shows dose-response curves for four 5-HT (serotonin) agonists in rat jugular vein. It can be seen that if response is plotted as a function of the percent receptor occupancy different receptor occupancies for the different agonists lead to different levels of response. For example, while 0.6 g force can be generated by 5-HT by occupying 30% of the receptors, the agonist 5-cyanotryptamine requires twice the receptor occupancy to generate the same response (i.e., the capability of 5-cyanotryptamine to induce response is half that of 5-HT [14]). These agonists are then said to possess different magnitudes of intrinsic efficacy.

It is important to consider affinity and efficacy as separately manipulatable properties. Thus, there are chemical features of agonists that pertain especially to affinity and other features that pertain to efficacy. Figure 1.7 shows a series of key chemical compounds made en route to the histamine H_2 receptor antagonist cimetidine (used for healing gastric ulcers). The starting point for this discovery program was the knowledge that histamine, a naturally occurring autacoid, activates histamine H_2 receptors in the stomach to cause acid secretion. This constant acid secretion is what prevents healing of lesions and ulcers. The task was then to design a molecule that would antagonize the histamine receptors mediating acid secretion and prevent histamine H_2 receptor activation to allow the ulcers to heal. This task was approached with the knowledge that molecules, theoretically, could be made that retained or even enhanced affinity but decreased the

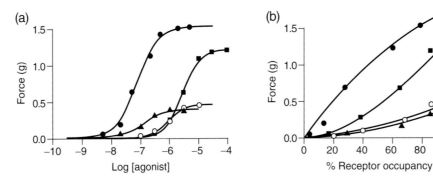

FIGURE 1.6 Differences between agonists producing contraction of rat jugular vein through activation of 5-HT receptors. (a) Dose-response curves to 5-HT receptor agonists, 5-HT (filled circles), 5-cyanotryptamine (filled squares), N,N-dimethyltryptamine (open circles), and N-benzyl-5-methoxytryptamine (filled triangles). Abscissae: logarithms of molar concentrations of agonist. (b) Occupancy response curves for curves shown in panel a. Abscissae: percent receptor occupancy by the agonist as calculated by mass action and the equilibrium dissociation constant of the agonist-receptor complex. Ordinates: force of contraction in g. Data drawn from [14].

efficacy of histamine (i.e., these were separate properties). As can be seen in Figure 1.7, molecules were consecutively synthesized with reduced values of efficacy and enhanced affinity until the target histamine H_2 antagonist cimetidine was made. This was a clear demonstration of the power of medicinal chemistry to separately manipulate affinity and efficacy for which, in part, the Nobel prize in medicine was awarded in 1988.

1.8 What Is Affinity?

The affinity of a drug for a receptor defines the strength of interaction between the two species. The forces controlling the affinity of a drug for the receptor are thermodynamic (enthalpy as changes in heat and entropy as changes in the state of disorder). The chemical forces between the components of the drug and the receptor vary in importance in relation to the distance the drug is away from the receptor binding surface. Thus, the strength of electrostatic forces (attraction due to positive and negative charges and/or complex interactions between polar groups) varies as a function of the reciprocal of the distance between the drug and the receptor. Hydrogen bonding (the sharing of a hydrogen atom between an acidic and basic group) varies in strength as a function of the fourth power of the reciprocal of the distance. Also involved are Van der Waals forces (weak attraction between polar and nonpolar molecules) and hydrophobic bonds (interaction of nonpolar surfaces to avoid interaction with water). The combination of all of these forces causes the drug to reside in a certain position within the protein binding pocket. This is a position of minimal free energy. It is important to note that drugs do not statically reside in one uniform position. As thermal energy varies in the system, drugs approach and dissociate from the protein surface. This is an important concept in pharmacology as it

sets the stage for competition between two drugs for a single binding domain on the receptor protein. The probability that a given molecule will be at the point of minimal free energy within the protein binding pocket thus depends on the concentration of the drug available to fuel the binding process and also the strength of the interactions for the complementary regions in the binding pocket (affinity). Affinity can be thought of as a force of attraction and can be quantified with a very simple tool first used to study the adsorption of molecules onto a surface; namely the Langmuir adsorption isotherm.

1.9 The Langmuir Adsorption Isotherm

Defined by the chemist Irving Langmuir (1881–1957, Figure 1.8), the model for affinity is referred to as the Langmuir adsorption isotherm. Langmuir reasoned that molecules had a characteristic rate of diffusion toward a surface (referred to as condensation and denoted α in his nomenclature) and also a characteristic rate of dissociation (referred to as evaporation and denoted as V_1; see Figure 1.8). He assumed that the amount of surface that already has a molecule bound is not available to bind another molecule. The surface area bound by molecule is denoted θ_1, expressed as a fraction of the total area. The amount of free area open for the binding of molecule, expressed as a fraction of the total area, is denoted as $1 - \theta_1$. The rate of adsorption toward the surface therefore is controlled by the concentration of drug in the medium (denoted μ in Langmuir's nomenclature) multiplied by the rate of condensation on the surface and the amount of free area available for binding:

$$\text{Rate of diffusion toward surface} = \alpha\mu(1 - \theta_1). \quad (1.1)$$

The rate of evaporation is given by the intrinsic rate of dissociation of bound molecules from the surface

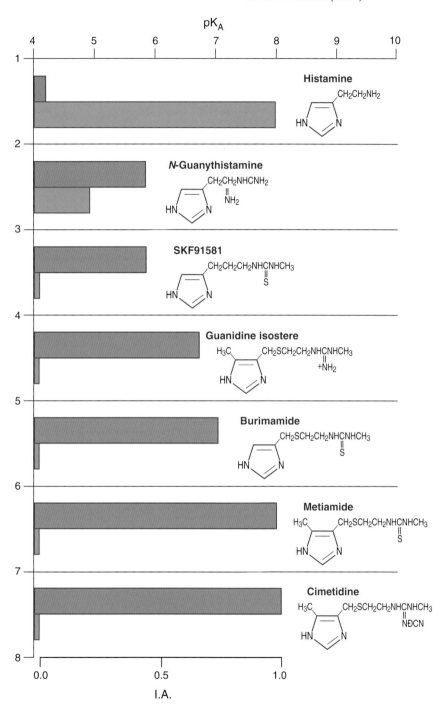

*"...we knew the receptor bound histamine, so it was
a matter of keeping affinity and losing efficacy..."*
—*Sir James Black (1996)*

FIGURE 1.7 Key compounds synthesized to eliminate the efficacy (burgundy red) and enhance the affinity (green) of histamine for histamine H_2 receptors to make cimetidine, one of the first histamine H_2 antagonists of use in the treatment of peptic ulcers. Quotation from James Black [10].

$$\theta_1 = \frac{\alpha\mu}{\alpha\mu + V_1}$$

FIGURE 1.8 The Langmuir adsorption isotherm representing the binding of a molecule to a surface. Photo shows Irving Langmuir (1881–1957), a chemist interested in the adsorption of molecules to metal filaments for the production of light. Langmuir devised the simple equation still in use today for quantifying the binding of molecules to surfaces. The equilibrium is described by condensation and evaporation to yield the fraction of surface bound (θ_1) by a concentration μ.

multiplied by the amount already bound:

$$\text{Rate of evaporation} = V_1\theta_1. \qquad (1.2)$$

Once equilibrium has been reached, the rate of adsorption equals the rate of evaporation. Equating [1.1] and [1.2] and rearranging yields:

$$\theta_1 = \frac{\alpha\mu}{\alpha\mu + V_1}. \qquad (1.3)$$

This is the Langmuir adsorption isotherm in its original form. In pharmacological nomenclature, it is rewritten in the convention

$$\rho = \frac{[AR]}{[R_t]} = \frac{[A]}{[A] + K_A}, \qquad (1.4)$$

where [AR] is the amount of complex formed between the ligand and the receptor and [R_t] is the total number of receptor sites. The ratio ρ refers to the fraction of maximal binding by a molar concentration of drug [A] with an equilibrium dissociation constant of K_A. This latter term is the ratio of the rate of offset (in Langmuir's terms V_1 and referred to as k_2 in receptor pharmacology) divided by the rate of onset (in Langmuir's terms α denoted k_1 in receptor pharmacology).

It is amazing to note that complex processes such as drug binding to protein, activation of cells, and observation of syncytial cellular response should apparently so closely follow a model based on these simple concepts. This was not lost on A. J. Clark in his treatise on drug receptor theory *The Mode of Action of Drugs on Cells* [4]:

It is an interesting and significant fact that the author in 1926 found that the quantitative relations between the concentration of acetylcholine and its action on muscle cells, an action the nature of which is wholly unknown, could be most accurately expressed by the formulae devised by Langmuir to express the adsorption of gases on metal filaments

—A. J. Clark (1937)

The term K_A is a concentration and it quantifies affinity. Specifically, it is the concentration that binds to 50% of the total receptor population (see Equation 1.4 when [A] = K_A). Therefore, the smaller the K_A, the higher is the affinity. Affinity is the reciprocal of K_A. For example, if $K_A = 10^{-8}$ M, then 10^{-8} M binds to 50% of the receptors. If $K_A = 10^{-4}$ M, a 10,000-fold higher concentration of the drug is needed to bind to 50% of the receptors (i.e., it is of lower affinity).

It is instructive to discuss affinity in terms of the adsorption isotherm in the context of measuring the amount of receptor bound for given concentrations of drug. Assume that values of fractional receptor occupancy can be visualized for various drug concentrations. The kinetics of such binding are shown in Figure 1.9. It can be seen that initially the binding is rapid in accordance with the fact that there are many unbound sites for the drug to choose. As the sites become occupied, there is a temporal reduction in binding until a maximal value for that concentration is attained. Figure 1.9 also shows that the binding of higher concentrations of drug is correspondingly increased. In keeping with the fact that this is first-order binding kinetics (where the rate is dependent on a rate constant multiplied by the concentration of reactant), the time to equilibrium is shorter for higher concentrations than for lower concentrations. The various values for receptor occupancy at different concentrations constitute a

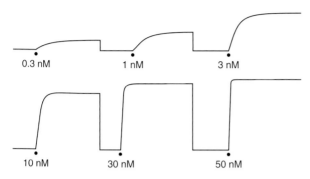

FIGURE 1.9 Time course for increasing concentrations of a ligand with a K_A of 2 nM. Initially the binding is rapid but slows as the sites become occupied. The maximal binding increases with increasing concentrations as does the rate of binding.

concentration-binding curve (shown in Figure 1.10a). There are two areas in this curve of particular interest to pharmacologists. The first is the maximal asymptote for binding. This defines the maximal number of receptive binding sites in the preparation. The binding isotherm Equation 1.4 defines the ordinate axis as the fraction of the maximal binding. Thus, by definition the maximal value is unity. However, in experimental studies real values of capacity are used since the maximum is not known. When the complete curve is defined, the maximal value of binding can be used to define fractional binding at various concentrations and thus define the concentration at which half maximal binding (binding to 50% of the receptor population) occurs. This is the equilibrium dissociation constant of the drug-receptor complex (K_A), the important measure of drug affinity. This comes from the other important region of the curve; namely, the midpoint. It can be seen from Figure 1.10a that graphical estimation of both the maximal asymptote and the midpoint is difficult to visualize from the graph in the form shown. A much easier format to present binding, or any concentration response data, is a semi-logarithmic form of the isotherm. This allows better estimation of the maximal asymptote and places the midpoint in a linear portion of the graph where

intrapolation can be done (see Figure 1.10b). Dose-response curves for binding are not often visualized as they require a means to detect bound (over unbound) drug. However, for drugs that produce pharmacological response (i.e., agonists) a signal proportional to bound drug can be observed. The true definition of dose-response curve is the observed in vivo effect of a drug given as a dose to a whole animal or human. However, it has entered into the common pharmacological jargon as a general depiction of drug and effect. Thus, a dose-response curve for binding is actually a binding concentration curve, and an in vitro effect of an agonist in a receptor system is a *concentration-response curve*.

1.10 What Is Efficacy?

The property that gives a molecule the ability to change a receptor, such that it produces a cellular response, is termed *efficacy*. Early concepts of receptors likened them to locks and drugs as keys. As stated by Paul Ehrlich, "Substances can only be anchored at any particular part of the organism if they fit into the molecule of the recipient complex like a piece of mosaic finds it's place in a pattern." This historically useful but inaccurate view of receptor function has in some ways hindered development models of efficacy. Specifically, the lock-and-key model implies a static system with no moving parts. However, a feature of proteins is their malleability. While they have structure, they do not have a single structure but rather many potential shapes referred to as *conformations*. A protein stays in a particular conformation because it is energetically favorable to do so (i.e., there is a minimal free energy for that conformation). If thermal energy enters the system, the protein may adopt another shape in response. Stated by Lindstrom-Lang and Schellman [15]:

> ...a protein cannot be said to have "a" secondary structure but exists mainly as a group of structures not too different from one another in free energy...In fact, the molecule must be conceived as trying every possible structure...
> —Lindstrom and Schellman (1959)

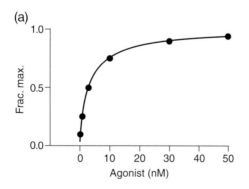

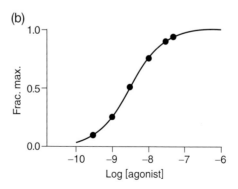

FIGURE 1.10 Dose-response relationship for ligand binding according to the Langmuir adsorption isotherm. (a) Fraction of maximal binding as a function of concentration of agonist. (b) Semi-logarithmic form of curve shown in panel a.

Not only are a number of conformations for a given protein possible, the protein samples these various conformations constantly. It is a dynamic not a static entity. Receptor proteins can spontaneously change conformation in response to the energy of the system. An important concept here is that small molecules, by interacting with the receptor protein, can bias the conformations that are sampled. It is in this way that drugs can produce active effects on receptor proteins (i.e., demonstrate efficacy). A thermodynamic mechanism by which this can occur is through what is known as conformational selection [16]. A simple illustration can be made by reducing the possible conformations of a given receptor protein to just two. These will be referred to as the "active" (denoted [R_a]) and "inactive" (denoted [R_i]) conformation.

Thermodynamically it would be expected that a ligand may not have identical affinity for both receptor conformations. This was an assumption in early formulations of conformational selection. For example, differential affinity for protein conformations was proposed for oxygen binding to hemoglobin [17] and for choline derivatives and nicotinic receptors [18]. Furthermore, assume that these conformations exist in an equilibrium defined by an allosteric constant L (defined as [R_a]/[R_i]) and that a ligand [A] has affinity for both conformations defined by equilibrium association constants K_a and αK_a, respectively, for the inactive and active states:

$$R_i \xrightarrow{\quad L \quad} R_a$$
$$\uparrow K_a \qquad \uparrow \alpha K_a. \qquad (1.5)$$
$$A \qquad\qquad A$$

It can be shown that the ratio of the active species Ra in the presence of a saturating concentration (ρ_∞) of the ligand versus in the absence of the ligand (ρ_0) is given by (see Section 1.13):

$$\frac{\rho_\infty}{\rho_0} = \frac{\alpha(1 + L)}{(1 + \alpha L)}. \qquad (1.6)$$

It can be seen that if the factor α is unity (i.e., the affinity of the ligand for R_a and R_i is equal [$K_a = \alpha K_a$]), then there will be no change in the amount of R_a when the ligand is present. However, if α is not unity (i.e., if the affinity of the ligand differs for the two species), then the ratio necessarily will change when the ligand is present. Therefore, the differential affinity for the two protein species will alter their relative amounts. If the affinity of the ligand is higher for R_a, then the ratio will be > 1 and the ligand will enrich the R_a species. If the affinity for the ligand for R_a is less than for R_i, then the ligand (by its presence in the system) will reduce the amount of R_a. For example, if the affinity of the ligand is 30-fold greater for the R_a state, then in a system where 16.7% of the receptors are spontaneously in the R_a state the saturation of the receptors with this

agonist will increase the amount of R_a by a factor of 5.14 (16.7 to 85%).

This concept is demonstrated schematically in Figure 1.11. It can be seen that the initial bias in a system of proteins containing two conformations (square and spherical) lies far toward the square conformation. When a ligand (filled circles) enters the system and selectively binds to the circular conformations, this binding process removes the circles driving the backward reaction from circles back to squares. In the absence of this backward pressure, more square conformations flow into the circular state to fill the gap. Overall, there is an enrichment of the circular conformations when unbound and ligand-bound circular conformations are totaled.

This also can be described in terms of the Gibbs free energy of the receptor-ligand system. Receptor conformations are adopted as a result of attainment of minimal free energy. Therefore, if the free energy of the collection of receptors changes so too will the conformational makeup of the system. The free energy of a system comprised of two conformations a_i and a_0 is given by [19]:

$$\sum \Delta G_i = \sum \Delta G_i^0 - RT \\ \times \sum \ln(1 + K_{a,i}[A])/\ln(1 + K_{a,0}[A]), \qquad (1.7)$$

where $K_{a,i}$ and $K_{a,0}$ are the respective affinities of the ligand for states i and 0. It can be seen that unless $K_{a,i} = K_{a,0}$ the logarithmic term will not equal zero and the free energy of the system will change ($\sum \Delta G_i \neq \sum \Delta G_i^0$). Thus, if a ligand has differential affinity for either state, then the free energy of the system will change in the presence of the ligand. Under these circumstances, a different conformational bias will be formed by the differential affinity of the ligand. From these models comes the concept that binding is not a passive process whereby a ligand simply adheres to a protein without changing it. The act of binding can itself bias the behavior of the protein. This is the thermodynamic basis of efficacy.

1.11 Dose-response Curves

The concept of "dose response" in pharmacology has been known and discussed for some time. A prescription written in 1562 for hyoscyamus and opium for sleep clearly states, "If you want him to sleep less, give him less" [13]. It was recognized by one of the earliest physicians, Paracelsus (1493–1541), that it is only the dose that makes something beneficial or harmful: "All things are poison, and nothing is without poison. The Dosis alone makes a thing not poison."

Dose-response curves depict the response to an agonist in a cellular or subcellular system as a function of the agonist concentration. Specifically, they plot response as a function of the logarithm of the concentration. They can be defined completely by three parameters; namely, location along the concentration axis, slope, and maximal asymptote

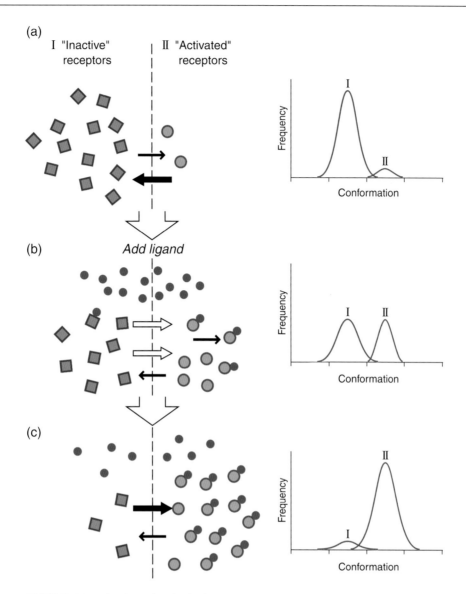

FIGURE 1.11 Conformational selection as a thermodynamic process to bias mixtures of protein conformations. (a) The two forms of the protein are depicted as circular and square shapes. The system initially is predominantly square. Gaussian curves to the right show the relative frequency of occurrence of the two conformations. (b) As a ligand (black dots) enters the system and prefers the circular conformations, these are selectively removed from the equilibrium between the two protein states. The distributions show the enrichment of the circular conformations at the expense of the square one. (c) A new equilibrium is attained in the presence of the ligand favoring the circular conformation because of the selective pressure of affinity between the ligand and this conformation. The distribution reflects the presence of the ligand and the enrichment of the circular conformation.

(Figure 1.12). At first glance, the shapes of dose-response curves appear to closely mimic the line predicted by the Langmuir adsorption isotherm and it is tempting to assume that dose-response curves reflect the first-order binding and activation of receptors on the cell surface. However, in most cases this resemblance is happenstance and dose-response curves reflect a far more complex amalgam of binding, activation, and recruitment of cellular elements of response. In the end, these may yield a sigmoidal curve but in reality they are far removed from the initial binding of drug and receptor. For example, in a cell culture with a collection of cells of varying threshold for depolarization the single-cell response to an agonist may be complete depolarization (in an all-or-none fashion). Taken as a complete collection,

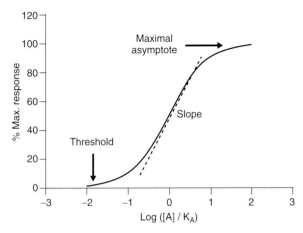

FIGURE 1.12 Dose-response curves. Any dose-response curve can be defined by the threshold (where response begins along the concentration axis), the slope (the rise in response with changes in concentration), and the maximal asymptote (the maximal response).

the depolarization profile of the culture where the cells all have differing thresholds for depolarization would have a Gaussian distribution of depolarization thresholds—some cells being more sensitive than others (Figure 1.13a). The relationship of depolarization of the complete culture to the concentration of a depolarizing agonist is the area under the Gaussian curve. This yields a sigmoidal dose-response curve (Figure 1.13b), which resembles the Langmuirian binding curve for drug-receptor binding. The slope of the latter curve reflects the molecularity of the drug-receptor interaction (i.e., one ligand binding to one receptor yields a slope for the curve of unity). In the case of the sequential

depolarization of a collection of cells, it can be seen that a more narrow range of depolarization thresholds yields a steeper dose-response curve, indicating that the actual numerical value of the slope for a dose-response curve cannot be equated to the molecularity of the binding between agonist and receptor. In general, shapes of dose-response curves are completely controlled by cellular factors and cannot be used to discern drug-receptor mechanisms. These must be determined indirectly by null methods.

1.11.1 Potency and Maximal Response

There are certain features of agonist dose-response curves that are generally true for all agonists. The first is that the magnitude of the maximal asymptote is totally dependent on the efficacy of the agonist and the efficiency of the biological system to convert receptor stimulus into tissue response (Figure 1.14a). This can be an extremely useful observation in the drug discovery process when attempting to affect the efficacy of a molecule. Changes in chemical structure that affect only the affinity of the agonist will have no effect on the maximal asymptote of the dose-response curve for that agonist. Therefore, if chemists wish to optimize or minimize efficacy in a molecule they can track the maximal response to do so. Second, the location, along the concentration axis of dose-response curves quantifies the *potency* of the agonist (Figure 1.14b). The potency is the molar concentration required to produce a given response. Potencies vary with the type of cellular system used to make the measurement and the level of response at which the measurement is made. A common measurement used to quantify potency is the EC_{50}; namely the molar concentration of an agonist

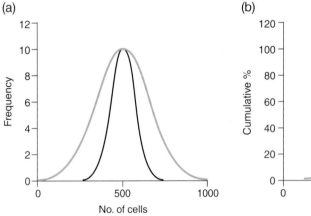

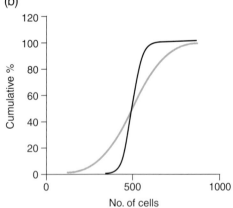

FIGURE 1.13 Factors affecting the slope of dose-response curves. (a) Gaussian distributions of the thresholds for depolarization of cells to an agonist in a cell culture. Solid line shows a narrow range of threshold and the lighter line a wider range. (b) Area under the curve of the Gaussian distributions shown in panel a. These would represent the relative depolarization of the entire cell culture as a function of the concentration of agonist. The more narrow range of threshold values corresponds to the dose-response curve of steeper slope. Note how the more narrow distribution in panel a leads to a steeper slope for the curve in panel b.

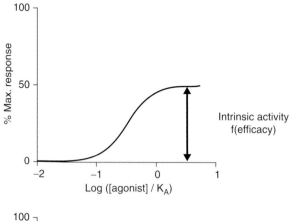

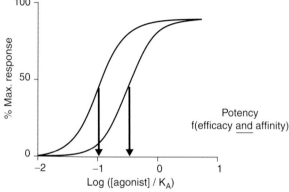

FIGURE 1.14 Major attributes of agonist dose-response curves. Maximal responses solely reflect efficacy while the potency (location along the concentration axis) reflects a complex function of both efficacy and affinity.

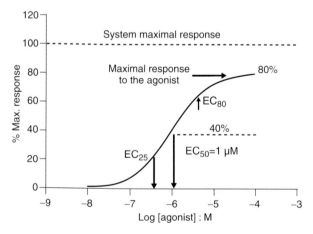

FIGURE 1.15 Dose-response curves. Dose-response curve to an agonist that produces 80% of the system maximal response. The EC_{50} (concentration producing 40% response) is $1\,\mu M$, the EC_{25}(20%) is $0.5\,\mu M$ and the EC_{80} (64%) is $5\,\mu M$.

required to produce 50% of the maximal response to the agonist. Thus, an EC_{50} value of $1\,\mu M$ indicates that 50% of the maximal response to the agonist is produced by a concentration of $1\,\mu M$ of the agonist (Figure 1.15). If the

agonist produces a maximal response of 80% of the system maximal response, then 40% of the system maximal response will be produced by $1\,\mu M$ of this agonist (Figure 1.15). Similarly, an EC_{25} will be produced by a lower concentration of this same agonist; in this case, the EC_{25} is $0.5\,\mu M$.

1.11.2 p-Scales and the Representation of Potency

Agonist potency is an extremely important parameter in drug receptor pharmacology. Invariably it is determined from log-dose response curves. It should be noted that since these curves are generated from semi-logarithmic plots the location parameter of these curves are *log normally distributed*. This means that the *logarithms* of the sensitivities (EC_{50}) and *not* the EC_{50} values themselves are normally distributed (Figure 1.16a). Since all statistical parametric tests must be done on data that come from normal distributions, all statistics (including comparisons of potency and estimates of errors of potency) must come from logarithmically expressed potency data. When log normally distributed EC_{50} data (Figure 1.16b) is converted to EC_{50} data, the resulting distribution is seriously skewed (Figure 1.16c). It can be seen that error limits on the mean of such a distribution are not equal (i.e., 1 standard error of the mean unit [see Chapter 8] either side of the mean gives different values on the skewed distribution [Figure 1.16c]). This is not true of the symmetrical normal distribution (Figure 1.16b).

One representation of numbers such as potency estimates is with the p-scale. The p-scale is the negative logarithm of number. For example, the pH is the negative logarithm of a hydrogen ion concentration ($10^{-5}\,Molar = pH = 5$). It is essential to express dose-response parameters as p-values ($-\log$ of the value, as in the pEC_{50}) since these are log normal. However, it sometimes is useful on an intuitive level to express potency as a concentration (i.e., the antilog value). One way this can be done and still preserve the error estimate is to make the calculation as p-values and then convert to concentration as the last step. For example, Table 1.2 shows five pEC_{50} values giving a mean pEC_{50} of 8.46 and a standard error of 0.21. It can be seen that the calculation of the mean as a converted concentration (EC_{50} value) leads to an apparently reasonable mean value of $3.8\,nM$, with a standard error of $1.81\,nM$. However, the 95% confidence limits (range of values that will include the true value) of the concentration value is meaningless in that one of them (the lower limit) is a negative number. The true value of the EC_{50} lies within the 95% confidence limits given by the mean $+ 2.57 \times$ the standard error, which leads to the values $8.4\,nM$ and $-0.85\,nM$. However, when pEC_{50} values are used for the calculations this does not occur. Specifically, the mean of 8.46 yields a mean EC_{50} of $3.47\,nM$. The 95% confidence limits on the pEC50 are 7.8 to 9.0. Conversion of these limits to EC50 values yields 95% confidence limits of $1\,nM$ to $11.8\,nM$. Thus, the true potency lies between the values of 1 and $11.8\,nM$ 95% of the time.

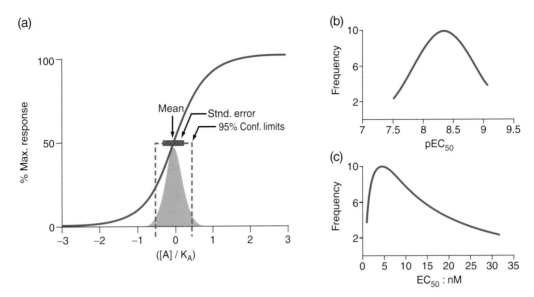

FIGURE 1.16 Log normal distributions of sensitivity of a pharmacological preparation to an agonist. (a) Dose-response curve showing the distribution of the EC_{50} values along the log concentration axis. This distribution is normal only on a log scale. (b) Log normal distribution of pEC_{50} values ($-\log EC_{50}$ values). (c) Skewed distribution of EC_{50} values converted from the pEC_{50} values shown in panel b.

TABLE 1.2

Expressing mean agonist potencies with error.

pEC_{50}[1]	EC_{50} (nM)[2]
8.5	3.16
8.7	2
8.3	5.01
8.2	6.31
8.6	2.51
Mean = 8.46	Mean = 3.8
SE = 0.21	SE = 1.81

[1]Replicate values of $-1/N\log EC_{50}$'s.

[2]Replicate EC_{50} values in nM.

1.12 Chapter Summary and Conclusions

- Some ideas on the origins and relevance of pharmacology and the concept of biological "receptors" are discussed.
- Currently there are drugs for only a fraction of the druggable targets present in the human genome.
- While recombinant systems have greatly improved the drug discovery process, pathological phenotypes still are a step away from these drug testing systems.
- Because of the previously cited, system-independent measures of drug activity (namely, affinity and efficacy) must be measured in drug discovery.
- Affinity is the strength of binding of a drug to a receptor. It is quantified by an equilibrium dissociation constant.

- Affinity can be depicted and quantified with the Langmuir adsorption isotherm.
- Efficacy is measured in relative terms (having no absolute scale) and quantifies the ability of a molecule to produce a change in the receptor (most often leading to a physiological response).
- Dose-response curves quantify drug activity. The maximal asymptote is totally dependent on efficacy, while potency is due to an amalgam of affinity and efficacy.
- Measures of potency are log normally distributed. Only p-scale values (i.e., pEC_{50}) should be used for statistical tests.

1.13 Derivations: Conformational Selections as a Mechanism of Efficacy

Consider a system containing two receptor conformations R_i and R_a that coexist in the system according to an allosteric constant denoted L:

$$R_i \xleftrightarrow{\quad L \quad} R_a$$
$$\uparrow K_a \qquad\qquad \uparrow \alpha K_a\cdot$$
$$A \qquad\qquad\quad A$$

Assume that ligand A binds to R_i with an equilibrium association constant K_a and R_a by an equilibrium association constant αK_a. The factor α denotes the differential affinity of the agonist for R_a (i.e., $\alpha = 10$ denotes a tenfold greater affinity of the ligand for the R_a state). The effect of α on the ability of the ligand to alter the equilibrium between R_i and R_a can be calculated by examining the amount of R_a species (both as R_a and AR_a) present in the system in the absence of ligand and in the presence of ligand. The equilibrium expression for $[R_a]+[AR_a])/[R_{tot}]$ where $[R_{tot}]$ is the total receptor concentration given by the conservation equation $[R_{tot}]=[R_i]+[AR_i]+[R_a]+[AR_a])$ is:

$$\rho = \frac{L(1 + \alpha[A]/K_A)}{[A]/K_A(1 + \alpha L) + 1 + L}, \tag{1.8}$$

where L is the allosteric constant, [A] is the concentration of ligand, K_A is the equilibrium dissociation constant of the agonist-receptor complex $(K_A = 1/K_a)$, and α is the differential affinity of the ligand for the R_a state. It can be seen that in the absence of agonist $([A]=0)$, $\rho_0 = L/(1+L)$ and in the presence of a maximal concentration of ligand (saturating the receptors; $[A] \to \infty$) $\rho_\infty = (\alpha(1+L))/(1+\alpha L)$. The effect of the ligand on changing the proportion of the R_a state is given by the ratio ρ/ρ_0. This ratio is given by

$$\frac{\rho_\infty}{\rho_0} = \frac{\alpha(1+L)}{(1+\alpha L)}. \tag{1.9}$$

Equation 1.9 indicates that if the ligand has an equal affinity for both the R_i and R_a states $(\alpha = 1)$ then ρ_∞/ρ_0 will equal unity and no change in the proportion of R_a will result from maximal ligand binding. However, if $\alpha > 1$, then the presence of the conformationally selective ligand will cause the ratio ρ_∞/ρ_0 to be > 1 and the R_a state will be enriched by presence of the ligand.

References

1. Maehle, A.-H., Prull, C.-R., and Halliwell, R. F. (2002). The emergence of the drug-receptor theory. *Nature Rev. Drug. Disc.* 1637–1642.
2. Paton, W. D. M. (1986). On becoming a pharmacologist. *Ann. Rev. Pharmacol. and Toxicol.* 26:1–22.
3. Drews, J. (2000). Drug discovery: A historical perspective. *Science* 287:1960–1964.
4. Clark, A. J. (1933). The mode of action of drugs on cells. Edward Arnold, London.
5. Clark, A. J. General pharmacology. In: *Heffner's Handbuch der Experimentaellen Pharmacokogie, Erganzungswerk Band* 4. Springer-Verlag, 1937.
6. Holmstedt, B., and Liljestrand, G. (1981). Readings in pharmacology, Raven Press, New York.
7. Marchese, A., George, S. R., Kolakowski, L. F., Lynch, K. R., and O'Dowd, B. F. (1999). Novel GPCR's and their endogenous ligands: Expanding the boundaries of physiology and pharmacology. *Trends Pharmacol. Sci.* 20:370–375.
8. Venter, J. C., et al. (2001). The sequence of the human genome. *Science* 291:1304–1351.
9. Link, R., Daunt, D., Barsh, G., Chruscinski, A., and Kobilka, B. (1992). Cloning of two mouse genes encoding α_2-adrenergic receptor subtypes and identification of a single amino acid in the mouse α_2-C10 homolog responsible for an interspecies variation in antagonist binding. *Mol. Pharmacol.* 42:16–17.
10. Black, J. W. (1996). A personal view of pharmacology. *Ann. Rev. Pharmacol. Toxicol.* 36:1–33.
11. Buscher, R., Hermann, V., and Insel, P. A. (1999). Human adrenoceptor polymorphisms: Evolving recognition of clinical importance. *Trends Pharmacol. Sci.* 20:94–99.
12. Stephenson, R. P. (1956). A modification of receptor theory. *Br. J. Pharmacol.* 11:379–393.
13. Norton, S. (2005). Origins of Pharmacology. *Mol. Interventions* 5:144–149.
14. Leff, P., Martin, G. R., and Morse, J. M. (1986). *Br. J. Pharmacol.* 89:493–499.
15. Linderstrom-Lang, A., and Schellman, P. (1959). Protein conformation. *Enzymes* 1:443–471.
16. Burgen, A. S. V. (1966). Conformational changes and drug action. *Fed. Proc.* 40:2723–2728.
17. Wyman, J. J., and Allen, D. W. (1951). The problem of the haem interaction in haemoglobin and the basis for the Bohr effect. *J. Polymer Sci.* 7:499–518.
18. Del Castillo, J., and Katz, B. (1957). Interaction at end-plate receptors between different choline derivatives. *Proc. Roy. Soc. Lond. B.* 146:369–381.
19. Freire, E. (2000). Can allosteric regulation be predicted from structure? *Proc. Natl. Acad. Sci. U.S.A.* 97:11680–11682.

2

How Different Tissues Process Drug Response

[Nature] can refuse to speak but she cannot give a wrong answer.
— **DR. CHARLES BRENTON HUGINS (1966)**
We have to remember that what we observe is not nature in itself, but nature exposed to our method of questioning . . .
— **WERNER HEISENBERG (1901–1976)**

2.1 Drug Response as Seen Through the "Cellular Veil"

If a drug possesses the molecular property of efficacy, then it produces a change in the receptor that may be detected by the cell. However, this can only occur if the stimulus is of sufficient strength and if the cell has the amplification machinery necessary to convert the stimulus into an observable response. In this sense, the cellular host system completely controls what the experimenter observes regarding the events taking place at the drug receptor. Drug activity is thus revealed through a "cellular veil" that can, in many cases, obscure or substantially modify drug-receptor activity (Figure 2.1). Minute signals, initiated either at the cell surface or within the cytoplasm of the cell, are interpreted, transformed, amplified, and otherwise altered by the cell to tailor that signal to its own particular needs. In receptor systems where a drug does produce a response, the relationship between the binding reaction (drug + receptor protein) and the observed response can be studied indirectly through observation of the cellular response as a function of drug concentration (dose-response curve). A general phenomenon observed experimentally is that cellular response most often is not linearly related to receptor occupancy (i.e., it does not require 100% occupation of all of the receptors to produce the maximal cellular response). Figure 2.2a shows a functional dose-response curve to human calcitonin in human embryonic kidney (HEK) cells transfected with cDNA for human calcitonin receptor type 2. The response being measured here is hydrogen ion release by the cells, a sensitive measure of cellular metabolism. Also shown (dotted line) is a curve for calcitonin binding to the receptors (as measured with radioligand binding). A striking feature of these curves is that the curve for function is shifted considerably to the left of the binding curve. Calculation of the receptor occupancy required for 50% maximal tissue response indicates that less than 50% occupancy; namely, more on the order of 3 to 4% is needed. In fact, a regression of tissue response upon the receptor occupancy is hyperbolic in nature (Figure 2.2b), showing a skewed relationship between receptor occupancy and cellular response. This skewed relationship indicates that the stimulation of the receptor initiated by binding is amplified by the cell in the process of response production.

The ability of a given agonist to produce a maximal system response can be quantified as a *receptor reserve*. The reserve refers to the percentage of receptors not required for production of maximal response (i.e., sometimes referred to as *spare receptors*). For example, a receptor reserve of 80% for an agonist means that the system maximal response is produced by activation of 20% of the receptor population by that agonist. Receptor reserves can be quite striking. Figure 2.3 shows guinea pig ileal smooth muscle contractions to the agonist histamine before and after irreversible inactivation of a large fraction of the receptors with the protein alkylating agent phenoxybenzamine. The fact that the depressed maximum dose-response curve is observed so far to the right of the control dose-response curve indicates a receptor reserve of 98% (i.e., only 2% of the receptors must be activated by histamine to produce the tissue maximal response [Figure 2.3b]). In teleological terms, this may be useful since it allows neurotransmitters to produce rapid activation of organs with minimal receptor occupancy leading to optimal and rapid control of function.

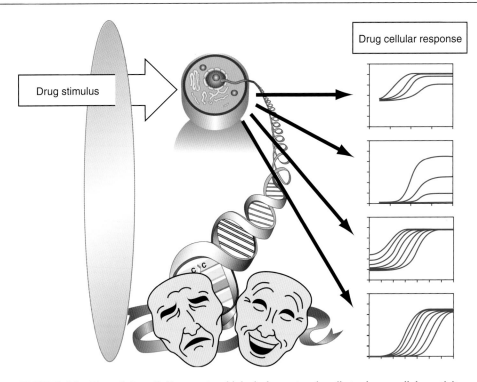

FIGURE 2.1 The cellular veil. Drugs act on biological receptors in cells to change cellular activity. The initial receptor stimulus usually alters a complicated system of interconnected metabolic biochemical reactions and the outcome of the drug effect is modified by the extent of these interconnections, the basal state of the cell, and the threshold sensitivity of the various processes involved. This can lead to a variety of apparently different effects for the same drug in different cells. Receptor pharmacology strives to identify the basic mechanism initiating these complex events.

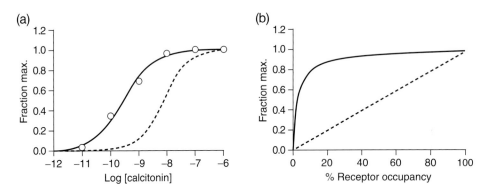

FIGURE 2.2 Binding and dose-response curves for human calcitonin on human calcitonin receptors type 2. (a) Dose-response curves for microphysiometry responses to human calcitonin in HEK cells (open circles) and binding in membranes from HEK cells (displacement of [^{125}I]-human calcitonin). Data from [1]. (b) Regression of microphysiometry responses to human calcitonin (ordinates) upon human calcitonin fractional receptor occupancy (abscissae). Dotted line shows a direct correlation between receptor occupancy and cellular response.

Receptor reserve is a property of the tissue (i.e., the strength of amplification of receptor stimulus inherent to the cells) *and* it is a property of the agonist (i.e., how much stimulus is imparted to the system by a given agonist receptor occupancy). This latter factor is quantified as the efficacy of the agonist. A high-efficacy agonist need occupy a smaller fraction of the receptor population than a lower-efficacy agonist to produce a comparable stimulus. Therefore, it is incorrect to ascribe a given tissue or cellular response system with a characteristic receptor reserve. The actual value of the receptor reserve will be unique to each agonist in that system. For example, Figure 2.4 shows the different amplification hyperbolae of CHO cells transfected with β-adrenoceptors in producing cyclic AMP responses to

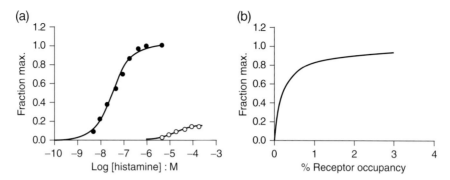

FIGURE 2.3 Guinea pig ileal responses to histamine. (a) Contraction of guinea pig ileal longitudinal smooth muscle (ordinates as a percentage of maximum) to histamine (abscissae, logarithmic scale). Responses obtained before (filled circles) and after treatment with the irreversible histamine receptor antagonist phenoxybenzamine (50 μM for 3 minutes; open circles). (b) Occupancy response curve for data shown in (a). Ordinates are percentage of maximal response. Abscissae are calculated receptor occupancy values from an estimated affinity of 20 μM for histamine. Note that maximal response is essentially observed after only 2% receptor occupancy by the agonist (i.e., a 98% receptor reserve for this agonist in this system). Data redrawn from [2].

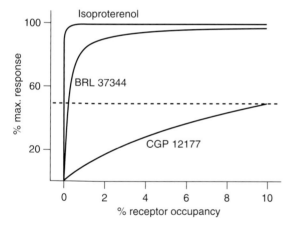

FIGURE 2.4 Occupancy-response curves for β-adrenoceptor agonists in transfected CHO cells. Occupancy (abscissae) calculated from binding affinity measured by displacement of [^{125}I]iodocyanopindolol. Response measured as increases in cyclic AMP. Drawn from [3].

three different β-adrenoceptor agonists. It can be seen that isoproterenol requires many times less receptors to produce 50% response than do both the agonists BRL 37344 and CGP 12177. This underscores the idea that the magnitude of receptor reserves are very much dependent on the efficacy of the agonist (i.e., one agonist's spare receptor is another agonist's essential one).

2.2 The Biochemical Nature of Stimulus-response Cascades

Cellular amplification of receptor signals occurs through a succession of saturable biochemical reactions. Different

receptors are coupled to different stimulus-response mechanisms in the cell. Each has its own function and operates on its own timescale. For example, receptor tyrosine kinases (activated by growth factors) phosphorylate target proteins on tyrosine residues to activate protein phosphorylation cascades such as MAP kinase pathways. This process, on a timescale on the order of seconds to days, leads to protein synthesis from gene transcription with resulting cell differentiation and/or cell proliferation. Nuclear receptors, activated by steroids, operate on a timescale of minutes to days and mediate gene transcription and protein synthesis. This leads to homeostatic, metabolic, and immunosuppression effects. Ligand gated ion channels, activated by neurotransmitters, operate on the order of milliseconds to increase the permeability of plasma membranes to ions. This leads to increases in cytosolic Ca^{2+}, depolarization or hyperolarization of cells. This results in muscle contraction, release of neurotransmitters or inhibition of these processes.

G-protein coupled receptors (GPCRs) react to a wide variety of molecules from some as small as acetylcholine to as large as the protein SDF-1α. Operating on a timescale of minutes to hours, these receptors mediate a plethora of cellular processes. The first reaction in the activation cascade for GPCRs is the binding of the activated receptor to a trimeric complex of proteins called G-proteins (Figure 2.5). These proteins—comprised of three subunits named α, β, and γ—act as molecular switches to a number of other effectors in the cell. The binding of activated receptors to the G-protein initiates the dissociation of GDP from the α-subunit of the G-protein complex, the binding of GTP, and the dissociation of the complex into α- and βγ-subunits. The separated subunits of the G-protein can activate effectors in the cell such as adenylate cyclase and ion channels. Amplification can occur at these early stages

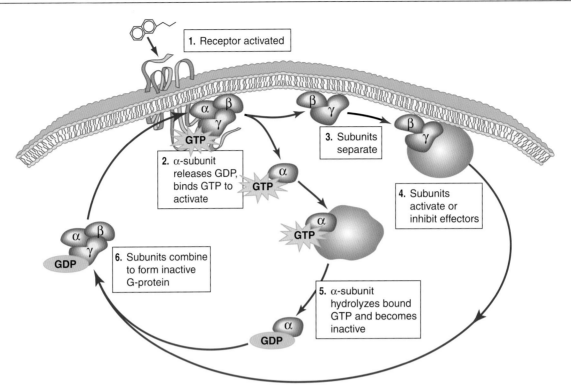

FIGURE 2.5 Activation of trimeric G-proteins by activated receptors. An agonist produces a receptor active state that goes on to interact with the G-protein. A conformational change in the G-protein causes bound GDP to exchange with GTP. This triggers dissociation of the G-protein complex into α- and βγ- subunits. These go on to interact with effectors such as adenylate cyclase and calcium channels. The intrinsic GTPase activity of the α-subunit hydolyzes bound GTP back to GDP and the inactived α-subunit reassociates with the βγ- subunits to repeat the cycle.

if one receptor activates more than one G-protein. The α-subunit also is a GTPase, which hydrolyzes the bound GTP to produce its own deactivation. This terminates the action of the α-subunit on the effector. It can be seen that the length of time that the α-subunit is active can control the amount of stimulus given to the effector and that this also can be a means of amplification (i.e., one α-subunit could activate many effectors). The α- and βγ- subunits then reassociate to complete the regulatory cycle (Figure 2.5). Such receptor-mediated reactions generate cellular molecules called second messengers. These molecules go on to activate or inhibit other components of the cellular machinery to change cellular metabolism and state of activation. For example, the second messenger (cyclic AMP) is generated by the enzyme adenylate cyclase from ATP. This second messenger furnishes fuel, through protein kinases, for phosphorylation of serine and threonine residues on a number of proteins such as other protein kinases, receptors, metabolic enzymes, ion channels, and transcription factors (see Figure 2.6). Activation of other G-proteins leads to activation of phospholipase C. These enzymes catalyze the hydrolysis of phosphatidylinositol 4,5-bisphosphate (PIP_2) to 1,2 diacylglycerol (DAG) and inositol1,4,5-triphosphate (IP_3) (see Figure 2.7). This latter second messenger interacts with receptors on intracellular

calcium stores resulting in the release of calcium into the cytosol. This calcium binds to calcium sensor proteins such as calmodulin or troponin C, which then go on to regulate the activity of proteins such as protein kinases, phosphatases, phosphodiesterase, nitric oxide synthase, ion channels, and adenylate cyclase. The second messenger DAG diffuses in the plane of the membrane to activate protein kinase C isoforms, which phosphorylate protein kinases, transcription factors, ion channels, and receptors. DAG also functions as the source of arachidonic acid which goes on to be the source of eicosanoid mediators such as prostanoids and leukotrienes. In general, all of these processes can lead to a case where a relatively small amount of receptor stimulation can result in a large biochemical signal. An example of a complete stimulus-response cascade for the β-adrenoceptor production of blood glucose is shown in Figure 2.8.

There are numerous second messenger systems such as those utilizing cyclic AMP and cyclic GMP, calcium and calmodulin, phosphoinositides, and diacylglerol with accompanying modulatory mechanisms. Each receptor is coupled to these in a variety of ways in different cell types. Therefore, it can be seen that it is impractical to attempt to quantitatively define each stimulus-response mechanism for each receptor system. Fortunately, this is not an

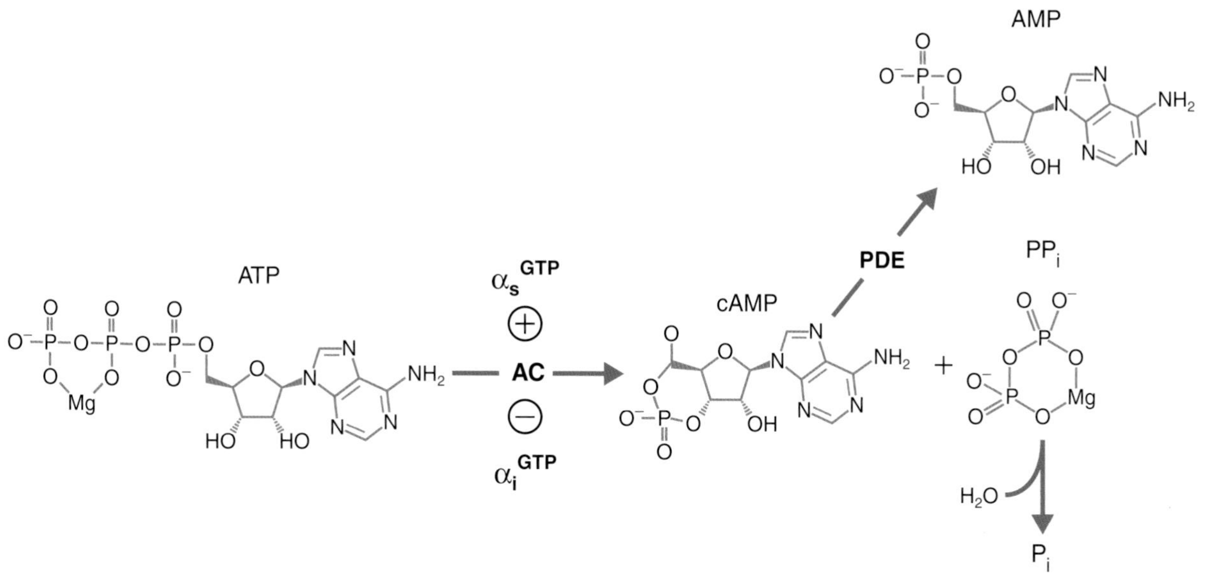

FIGURE 2.6 Production of cyclic AMP from ATP by the enzyme adenylate cyclase. Cyclic AMP is a ubiquitous second messenger in cells activating numerous cellular pathways. The adenylate cyclase is activated by the a subunit of G_s-protein and inhibited by the α-subunit of G_i-protein. Cyclic AMP is degraded by phosphodiesterases in the cell.

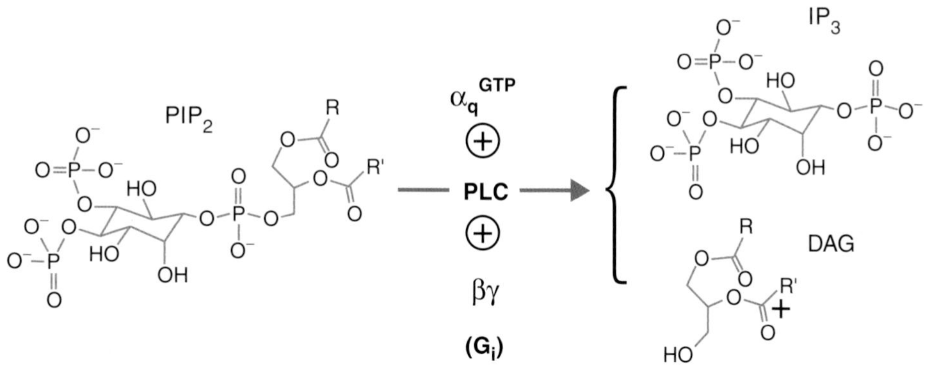

FIGURE 2.7 Production of second messengers inositol1,4,5-triphosphate (IP_3) and diacylglycerol (DAG) through activation of the enzyme phospholipase C. This enzyme is activated by the α- subunit of G_q-protein and also by $\beta\gamma$ subunits of G_i-protein. IP_3 stimulates the release of Ca^{2+} from intracellular stores while DAG is a potent activator of protein kinase C.

important prerequisite in the pharmacological process of classifying agonists since these complex mechanisms can be approximated by simple mathematical functions.

2.3 The Mathematical Approximation of Stimulus-response Mechanisms

Each of the processes shown in Figure 2.8 can be described by a Michaelis-Menten type of biochemical reaction, a standard generalized mathematical equation describing the interaction of a substrate with an enzyme. Michaelis and Menten realized in 1913 that the kinetics of enzyme reactions differed from the kinetics of conventional

chemical reactions. They visualized the reaction of substrate and an enzyme yielding enzyme plus substrate as a form of the equation: reaction velocity = (maximal velocity of the reaction × substrate concentration)/(concentration of substrate + a fitting constant K_m). The constant K_m (referred to as the Michaelis-Menten constant) characterizes the tightness of the binding of the reaction between substrate and enzyme, essentially a quantification of the coupling efficiency of the reaction. The K_m is the concentration at which the reaction is half the maximal value, or in terms of kinetics the concentration at which the reaction runs at half its maximal rate. This model forms the basis of enzymatic biochemical reactions and can

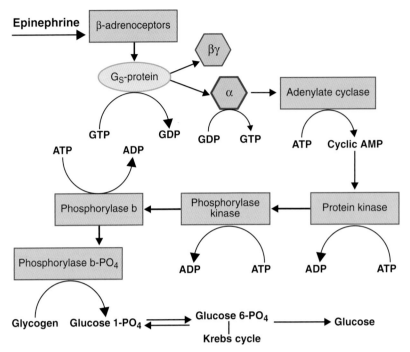

FIGURE 2.8 Stimulus response cascade for the production of blood glucose by activation of β-adrenoceptors. Redrawn from [4].

be used as a mathematical approximation of such functions.

As with the Langmuir adsorption isotherm, which in shape closely resembles Michaelis-Menten type biochemical kinetics, the two notable features of such reactions are the location parameter of the curve along the concentration axis (the value of K_m or the magnitude of the coupling efficiency factor) and the maximal rate of the reaction (V_{max}). In generic terms, Michaelis-Menten reactions can be written in the form

$$\text{Velocity} = \frac{[\text{substract}] \cdot V_{max}}{[\text{substract}] + K_m} = \frac{[\text{input}] \cdot \text{MAX}}{[\text{input}] + \beta}, \quad (2.1)$$

where β is a generic coupling efficiency factor. It can be seen that the velocity of the reaction is inversely proportional to the magnitude of β (i.e., the lower the value of β the more efficiently is the reaction coupled). If it is assumed that the stimulus-response cascade of any given cell is a series succession of such reactions, there are two general features of the resultant that can be predicted mathematically. The first is that the resultant of the total series of reactions will itself be of the form of the same hyperbolic shape (see Section 2.11.1). The second is that the location parameter along the input axis (magnitude of the coupling efficiency parameter) will reflect a general amplification of any single reaction within the cascade (i.e., the magnitude of the coupling parameter for the complete series will be lower than the coupling parameter of any single reaction; see Figure 2.9). The magnitude of β_{total} for the series sum of

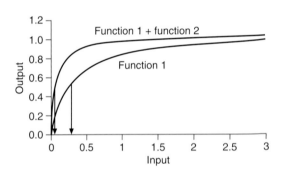

FIGURE 2.9 Amplification of stimulus through successive rectangular hyperbolae. The output from the first function ($\beta = 0.3$) becomes the input of a second function with the same coupling efficiency ($\beta = 0.3$), to yield a more efficiently coupled overall function ($\beta = 0.069$). Arrows indicate the potency for input to yield 50% maximal output for the first function and the series functions.

two reactions (characterized by β_1 and β_2) is given by (see Section 2.11.2)

$$\beta_{total} = \frac{\beta_1 \beta_2}{1 + \beta_2}. \quad (2.2)$$

It can be seen from Equation 2.2 that for positive non-zero values of β_2, $\beta_{total} < \beta_1$. Therefore, the location parameter of the rectangular hyperbola of the composite set of reactions in series is shifted to the left (increased

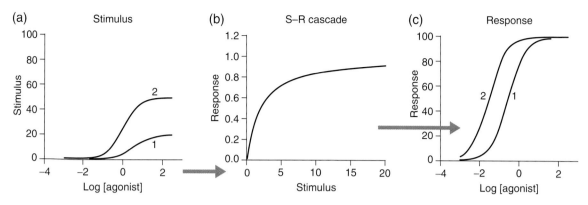

FIGURE 2.10 The monotonic nature of stimulus-response mechanisms. (a) Receptor stimulus generated by two agonists designated 1 and 2 as a function of agonist concentration. (b) Rectangular hyperbola characterizing the transformation of receptor stimulus (abscissae) into cellular response (ordinates) for the tissue. (c) The resulting relationship between tissue response to the agonists as a function of agonist concentration. The general rank order of activity (2 > 1) is preserved in the response as a reflection of the monotonic nature of the stimulus-response hyperbola.

potency) of that for the first reaction in the sequence (i.e., there is amplification inherent in the series of reactions).

The fact that the total stimulus-response chain can be approximated by a single rectangular hyperbola furnishes the basis of using end organ response to quantify agonist effect in a non-system-dependent manner. An important feature of such a relationship is that it is monotonic (i.e., there is only one value of y for each value of x). Therefore, the relationship between the strength of signal imparted to the receptor between two agonists is accurately reflected by the end organ response (Figure 2.10). This is the primary reason pharmacologists can circumvent the effects of the cellular veil and discern system-independent receptor events from translated cellular events.

2.4 System Effects on Agonist Response: Full and Partial Agonists

For any given receptor type, different cellular hosts should have characteristic efficiencies of coupling and these should characterize all agonists for that same receptor irrespective of the magnitude of the efficacy of the agonists. Different cellular backgrounds have different capabilities for amplification of receptor stimuli. This is illustrated by the strikingly different magnitudes of the receptor reserves for calcitonin and histamine receptors shown in Figures 2.2 and 2.3. Figure 2.11 shows the response produced by human calcitonin activation of the human calcitonin receptor type 2 when it is expressed in three different cell formats (human embryonic kidney cells [HEK 293 cells], Chinese hamster ovary cells [CHO cells], and *Xenopus laevis* melanophores). From this figure it can be seen that while only 3% receptor activation by this agonist is required for 50% response in melanophores this same occupancy in CHO

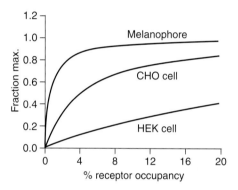

FIGURE 2.11 Receptor-occupancy curves for activation of human calcitonin type 2 receptors by the agonist human calcitonin. Ordinates (response as a fraction of the maximal response to human calcitonin). Abscissae (fractional receptor occupancy by human calcitonin). Curves shown for receptors transfected into three cell types: human embryonic kidney cells (HEK), Chinese hamster ovary cells (CHO), and *Xenopus laevis* melanophores. It can be seen that the different cell types lead to differing amplification factors for the conversion from agonist receptor occupancy to tissue response.

cells produces only 10% response and even less in HEK cells.

One operational view of differing efficiencies of receptor coupling is to consider the efficacy of a given agonist as a certain mass characteristic of the agonist. If this mass were to be placed on one end of a balance, it would depress that end by an amount dependent on the weight. The amount that the end is depressed would be the stimulus (see Figure 2.12). Consider the other end of the scale as reflecting the placement of the weight on the scale (i.e., the displacement of the other end is the response of the cell). Where along the arm this displacement is viewed

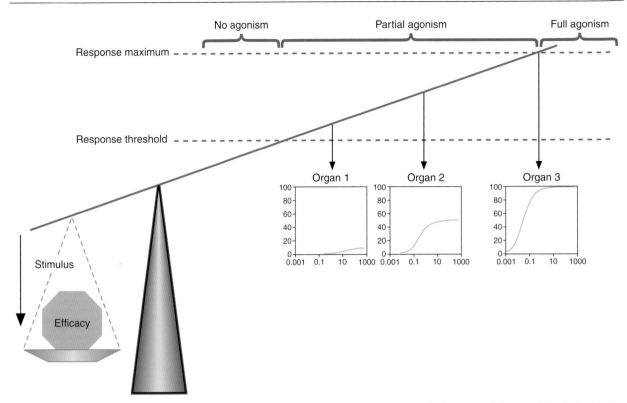

FIGURE 2.12 Depiction of agonist efficacy as a weight placed on a balance to produce displacement of the arm (stimulus) and the observation of the displacement of the other end of the arm as tissue response. The vantage point determines the amplitude of the displacement. Where no displacement is observed, no agonism is seen. Where the displacement is between the limits of travel of the arm (threshold and maximum), partial agonism is seen. Where displacement goes beyond the maximal limit of travel of the arm, uniform full agonism is observed.

reflects the relative amplification of the original stimulus (i.e., the closer to the fulcrum the less the amplification). Therefore, different vantage points along the displaced end of the balance arm reflect different tissues with different amplification factors (different magnitudes of coupling parameters). The response features of cells have limits (i.e., a threshold for detecting the response and a maximal response characteristic of the tissue). Depending on the efficiency of stimulus-response coupling apparatus of the cell, a given agonist could produce no response, a partially maximal response, or the system maximal response (see Figure 2.12). The observed response to a given drug gives a label to the drug in that system. Thus, a drug that binds to the receptor but produces no response is an *antagonist*, a drug that produces a submaximal response is a *partial agonist*, and a drug that produces the tissue maximal response is termed a *full agonist* (see Figure 2.13). It should be noted that while these labels often are given to a drug and used across different systems as identifying labels for the drug they are in fact dependent on the system. Therefore, the magnitude of the response can completely change with changes in the coupling efficiency of the system. For example, the low-efficacy β-adrenoceptor agonist prenalterol can be an antagonist in guinea pig extensor digitorum longus muscle, a

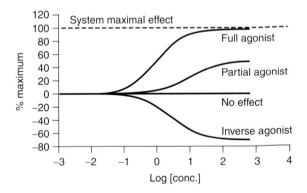

FIGURE 2.13 The expression of different types of drug activities in cells. A drug that produces the full maximal response of the biological system is termed a full agonist. A drug that produces a submaximal response is a partial agonist. Drugs also may produce no overt response or may actively reduce basal response. This latter class of drug is known as inverse agonist. These ligands have negative efficacy. This is discussed specifically in Chapter 3.

partial agonist in guinea pig left atria, and nearly a full agonist in right atria from thyroxine-treated guinea pigs (Figure 2.14).

As noted previously, the efficacy of the agonist determines the magnitude of the initial stimulus given to the

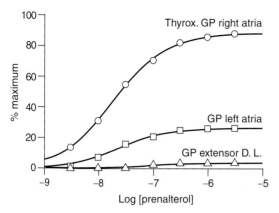

FIGURE 2.14 Dose-response curves to the β-adrenoceptor low-efficacy agonist prenalterol in three different tissues from guinea pigs. Responses all mediated by β_1-adrenoceptors. Depending on the tissue, this drug can function as nearly a full agonist, a partial agonist, or a full antagonist. Redrawn from [5].

receptor, and therefore the starting point for the input into the stimulus-response cascade. As agonists are tested in systems of varying coupling efficiency, it will be seen that the point at which system saturation of the stimulus-response cascade is reached differs for different agonists. Figure 2.15 shows two agonists, one of higher efficacy than the other. It can be seen that both are partial agonists in tissue A but that agonist 2 saturates the maximal response producing capabilities of tissue B and is a full agonist. The same is not true for agonist 1. In a yet more efficiently coupled system (tissue C), both agonists are full agonists. This illustrates the obvious error in assuming that all agonists that produce the system maximal response have equal efficacy. All full agonists in a given system may not have equal efficacy.

The more efficiently coupled is a given system the more likely agonists will produce the system maximum response (i.e., be full agonists). It can be shown also that if an

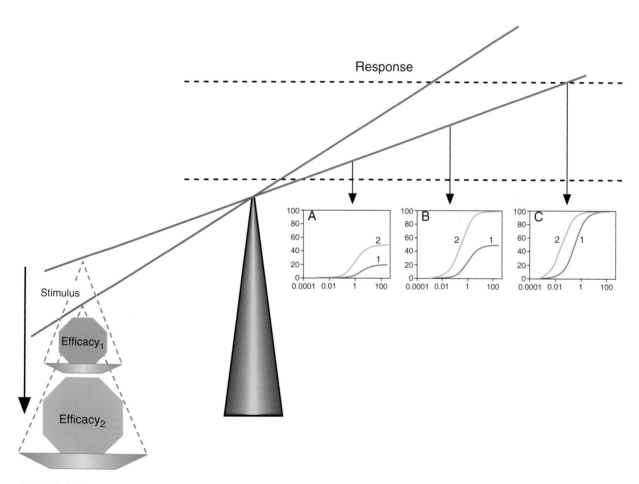

FIGURE 2.15 Depiction of agonist efficacy as a weight placed on a balance to produce displacement of the arm (stimulus) and the observation of the displacement of the other end of the arm as tissue response for two agonists, one of higher efficacy (Efficacy$_2$) than the other (Efficacy$_1$). The vantage point determines the amplitude of the displacement. In system A, both agonists are partial agonists. In system B, agonist 2 is a full agonist and agonist 1 a partial agonist. In system C, both are full agonists. It can be seen that the tissue determines the extent of agonism observed for both agonists and that system C does not differentiate the two agonists on the basis of efficacy.

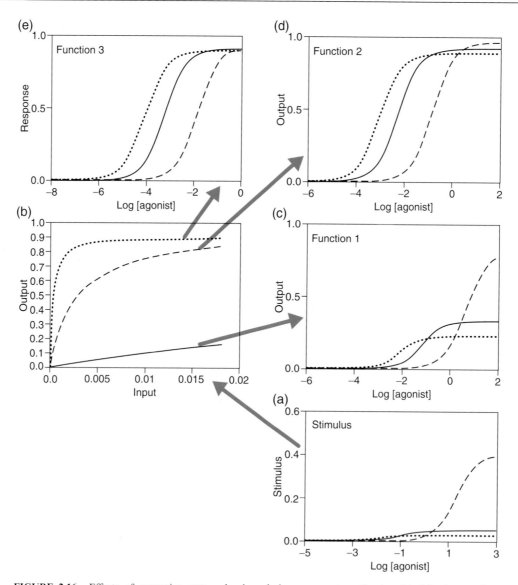

FIGURE 2.16 Effects of successive rectangular hyperbolae on receptor stimulus. (a) Stimulus to three agonists. (b) Three rectangular hyperbolic stimulus-response functions in series. Function 1 ($\beta = 0.1$) feeds function 2 ($\beta = 0.03$), which in turn feeds function 3 ($\beta = 0.1$). (c) Output from function 1. (d) Output from function 2 (functions 1 and 2 in series). (e) Final response: output from function 3 (all three functions in series). Note how all three are full agonists when observed as final response.

agonist saturates any biochemical reaction within the stimulus-response cascade it will produce full agonism (see Section 2.11.3). This also means that there will be an increasing tendency for an agonist to produce the full system maximal response the further down the stimulus-response cascade the response is measured. Figure 2.16 shows three agonists all producing different amounts of initial receptor stimulus. These stimuli are then passed through three successive rectangular hyperbolae simulating the stimulus-response cascade. As can be seen from the figure, by the last step all of the agonists are full agonists.

Viewing response at this point gives no indication of differences in efficacy.

2.5 Differential Cellular Response to Receptor Stimulus

As noted in the previous discussion, different tissues have varying efficiencies of stimulus-response coupling. However, within a given tissue there may be the capability of choosing or altering the responsiveness of the system to agonists. This can be a useful technique in the study of agonists. Specifically, the ability to observe full agonists as partial agonists enables the experimenter to compare relative efficacies (see previous material). Also, if stimulus-response capability can be reduced weak partial agonists can be studied as antagonists to gain measures of affinity. There are three general approaches to add texture to agonism: (1) choice of response pathway, (2) augmentation or modulation of pathway stimulus, and (3) manipulation of receptor density. This latter technique is operable only in recombinant systems where receptors are actively expressed in surrogate systems.

2.5.1 Choice of Response Pathway

The production of second messengers in cells by receptor stimulation leads to a wide range of biochemical reactions. As noted in the previous discussion, these can be approximately described by Michaelis-Menten type reaction curves and each will have unique values of maximal rates of reaction and sensitivities to substrate. There are occasions where experimenters have access to different end points of these cascades, and with them different amplification factors for agonist response. One such case is the stimulation of cardiac β-adrenoceptors. In general, this leads to a general excitation of cardiac response comprised of an increase in heart rate (for right atria), an increased force of contraction (inotropy), and an increase in the rate of muscle relaxation (lusitropy). These latter two cardiac functions can be accessed simultaneously from measurement of isometric cardiac contraction and each has its own sensitivity to β-adrenoceptor excitation (lusitropic responses being more efficiently coupled to elevation of cyclic AMP than inotropic responses). Figure 2.17a shows the relative sensitivity of cardiac lusitropy and intropy to elevations in cyclic AMP in guinea pig left atria. It can be seen that the coupling of lusitropic response is fourfold more efficiently coupled to cyclic AMP elevation than is inotropic response. Such differential efficiency of coupling

can be used to dissect agonist response. For example, the inotropic and lusitropic responses of the β-adrenoceptor agonists isoproterenol and prenalterol can be divided into different degrees of full and partial agonism (Figure 2.18). It can be seen from Figure 2.18a that there are concentrations of isoproterenol that increase the rate of myocardial relaxation (i.e., 0.3 nM) without changing inotropic state. As the concentration of isoproterenol increases the inotropic response appears (Figure 2.18b and c). Thus, the dose-response curve for myocardial relaxation for this full agonist is shifted to the left of the dose-response curve for inotropy in this preparation (Figure 2.18d). For a partial agonist such as prenalterol, there is nearly a complete dissociation between cardiac lusitropy and inotropy (Figure 2.18e). Theoretically, an agonist of low efficacy can be used as an antagonist of isoproterenol response in the more poorly coupled system (inotropy) and then compared with respect to efficacy (observation of visible response) in the more highly coupled system.

2.5.2 Augmentation or Modulation of Stimulus Pathway

The biochemical pathways making up the cellular stimulus-response cascade are complex systems with feedback and modulation mechanisms. Many of these are mechanisms to protect against overstimulation. For example, cells contain phosphodiesterase enzymes to degrade cyclic AMP to provide a fine control of stimulus strength and duration. Inhibition of phosphodiesterase therefore can remove this control and increase cellular levels of cyclic AMP. Figure 2.19a shows the effect of phosphodiesterase inhibition on the inotropic response of guinea pig papillary muscle. It can be seen from this figure that whereas 4.5% receptor stimulation by isoproterenol is required for 50% inotropic response in the natural system (where phosphodiesterase modulated intracellular cyclic AMP response) this is reduced to only 0.2% required receptor stimulation after inhibition of phosphodiesterase degradation of intracellular cyclic AMP. This technique can be used to modulate responses as well. Smooth muscle contraction requires extracellular calcium ion (calcium entry mediates contraction). Therefore, reduction of the calcium concentration in the extracellular space causes a modulation of the contractile responses (see example for the muscarinic contractile agonist carbachol, Figure 2.19b). In general the sensitivity of functional systems can be manipulated by antagonism of modulating mechanisms and control of cofactors needed for cellular response.

2.5.3 Differences in Receptor Density

The number of functioning receptors controls the magnitude of the initial stimulus given to the cell by an agonist. Number of receptors on the cell surface is one means the cell can control its stimulatory environment. Thus, it is not surprising that receptor density varies with different cell types. Potentially, this can be used to control

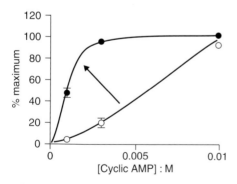

FIGURE 2.17 Differential efficiency of receptor coupling for cardiac function. (a) Guinea pig left atrial force of contraction (inotropy, open circles) and rate of relaxation (lusitropy, filled circles) as a function (ordinates) of elevated intracellular cyclic AMP concentration (abscissae). Redrawn from [6].

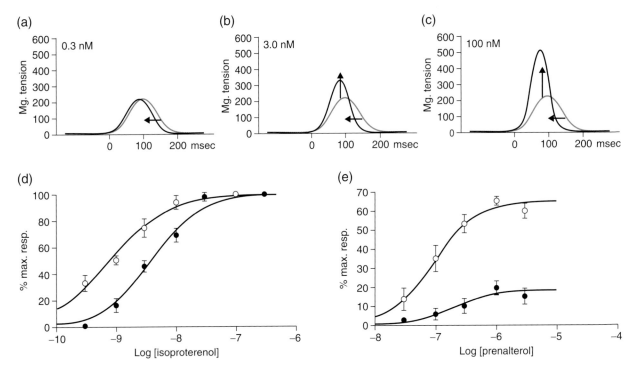

FIGURE 2.18 Inotropic and lusitropic responses of guinea pig left atria to β-adrenoceptor stimulation. Panels A to C: isometric tension waveforms of cardiac contraction (ordinates are mg tension; abscissae are msec). (a) Effect of 0.3 nM isoproterenol on the waveform. The wave is shortened due to an increase in the rate of diastolic relaxation, whereas no inotropic response (change in peak tension) is observed at this concentration. (b) A further shortening of waveform duration (lusitropic response) is observed with 3 nM isoproterenol. This is concomitant with positive inotropic response (increase maximal tension). (c) This trend continues with 100 nM isoproterenol. (d) Dose-response curves for inotropy (filled circles) and lusitropy (open circles) in guinea pig atria for isoproterenol. (e) Dose-response curves for inotropy (filled circles) and lusitropy (open circles) in guinea pig atria for the β-adrenoceptor partial agonist prenalterol. Data redrawn from [6].

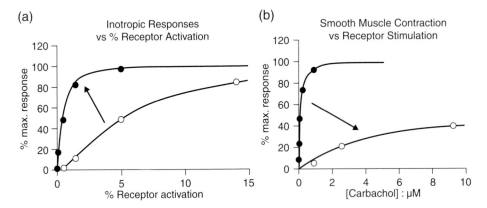

FIGURE 2.19 Potentiation and modulation of response through control of cellular processes. (a) Potentiation of inotropic response to isoproterenol in guinea pig papillary muscle by the phosphodiesterase inhibitor isobutylmethylxanthine (IBMX). Ordinates: percent of maximal response to isoproterenol. Abscissa: percent receptor occupancy by isoproterenol (log scale). Responses shown in absence (open circles) and presence (filled circles) of IBMX. Data redrawn from [7]. (b) Effect of reduction in calcium ion concentration on carbachol contraction of guinea pig ileum. Responses in the presence of 2.5 mM (filled circles) and 1.5 mM (open circles) calcium ion in physiological media bathing the tissue. Data redrawn from [8].

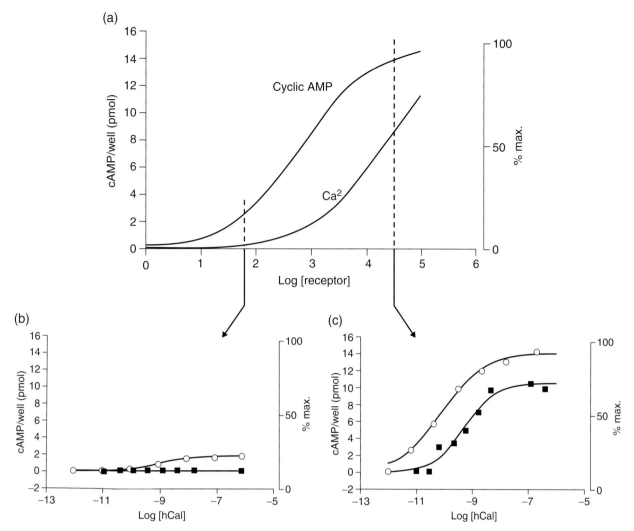

FIGURE 2.20 Effect of receptor expression level on responses of human calcitonin receptor type 2 to human calcitonin. (a) Cyclic AMP and calcium responses for human calcitonin activation of the receptor. Abscissae: logarithm of receptor density in fmole/mg protein. Ordinates: pmole cyclic AMP (left-hand axis) or calcium entry as a percentage maximum of response to human calcitonin. Two receptor expression levels are shown: at 65 fmole/mg, there is sufficient receptor to only produce a cyclic AMP response. At 30,000 fmole/mg receptor, more cyclic AMP is produced, but there is also sufficient receptor to couple to Gq protein and produce a calcium response. (b and c) Dose-response curves to human calcitonin for the two responses in cell lines expressing the two different levels of receptor. Effects on cyclic AMP levels (open circles; left-hand ordinal axes) and calcium entry (filled squares; right-hand ordinal axes) for HEK cells expressing calcitonin receptors at 65 fmole/mg (panel b) and 30,000 fmole/mg (panel c). Data redrawn from [10].

the responses to agonists since low receptor densities will produce less response than higher densities. Experimental control of this factor can be achieved in recombinant systems. The methods of doing this are discussed more fully in Chapter 5. Figure 2.20 shows the cyclic AMP and calcium responses to human calcitonin activating calcitonin receptors in human embryonic kidney cells. Shown are responses from two different recombinant stable recombinant cell lines of differing receptor density. It can be seen that not only does the quantity of response change with increasing receptor number response (note ordinate scales for cyclic AMP production in Figure 2.20b and c) but the

quality of the response changes. Specifically, calcitonin is a pleiotropic receptor with respect to the G-proteins with which it interacts (this receptor can couple to Gs, Gi, and Gq proteins). In cells containing a low number of receptors, there is an insufficient density to activate Gq proteins and thus no Gq response (calcium signaling) is observed (see Figure 2.20b). However, in cells with a higher receptor density both a cyclic AMP and a calcium response (indicative of concomitant Gs and Gq protein activation) is observed (Figure 2.20c). In this way, the receptor density controls the overall composition of the cellular response to the agonist.

2.6 Receptor Desensitization and Tachyphylaxis

There is a temporal effect that must be considered in functional experiments; namely, the *desensitization* of the system to sustained or repeated stimulation. Receptor response is regulated by processes of phosphorylation and internalization, which can prevent overstimulation of physiological function in cells. This desensitization can be specific for a receptor, in which case it is referred to as *homologous desensitization*, or it can be related to modulation of a pathway common to more than one receptor and thus be *heterologous desensitization*. In this latter case, repeated stimulation of one receptor may cause the reduction in responsiveness of a number of receptors. The effects of desensitization on agonist dose-response curves are not uniform. Thus, for powerful highly efficacious agonists desensitization can cause a dextral displacement of the dose-response with no diminution of maximal response (see Figure 2.21a). In contrast, desensitization can cause a depression of the maximal response to weak partial agonists (see Figure 2.21b). The overall effects of desensitization on dose-response curves relate to the effective receptor reserve for the agonist in a particular system. If the desensitization process eliminates receptor responsiveness where it is essentially irreversible in terms of the timescale of response (i.e., response occurs in seconds whereas reversal from desensitization may require hours), then the desensitization process will mimic the removal of active receptors from the tissue. Therefore, for an agonist with a high receptor reserve (i.e., only a small portion of the receptors are required for production of maximal tissue response) desensitization will not depress the maximal response until a proportion greater than the reserve is affected. In contrast, for an agonist with no receptor reserve desensitization will produce an immediate decrease in the maximal response. These factors can be relevant to the choice of agonists for therapeutic application. This is discussed more fully in Chapter 10.

2.7 The Measurement of Drug Activity

In general there are two major formats for pharmacological experiments: cellular function and biochemical binding. Historically, function has been by far the more prevalent form of experiment. From the turn of the century, isolated tissues have been used to detect and quantify drug activity. Pioneers such as Rudolph Magnus (1873–1927) devised methods of preserving the physiological function of isolated tissues (i.e., isolated intestine) to allow the observation of drug-induced response. Such preparations formed the backbone of all in vitro pharmacological experimental observation and furnished the data to develop drug receptor theory. Isolated tissues were the workhorses of pharmacology and various laboratories had their favorite. As put by W. D. M Paton [9]:

> The guinea pig longitudinal muscle is a great gift to the pharmacologist. It has low spontaneous activity; nicely graded responses (not too many tight junctions); is highly sensitive to a very wide range of stimulants; is tough, if properly handled, and capable of hours of reproducible behavior.
>
> —W. D. M. Paton (1986)

All of drug discovery relied upon such functional assays until introduction of binding techniques. Aside from the obvious shortcoming of using animal tissue to predict human responsiveness to drugs, isolated tissue formats did not allow for high-throughput screening of compounds (i.e., the experiments were labor intensive). Therefore, the numbers of compounds that could be tested for potential activity were limited by the assay format. In the mid 1970s, a new technology (in the form of biochemical binding) was introduced and this quickly became a major approach to the study of drugs. Both binding and function are valuable and have unique application and it is worth considering the strengths and shortcomings of both approaches in the context of the study of drug-receptor interaction.

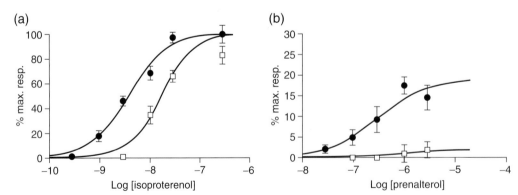

FIGURE 2.21   Effects of desensitization on inotropic responses of guinea pig atria to isoproterenol (panel a) and prenalterol (panel b). Ordinates: response as a percent of the maximal reaponse to isoproterenol. Abscissae: logarithms of molar concentrations of agonist (log scale). Responses shown after peak response attained (within 5 minutes, filled circles) and after 90 minutes of incubation with the agonist (open triangles). Data redrawn from [6].

2.8 Advantages and Disadvantages of Different Assay Formats

High-throughput volume was the major reason for the dominance of binding in the 1970s and 1980s. However, technology has now progressed to the point where the numbers of compounds tested in functional assays can equal or even exceed the volume that can be tested in binding studies. Therefore, this is an obsolete reason for choosing binding over function and the relative scientific merits of both assay formats can now be used to make the choice of assay for drug discovery. There are advantages and disadvantages to both formats. In general, binding assays allow the isolation of receptor systems by use of membrane preparations and selective radioligand (or other traceable ligands; see material following) probes. The interference with the binding of such a probe can be used as direct evidence of an interaction of the molecules with the receptor. In contrast, functional studies in cellular formats can be much more complex in that the interactions may not be confined to the receptor but rather extend further into the complexities of cellular functions. Since these may be cell-type dependent, some of this information may not be transferable across systems and therefore not useful for prediction of therapeutic effects. However, selectivity can be achieved in functional assays through the use of selective agonists. Thus, even in the presence of mixtures of functional receptors a judicious choice of agonist can be used to select the receptor of interest and reduce nonspecific signals.

In binding, the molecules detected are only those that interfere with the specific probe chosen to monitor receptor activity. There is a potential shortcoming of binding assays in that often the pharmacological probes used to monitor receptor binding are not the same probes that are relevant to receptor function in the cell. For example, there are molecules that may interfere with the physiological relevant receptor probe (the G-proteins that interact with the receptor and control cellular response to activation of that receptor) but not with the probe used for monitoring receptor binding. This is true for a number of interactions generally classified as *allosteric* (*vide infra*; see Chapters 4 and 7 for details) interactions. Specifically, allosteric ligands do not necessarily interact with the same binding site as the endogenous ligand (or the radioligand probe in binding) and therefore binding studies may not detect them.

Receptor levels in a given preparation may be insufficient to return a significant binding signal (i.e., functional responses are highly amplified and may reveal receptor presence in a more sensitive manner than binding). For example, CHO cells show a powerful 5-HT$_{1B}$ receptor-mediated agonist response to 5-HT that is blocked in nanomolar concentrations by the antagonist ($\pm$)-cyanopindolol [11]. However, no significant binding of the radioligand [^{125}I]-iodocyanopindolol is observed. Therefore, in this case the functional assay is a much more sensitive indicator of 5-HT responses. The physiological relevant probe (one that affects the cellular metabolism) can be monitored by observing cellular function.

Therefore, it can be argued that functional studies offer a broader scope for the study of receptors than do binding studies. Another major advantage of function over binding is the ability of the former, and not the latter, to directly observe ligand efficacy. Binding only registers the presence of the ligand bound to the receptor but does not return the amount of stimulation that the bound agonist imparts to the system.

In general, there are advantages and disadvantages to both assay formats and both are widely employed in pharmacological research. The specific strengths and weaknesses inherent in both approaches are discussed in more detail in Chapters 4 and 5. As a preface to the consideration of these two major formats, a potential issue with both of them should be considered; namely, dissimulations between the concentrations of drugs added to the experimentally accessible receptor compartment and the actual concentration producing the effect.

2.9 Drug Concentration as an Independent Variable

In pharmacological experiments the independent variable is drug concentration and the dependent (observed) variable is tissue response. Therefore, all measures of drug activity, potency, and efficacy are totally dependent on accurate knowledge of the concentration of drug at the receptor producing the observed effect. With no knowledge to the contrary, it is assumed that the concentration added to the receptor system by the experimenter is equal to the concentration acting at the receptor (i.e., there is no difference in the magnitude of the independent variable). However, there are potential factors in pharmacological experiments that can negate this assumption and thus lead to serious error in the measurement of drug activity. One is error in the concentration of the drug that is able to reach the receptor.

2.9.1 Dissimulation in Drug Concentration

The receptor compartment is defined as the aqueous volume containing the receptor and cellular system. It is assumed that free diffusion leads to ready access to this compartment (i.e., that the concentration within this compartment is the free concentration of drug at the receptor). However, there are factors that can cause differences between the experimentally accessible liquid compartment and the actual receptor compartment. One obvious potential problem is limited solubility of the drug being added to the medium. The assumption is made tacitly that the dissolved drug in the stock solution, when added to the medium bathing the pharmacological preparation, will stay in solution. There are cases where this may not be a valid assumption.

Many drug-like molecules have aromatic substituents and thus have limited aqueous solubility. A routine practice is to dissolve stock drugs in a solvent known to dissolve many types of molecular structures. One such solvent is

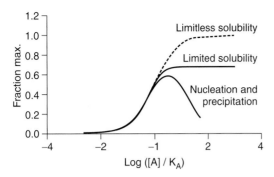

FIGURE 2.22 Theoretical effects of agonist insolubility on dose-response curves. Sigmoidal curve partially in dotted lines shows the theoretically ideal curve obtained when the agonist remains in solution throughout the course of the experiment determining the dose-response relationship. If a limit to the solubility is reached, then the responses will not increase beyond the point at which maximal solubility of the agonist is attained (labeled limited solubility). If the precipitation of the agonist in solution causes nucleation that subsequently causes precipitation of the amount already dissolved in solution, then a diminution of the previous response may be observed.

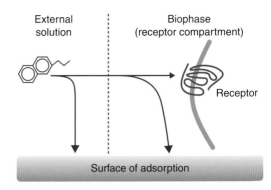

FIGURE 2.23 Schematic diagram showing the routes of possible removal of drug from the receptor compartment. Upon diffusion into the compartment, the drug may be removed by passive adsorption en route. This will cause a constant decrease in the steady-state concentration of the drug at the site of the receptor until the adsorption process is saturated.

dimethylsulfoxide (DMSO). This solvent is extremely useful because physiological preparations such as cells in culture or isolated tissues can tolerate relatively high concentrations of DMSO (i.e., 0.5 to 2%) with no change in function. When substances dissolved in one solvent are diluted into another solvent where the substance has different (less) solubility, local concentration gradients may exceed the solubility of the substance in the mixture. When this occurs, the substance may begin to come out of solution in these areas of limited solubility (i.e., microcrystals may form). This may in turn lead to a phenomenon known as nucleation, whereby the microcrystals form the seeds required for crystallization of the substance from the solution. The result of this process can be the complete crystallization of the substance from the entire mixture. For this reason, the dilution into the solution of questionable solubility (usually the aqueous physiological salt solution) should be done at the lowest concentration possible to ensure against nucleation and potential loss of solubility of the drug in the pharmacological medium. All dilutions of the stock drug solution should be carried out in the solution of maximal solubility, usually pure DMSO and the solution for pharmacological testing taken directly from these stocks. Even under these circumstances, the drug may precipitate out of the medium when added to the aqueous medium. Figure 2.22 shows the effects of limited solubility on a dose-response curve to an agonist. Solubility limits are absolute. Thus, once the limit is reached no further addition of stock solution will result in an increased soluble drug concentration. Therefore, the response at that solubility limit defines the maximal response for that preparation. If the solubility is below that required for the true maximal response to be observed (dotted line Figure 2.20), then an erroneously truncated response to the drug will be

observed. A further effect on the dose-response curve can be observed if the drug, upon entering the aqueous physiological solution, precipitates because of local supersaturated concentration gradients. This could lead to nucleation and subsequent crystallization of the drug previously dissolved in the medium. This would reduce the concentration below the previously dissolved concentration and lead to a decrease in the maximal response (bell-shaped dose-response curve, Figure 2.22).

Another potential problem causing differences in the concentration of drug added to the solution (and that reaching the receptors) is the sequestration of drug in regions other than the receptor compartment (Figure 2.23). Some of these effects can be due to active uptake or enzymatic degradation processes inherent in the biological preparation. These are primarily encountered in isolated whole tissues and are not a factor in in vitro assays comprised of cellular monolayers. However, another factor that is common to nearly all in vitro systems is the potential adsorption of drug molecules onto the surface of the vessel containing the biological system (i.e., well of a cell culture plate). The impact of these mechanisms depends on the drug and the nature of the surface, being more pronounced for some chemical structures and also more pronounced for some surfaces (i.e., non-silanized glass). Table 2.1 shows the striking differences in adsorption of [^{3}H]-endorphin with pretreatment of the surface with various agents. It can be seen that a difference of over 99.9% can be observed when the surface is treated with a substance that prevents adsorption such as myelin basic protein.

2.9.2 Free Concentration of Drug

If the adsorption process is not saturable within the concentration range of the experiment, it becomes a sink claiming a portion of the drug added to the medium—the magnitude of which is dependent on the maximal capacity of the sink ($[\Omega]$) and the affinity of the ligand for the site

TABLE 2.1

Effect of pretreatment of surface on adsorption of [³H]-endorphin.

Treatment	fmole Adsorbed	% Reduction over Lysine Treatment
Lysine	615	0
Arginine	511	16.9
Bovine serum albumin	383	38
Choline chloride	19.3	97
Polylysine	1.7	99.5
Myelin basic protein	1.5	99.9

Data from [12].

of adsorption ($1/K_{ad}$, where K_{ad} is the equilibrium dissociation constant of the ligand-adsorption site complex). The receptor then interacts with the remaining free concentration of drug in the compartment. The free concentration of drug, in the presence of an adsorption process, is given as (see Section 2.11.4):

$$[A_{free}] = [A_T] - \frac{1}{2}\left\{[A_T] + K_{ad} + \Omega \right.$$
$$\left. - \sqrt{([A_T^*] + K_{ad} + \Omega)^2 - 4[A_T]\Omega}\right\}. \quad (2.3)$$

The free concentration of a drug [A_{free}] in a system containing an adsorption process with maximal capacity ranging from 0.01 to 10 μM and for which the ligand has an affinity ($1/K_d$) is shown in Figure 2.24a. It can be seen that there is a constant ratio depletion of free ligand in the medium at low concentrations until the site of adsorption begins to be saturated. When this occurs, there is a curvilinear portion of the line reflecting the increase in the

free concentration of ligand in the receptor compartment due to cancellation of adsorption-mediated depletion (adsorption sites are fully bound and can no longer deplete ligand). It is useful to observe the effects such processes can have on dose-response curves to drugs. Figure 2.24b shows the effect of an adsorption process on the observed effects of an agonist in a system where an adsorption process becomes saturated at the higher concentrations of agonist. It can be seen that there is a change of shape of the dose-response curve (increase in Hill coefficient with increasing concentration). This is characteristic of the presence of an agonist removal process that is saturated at some point within the concentration range of agonist used in the experiment.

In general, it should be recognized that the most carefully designed experimental procedure can be completely derailed by processes causing differences in what is thought to be the concentration of drug at the receptor and the actual concentration producing the effect. Insofar as experiments can be done to indicate that these effects are not operative in a given experiment they should be.

2.10 Chapter Summary and Conclusions

- It is emphasized that drug activity is observed through a translation process controlled by cells. The aim of pharmacology is to derive system-independent constants characterizing drug activity from the indirect product of cellular response.
- Different drugs have different inherent capacities to induce response (intrinsic efficacy). Thus, equal cellular responses can be achieved by different fractional receptor occupancies of these drugs.

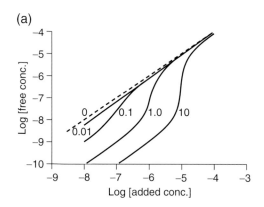
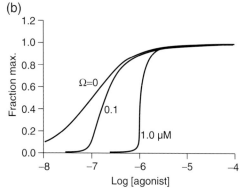

FIGURE 2.24 Effects of a saturable adsorption process on concentrations of agonist (panel a) and dose-response curves to agonists (panel b). (a) Concentrations of drug added to system (abscissae, log scale) versus free concentration in solution (ordinates, log scale). Numbers next to curves indicate the capacity of the adsorption process in μM. The equilibrium dissociation constant of the agonist/adsorption site is 10 nM. Dotted line indicates no difference between added concentrations and free concentration in solution. (b) Effect of a saturable adsorption process on agonist dose-response curves. Numbers next to curves refer to the maximal capability of the adsorption process. The equilibrium dissociation constant of the agonist/adsorption site is 0.1 μM. Curve furthest to the left is the curve with no adsorption taking place.

- Some cellular stimulus-response pathways and second messengers are briefly described. The overall efficiency of receptor coupling to these processes is defined as the stimulus-response capability of the cell.
- While individual stimulus-response pathways are extremely complicated, they all can be mathematically described with hyperbolic functions.
- The ability to reduce stimulus-response mechanisms to single monotonic functions allows relative cellular response to yield receptor-specific drug parameters.
- When the maximal stimulus-response capability of a given system is saturated by agonist stimulus, the agonist will be a full agonist (produce full system response). Not all full agonists are of equal efficacy; they only all saturate the system.
- In some cases, the stimulus-response characteristics of a system can be manipulated to provide a means to compare maximal responses of agonists (efficacy).
- Receptor desensitization can have differing overall effects on high- and low-efficacy agonists.
- All drug parameters are predicated on an accurate knowledge of the concentration of drug acting at the receptor. Errors in this independent variable negate all measures of dependent variables in the system.
- Adsorption and precipitation are two commonly encountered sources of error in drug concentration.

2.11 Derivations

- Series hyperbolae can be modeled by a single hyperbolic function (2.11.1)
- Successive rectangular hyperbolic equations necessarily lead to amplification (2.11.2)
- Saturation of any step in a stimulus cascade by two agonists leads to identical maximal final responses for the two agonists (2.11.3)
- Measurement of free drug concentration in the receptor compartment (2.11.4)

2.11.1 Series Hyperbolae Can Be Modeled by a Single Hyperbolic Function

Rectangular hyperbolae are of the general form

$$y = \frac{Ax}{x + B}. \tag{2.4}$$

Assume a function

$$y_1 = \frac{x}{x + \beta_1}, \tag{2.5}$$

where the output y_1 becomes the input for a second function of the form

$$y_2 = \frac{y_1}{y_1 + \beta_2}. \tag{2.6}$$

It can be shown that a series of such functions can be generalized to the form

$$y_n = \frac{x}{x(1 + \beta_n(1 + \beta_{n-1}(1 + \beta_{n-2}(1 + \beta_{n-3})..)..)...) + (\beta_n \bullet ..\beta_1)}, \tag{2.7}$$

which can be rewritten in the form of Equation 2.4, where $A = (1 + \beta_n(1 + \beta_{n-1}(1 + \beta_{n-2}(1 + \beta_{n-3})..)...)..)^{-1}$ and $B = (\beta_n...\bullet..\beta_1)/(1 + \beta_n(1 + \beta_{n-1}(1 + \beta_{n-2})..)...)..)$. Thus, it can be seen that the product of a succession of rectangular hyperbolae is itself a hyperbola.

2.11.2 Successive Rectangular Hyperbolic Equations Necessarily Lead to Amplification

Assume a rectangular hyperbola of the form

$$\rho_1 = \frac{[A]}{[A] + K_A}, \tag{2.8}$$

where $[A]$ is the molar concentration of drug and K_A is the location parameter of the dose-response curve along the concentration axis (the potency). Assume also a second rectangular hyperbola where the input function is defined by Equation 2.8:

$$\rho_2 = \frac{[A]/([A] + K_A)}{([A]/([A] + K_A)) + \beta}. \tag{2.9}$$

The term β is the coupling efficiency constant for the second function. The location parameter (potency) of the second function (denoted K_{obs}) is given by

$$K_{obs} = \frac{K_A \beta}{1 + \beta}. \tag{2.10}$$

It can be seen that for non-zero and positive values of β that $K_{obs} < K_A$ (i.e., the potency of the overall process will be greater than the potency for the initial process).

2.11.3 Saturation of Any Step in a Stimulus Cascade by Two Agonists Leads to Identical Maximal Final Responses for the Two Agonists

For a given agonist $[A]$, the product of any one reaction in the stimulus response cascade is given by

$$\text{Output}_1 = \frac{[A] \cdot M_1}{[A] + \beta_1}, \tag{2.11}$$

where M_1 is the maximal output of the reaction and β_1 is the coupling constant for the reaction. When this product becomes the substrate for the next reaction, the output becomes

$$\text{Output}_2 = \frac{[A] \cdot M_1 M_2}{[A](M_1 + \beta_2) + \beta_1 \beta_2}. \tag{2.12}$$

The maximal output from this second reaction (i.e., as $[A] \to \infty$) is

$$Max_2 = \frac{M_1 M_2}{M_1 + \beta_2}. \qquad (2.13)$$

By analogy, the maximal output from the second reaction for another agonist $[A']$ is

$$Max_2' = \frac{M_1' M_2}{M_1' + \beta_2}. \qquad (2.14)$$

The relative maximal responses for the two agonists are therefore

$$Relative\ Maxima = \frac{Max_2}{Max_2'} = \frac{1 + \beta_2/M_1'}{1 + \beta_2/M_1}. \qquad (2.15)$$

It can be seen from this equation that if $M_1 = M_1'$ (i.e., if the maximal response to two agonists in any previous reaction in the cascade is equal) the relative maxima of the two agonists in subsequent reactions will be equal ($Max_2/Max_2' = 1$).

2.11.4 Measurement of Free Drug Concentration in the Receptor Compartment

Assume that the total drug concentration $[A_T]$ is the sum of the free concentration $[A_{free}]$ and the concentration bound to a site of adsorption $[AD]$ (therefore, $[A_{free}] = [A_T] - [AD]$). The mass action equation for adsorption is

$$[AD] = \frac{([A_T] - [AD])\Omega}{[A_T] - [AD] + K_{ad}}, \qquad (2.16)$$

where the maximal number of adsorption sites is Ω and the equilibrium dissociation constant of the drug site of adsorption is K_{ad}. Equation 2.16 results in the quadratic equation

$$[AD]^2 - [AD](\Omega + [A_T] + K_{ad}) + [A_T]\Omega = 0, \qquad (2.17)$$

one solution for which is

$$\frac{1}{2}\left\{ [A_T] + K_{ad} + \Omega - \sqrt{\left([A_T^*] + K_{ad} + \Omega\right)^2 - 4[A_T]\Omega} \right\}. \qquad (2.18)$$

Since $[A_{free}] = [A_T] - [AD]$, then

$$[A_{free}] = [A_T] - \frac{1}{2}\left\{ [A_T] + K_{ad} + \Omega \right.$$
$$\left. - \sqrt{\left([A_T^*] + K_{ad} + \Omega\right)^2 - 4[A_T]\Omega} \right\}. \qquad (2.19)$$

References

1. Chen, W.-J., Armour, S., Way, J., Chen, G. C., Watson, C., Irving, P. E., Cobb, J., Kadwell, S., Beaumont, K., Rimele, T., and Kenakin, T. P. (1997). Expression cloning and receptor pharmacology of human calcitonin receptors from MCF-7 cells and their relationship to amylin receptors. *Mol. Pharmacol.* **52**:1164–1175.

2. Kenakin, T. P., and Cook, D. A. (1976). Blockade of histamine-induced contractions of intestinal smooth muscle by irreversibly acting agents. *Can. J. Physiol. Pharmacol.* **54**:386–392.

3. Wilson, S., Chambers, J. K., Park, J. E., Ladurner, A., Cronk, D. W., Chapman, C. G., Kallender, H., Browne, M. J., Murphy, G. J., and Young, P. W. (1996). Agonist potency at the cloned human beta-3 adrenoceptor depends on receptor expression level and nature of assay. *J. Pharmacol. Exp. Ther.* **279**:214–221.

4. Goldberg, N. D. (1975). Cyclic nucleotides and cell function. In: *Cell membranes, biochemistry, cell biology, and pathology.* edited by G. Weissman and R. Claiborne, pp. 185–202. H. P. Publishing, New York.

5. Kenakin, T. P., and Beek, D. (1980). Is prenalterol (H 133/80) really a selective beta-1 adrenoceptor agonist? Tissue selectivity resulting selective beta-1 adrenoceptor agonist? Tissue selectivity resulting from difference in stimulus-response relationships. *J. Pharmacol. Exp. Ther.* **213**:406–413.

6. Kenakin, T. P., Ambrose, J. R., and Irving, P. E. (1991). The relative efficiency of beta-adrenoceptor coupling to myocardial inotropy and diastolic relaxation: Organ-selective treatment of diastolic dysfunction. *J. Pharmacol. Exp. Ther.* **257**:1189–1197.

7. Kenakin, T. P., and Beek, D. (1984). The measurement of the relative efficacy of agonists by selective potentiation of tissue responses: Studies with isoprenaline and prenalterol in cardiac tissue. *J. Auton. Pharmacol.* **4**:153–159.

8. Burgen, A. S. V., and Spero, L. (1968). The action of acetylcholine and other drugs on the efflux of potassium and rubidium from smooth muscle of the guinea-pig intestine. *Br. J. Pharmacol.*, **34**:99–115.

9. Paton, W. D. M. (1986). On becoming a pharmacologist. *Ann. Rev. Pharmacol. and Toxicol.* **26**:1–22.

10. Kenakin, T. P. (1997). Differences between natural and recombinant G-protein coupled receptor systems with varying receptor/G-protein stoichiometry. *Trends Pharmacol. Sci.* **18**:456–464.

11. Giles, H., Lansdell, S. J., Bolofo, M.-L., Wilson, H. L., and Martin, G. R. (1996). Characterization of a 5-HT$_{1B}$ receptor on CHO cells: Functional responses in the absence of radioligand binding. *Br. J. Pharmacol.* **117**:1119–1126.

12. Ferrar, P., and Li, C. H. (1980). β-endorphin: Radioreceptor binding assay. *Int. J. Pept. Protein Res.* **16**:66–69.

3

Drug-Receptor Theory

What is it that breathes fire into the equations and makes a universe for them to describe?
— STEPHEN W. HAWKING

An equation is something for eternity...
— ALBERT EINSTEIN

Casual observation made in the course of a purely theoretical research has had the most important results in practical medicine....

Saul was not the last who, going forth to see his father's asses, found a kingdom.
— ARTHUR ROBSERTSON CUSHNY (1866–1926)

3.1 About This Chapter

This chapter discusses the various mathematical models that have been put forward to link the experimental observations (relating to drug-receptor interactions) and the events taking place on a molecular level between the drug and protein recognition sites. A major link between the data and the biological understanding of drug-receptor activity is the model. In general, experimental data is a sampling of a population of observations emanating from a system. The specific drug concentrations tested control the sample size and the resulting dependent variables reflect what is happening at the biological target. A model defines the complete relationship for the whole population (i.e., for an infinite number of concentrations). The choice of model, and how it fits into the biology of what is thought to be occurring, is critical to the assessment of the experiment. For example, Figure 3.1a shows a set of dose-response data fit to two mathematical functions. It can be seen that both

equations appear to adequately fit the data. The first curve is defined by

$$y = 78\left(1 - e^{-\left(0.76\left([A]^{0.75}\right)\right)}\right) - 2. \tag{3.1}$$

This is simply a collection of constants in an exponential function format. The constants cannot be related to the interactions at a molecular level. In contrast, the refit of the data to the Langmuir adsorption isotherm

$$y = \frac{80 \cdot [A]}{[A] + EC_{50}} \tag{3.2}$$

allows some measure of interpretation (i.e., the location parameter along the concentration axis may reflect affinity and efficacy while the maximal asymptote may reflect efficacy; Figure 3.1b). In this case, the model built on chemical concepts allows interpretation of the data in molecular terms. The fitting of experimental data to equations derived from models of receptor function are at least consistent with the testing and refinement of these models with the resulting further insight into biological behavior. An early proponent of using such models and laws to describe the very complex behavior of physiological systems was A. J. Clark, known as the originator of receptor pharmacology. As put by Clark in his monograph *The Mode of Action of Drugs on Cells* [1]:

> The general aim of this author in this monograph has been to determine the extent to which the effects produced by drugs on cells can be interpreted as processes following known laws of physical chemistry.
> — A. J. Clark (1937)

A classic example of where definitive experimental data necessitated refinement and extension of a model of drug-receptor interaction involved the discovery of constitutive receptor activity in GPCR systems. The state of the art model before this finding was the ternary complex model for GPCRs, a model that cannot accommodate ligand-independent (constitutive) receptor activity.

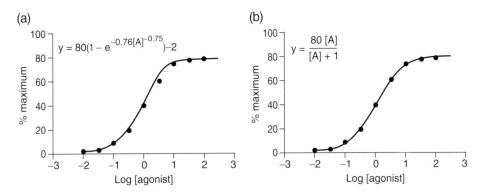

FIGURE 3.1 Data set fit to two functions of the same general shape. (a) Function fit to the exponential Equation 3.1. (b) Function fit to rectangular hyperbola of the form 80*[A]/([A] + 1).

With the experimental observation of constitutive activity for GPCRs by Costa and Herz [2], a modification was needed. Subsequently, Samama and colleagues [3] presented the extended ternary complex model to fill the void. This chapter discusses relevant mathematical models and generally offers a linkage between empirical measures of activity and molecular mechanisms.

3.2 Drug-Receptor Theory

The various equations used to describe the quantitative activity of drugs and the interaction of those drugs with receptors is generally given the name drug receptor theory. The models used within this theory originated from those used to describe enzyme kinetics. A. J. Clark is credited with applying quantitative models to drug action. His classic books *The Mode of Action of Drugs on Cells* [1] and *Handbook of Experimental Pharmacology* [4] served as the standard texts for quantitative receptor pharmacology for many years.

> A consideration of the more striking examples of specific drug antagonisms shows that these in many cases follow recognizable laws, both in the case of enzymes and cells.
> — A. J. Clark (1937)

With increasing experimental sophistication has come new knowledge of receptor function and insights into the ways in which drugs can affect that function. In this chapter, drug receptor theory is described in terms of what is referred to as "classical theory"; namely, the use and extension of concepts described by Clark and other researchers such as Stephenson [5], Ariens [6, 7], MacKay [8], and Furchgott [9, 10]. In this sense, classical theory is an amalgam of ideas linked chronologically. These theories were originated to describe the functional effects of drugs on isolated tissues and thus naturally involved functional physiological outputs. Another model used to describe functional drug activity, derived by Black and Leff [11], is termed the operational model. Unlike classical theory, this model makes no assumptions about the intrinsic ability of drugs to produce response. The operational model is a very important new tool in receptor pharmacology and is used throughout this book to illustrate receptor methods and concepts. Another model used primarily to describe the function of ion channels is termed two-state theory. This model contributed ideas essential to modern receptor theory, specifically in the description of drug efficacy in terms of the selective affinity for protein conformation. Finally, the idea that proteins translocate within cell membranes [12] and the observation that seven transmembrane receptors couple to separate G-proteins in the membrane led to the ternary complex model. This scheme was first described by De Lean and colleagues [13] and later modified to the extended ternary complex model by Samama and co-workers [3]. These are described separately as a background to discussion of drug-receptor activity and as context for the description of the quantitative tools and methods used in receptor pharmacology to quantify drug effect.

3.3 The Use of Mathematical Models in Pharmacology

Mathematical models are the link between what is observed experimentally and what is thought to occur at the molecular level. In physical sciences, such as chemistry, there is a direct correspondence between the experimental observation and the molecular world (i.e., a nuclear magnetic resonance spectrum directly reflects the interaction of hydrogen atoms on a molecule). In pharmacology the observations are much more indirect, leaving a much wider gap between the physical chemistry involved in drug-receptor interaction and what the cell does in response to those interactions (through the "cellular veil"). Hence, models become uniquely important.

There are different kinds of mathematical models, and they can be classified in two ways: by their complexity and by the number of estimatable parameters they use. The most simple models are cartoons with few very parameters.

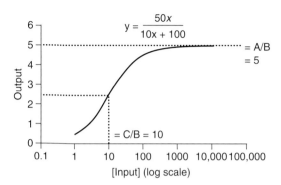

FIGURE 3.2 General curve for an input/output function of the rectangular hyperbolic form ($y = 50x/(10x + 100)$). The maximal asymptote is given by A/B and the location parameter (along the x axis) is given by C/B (see text).

These, such as the "black box" that was the receptor at the turn of the century, usually are simple input/output functions with no mechanistic description (i.e., the drug interacts with the receptor and a response ensues). Another type, termed the Parsimonious model, is also simple but has a greater number of estimatable parameters. These do not completely characterize the experimental situation completely but do offer insights into mechanism. Models can be more complex as well. For example, complex models with a large number of estimatable parameters can be used to simulate behavior under a variety of conditions (simulation models). Similarly, complex models for which the number of independently verifiable parameters is low (termed heuristic models) can still be used to describe complex behaviors not apparent by simple inspection of the system.

In general, a model will express a relationship between an independent variable (input by the operator) and one or more dependent variables (output, produced by the model). A ubiquitous form of equation for such input/output functions are curves of the rectangular hyperbolic form. It is worth illustrating some general points about models with such an example. Assume that a model takes on the general form

$$\text{Output} = \frac{[\text{Input}] \cdot A}{B \cdot [\text{Input}] + C}. \tag{3.3}$$

The form of that function is shown in Figure 3.2. There are two specific parameters that can be immediately observed from this function. The first is that the maximal asymptote of the function is given solely by the magnitude of A/B. The second is that the location parameter of the function (where it lies along the input axis) is given by C/B. It can be seen that when [Input] equals C/B the output necessarily will be 0.5. Therefore, whatever the function the midpoint of the curve will lie on a point at Input = C/B. These ideas are useful since they describe two essential behaviors of any drug-receptor model; namely, the maximal response (A/B) and the potency (concentration of input required for effect; C/B). Many of the complex equations

used to describe drug-receptor interaction can be reduced to these general forms and the maxima and midpoint values used to furnish general expressions for the dependence of efficacy and potency on the parameters of the mechanistic model used to furnish the equations.

3.4 Some Specific Uses of Models in Pharmacology

Models can be very useful in designing experiments, predicting drug effect and describing complex systems. Ideally, models should be comprised of species that can be independently quantified. Also, the characteristics of the processes that produce changes in the amounts of these species should be independently verifiable. The difference between a heuristic model and a simulation model is that the latter has independently verifiable constants for at least some of the processes. An ideal model also has internal checks that allow the researcher to determine that the calculation is or is not following predicted patterns set out by the model. A classic example of an internal check for a model is the linearity and slope of a Schild regression for simple competitive antagonism (see Chapter 6). In this case, the calculations must predict a linear regression of linear slope or the model of simple competitive antagonism is not operable. The internal check determines the applicability of the model.

Models can also predict apparently aberrant behaviors in systems that may appear to be artifactual (and therefore appear to denote experimental problems) but are in fact perfectly correct behaviors according to a given complex system. Simulation with modeling allows the researcher to determine if the data is erroneous or indicative of a correct system activity. For example, consider a system in which the receptors can form dimers and where the affinity of a radioligand (radioactive molecule with affinity for the receptor allowing measurement of ligand-receptor complex binding to be measured) is different for the single receptor and the dimer. It is not intuitively obvious how the system will behave when a nonradioactive ligand that also binds to the receptor is added. In a standard single receptor system, preincubation with a radioligand followed by addition of a nonradioactive ligand will produce displacement of the radioligand. This will cause a decrease in the bound radioactive signal. The result usually is a sigmoidal dose-response curve for displacement of the radioligand by the nonradioactive ligand (see Figure 3.3). This is discussed in some detail in Chapter 4. The point here is that addition of the same nonradioactive ligand to a system of pre-bound radioligand would be expected to produce a decrease in signal. However, in the case of dimerization if the combination of two receptors forms a "new" receptor of higher affinity for the radioligand addition of a nonradioligand may actually increase the amount of radioligand bound before it decreases it [14]. This is an apparent paradox (addition of a nonradioactive species actually increasing the binding of radioactivity to a receptor). The equation for the amount of radioactive ligand [A*] bound (signal denoted ω) in the presence of a range of concentrations of nonradioactive ligand [A] is (derivation 3.13.1) given as

$$\omega = \frac{\left([A^*]/K_d + \alpha[A^*][A]/K_d^2 + 2\alpha([A^*]/K_d)^2\right)\left(1 + [A^*]/K_d + \alpha([A^*]/K_d)^2\right)}{\left(1 + [A]/K_d + [A^*]/K_d + \alpha([A^*][A]/K_d^2 + \alpha([A^*]/K_d)^2 + \alpha([A]/K_d)^2\right)\left([A^*]/K_d + 2\alpha([A^*]/K_d)^2\right)} \cdot$$

(3.4)

As shown in Figure 3.3, addition of the nonradioactive ligand to the system can increase the amount of bound radioactivity for a system where the affinity of the ligand is higher for the dimer than it is for the single receptor. The prediction of this effect by the model changes the interpretation of a counterintuitive finding to one that conforms to the experimental system. Without the benefit of the modeling, an observation of an increased binding of radioligand with the addition of a nonradioactive ligand might have been interpreted erroneously.

Models also can assist in experimental design and the determination of the limits of experimental systems. For example, it is known that three proteins mediate the interaction of HIV with cells; namely, the chemokine receptor CCR5, the cellular protein CD4, and the viral coat protein gp120. An extremely useful experimental system to study this interaction is one in which radioactive CD4, prebound to soluble gp120, is allowed to bind to cellular receptor CCR5. This system can be used to screen for

potential drugs that may block this interaction and thus be useful as a treatment for AIDS. One of the practical problems with this approach is the availability and expense of purified gp120. This reagent can readily be prepared in crude broths but very pure samples are difficult to obtain. A practical question then is to what extent would uncertainty in the concentration of gp120 affect an assay that examines the binding of a complex of radioactive CD4 and gp120 with the CCR5 receptor in the presence of potential drugs that block the complex. It can be shown in this case that the model of interaction predicts the following equation for the relationship between the concentrations of radioactive CD4 [CD], crude gp120 [gp], [CCR5], and the ratio of the observed potency of a displacing ligand [B] to its true potency (i.e., to what extent errors in the potency estimation will be made with errors in the true concentration of gp120; see Section 3.13.2):

$$K_4 = \frac{[IC_{50}]}{([CD]/K_1)([gp]/K_2) + 1},$$

(3.5)

where K_4, K_1, and K_2 are the equilibrium dissociation constants of the ligand [B], CD4, and gp120 and the site of interaction with CCR5/CD4/gp120. The relationship between the concentration of radioligand used in the assay and the ratio of the observed potency of the ligand in blocking the binding to the true potency is shown in Figure 3.4. The gray lines indicate this ratio with a 50% error in the concentration of gp120 (crude gp120 preparation). It can be seen from this figure that as long as the concentration of radioligand is kept below $[CD_4]/K_1 = 0.1$ differences between the assumed concentration of gp120 in the assay and true concentrations make little difference to the estimation of ligand potency. In this case, the model delineates experimental parameters for the optimal performance of the assay.

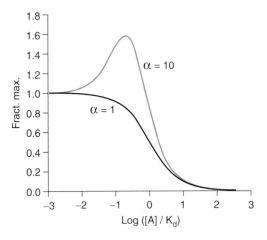

FIGURE 3.3 Displacement of prebound radioligand [A*] by non-radioactive concentrations of [A]. Curve for $\alpha = 1$ denotes no cooperativity in binding (i.e., formation of the receptor dimer does not lead to a change in the affinity of the receptor for either [A] or [A*]). The curve $\alpha = 10$ indicates a system whereby formation of the receptor dimer leads to a tenfold increase in the affinity for both [A*] and [A]. In this case, it can be seen that addition on the nonradioactive ligand [A] actually leads to an increase in the amount of radioligand [A*] bound before a decrease at higher concentrations of [A]. For this simulation $[A^*]/K_d = 0.1$.

3.5 Classical Model of Receptor Function

The binding of a ligand [A] to a receptor R is assumed to follow mass action according to the Langmuir adsorption isotherm (see Equation 1.4), as defined by Clark [1, 4]. No provision for different drugs of differing propensities to stimulate receptors was made until E. J. Ariens [6, 7] introduced a proportionality factor (termed *intrinsic activity* and denoted α in his terminology) was added to the binding function [5]. Intrinsic activity is the maximal response to an agonist expressed as a fraction of the

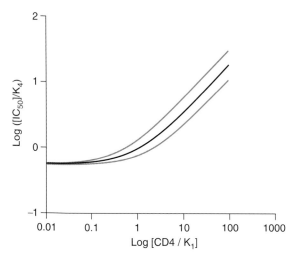

FIGURE 3.4 Errors in the estimation of ligand potency for displacement of radioactive CD4-gp120 complex (surrogate for HIV binding) as a function of the concentration of radioactive CD4 (expressed as a fraction of the equilibrium dissociation constant of the CD4 for its binding site). Gray lines indicate a 50% error in the concentration of gp120. It can be seen that very little error in the potency estimation of a displacing ligand is incurred at low concentrations of radioligand but that this error increases as the concentration of CD4 is increased.

maximal response for the entire system (i.e., $\alpha = 1$ indicates that the agonist produces the maximal response, $\alpha = 0.5$ indicates half the maximal response, and so on). An intrinsic activity of zero indicates no agonism. Within this framework, the equation for response is thus:

$$\text{Response} = \frac{[A]\alpha}{[A] + K_A}, \qquad (3.6)$$

where K_A is the equilibrium dissociation of the agonist-receptor complex. Note how in this scheme response is assumed to be a direct linear function of receptor occupancy multiplied by a constant. This latter requirement was seen to be a shortcoming of this approach since it was known that many nonlinear relationships between receptor occupancy and tissue response existed. This was rectified by Stephenson [5], who revolutionized receptor theory by introducing the abstract concept of *stimulus*. This is the amount of activation given to the receptor upon agonist binding. Stimulus is processed by the tissue to yield response. The magnitude of the stimulus is a function (denoted f in Equation 3.7) of another abstract quantity, referred to as *efficacy* (denoted e in Equation 3.7). Stephenson also assumed that the tissue response was some function (not direct) of stimulus. Thus, tissue response was given by

$$\text{Response} = \text{f(Stimulus)} = \text{f}\left[\frac{[A]e}{[A] + K_A}\right]. \qquad (3.7)$$

It can be seen that efficacy in this model is both an agonist and a tissue-specific term. Furchgott [9] separated the tissue

and agonist components of efficacy by defining a term *intrinsic efficacy* (denoted ε), which is a strictly agonist-specific term (i.e., this term defines the quantum stimulus given to a single receptor by the agonist). The product of receptor number ($[R_t]$) and intrinsic efficacy is then considered to be the agonist- and tissue-dependent element of agonism:

$$\text{Response} = \text{f}\left[\frac{[A] \cdot \varepsilon \cdot [R_t]}{[A] + K_A}\right]. \qquad (3.8)$$

The function f is usually hyperbolic, which introduces the nonlinearity between receptor occupancy and response. A common experimentally observed relationship between receptor stimulus and response is a rectangular hyperbola (see Chapter 2). Thus, response can be thought of as a hyperbolic function of stimulus:

$$\text{Response} = \frac{\text{Stimulus}}{\text{Stimulus} + \beta}, \qquad (3.9)$$

where β is a fitting factor representing the efficiency of coupling between stimulus and response. Substituting for stimulus from Equation 3.7 and rearranging, response in classical theory is given as

$$\text{Response} = \text{f}\left[\frac{[A][R_t]\varepsilon/\beta}{[A](([R_t]\varepsilon/\beta) + 1) + K_A}\right]. \qquad (3.10)$$

The various components of classical theory relating receptor occupancy to tissue response are shown schematically in Figure 3.5. It will be seen that this formally is identical to the equation for response derived in the operational model (see material following), where $\tau = [R_t]\varepsilon/\beta$.

It is worth exploring the effects of the various parameters on agonist response in terms of classical receptor theory. Figure 3.6 shows the effect of changing efficacy. It can be seen that increasing efficacy causes an increased maximal response with little shift to the left of the dose-response curves until the system maximal response is achieved. Once this occurs (i.e., the agonist is a full agonist in the system), increasing efficacy has no further effect on the maximal response but rather causes shifts to the left of the dose-response curves (Figure 3.6a). In contrast, changing K_A the equilibrium dissociation constant of the agonist-receptor complex has no effect on maximal response but only shifts the curves along the concentration axis (Figure 3.6b).

3.6 The Operational Model of Receptor Function

Black and Leff [11] presented a model, termed the operational model, that avoids the inclusion of ad hoc terms for efficacy. This model is based on the experimental observation that the relationship between agonist concentration and tissue response is most often hyperbolic. This allows for response to be expressed in terms of

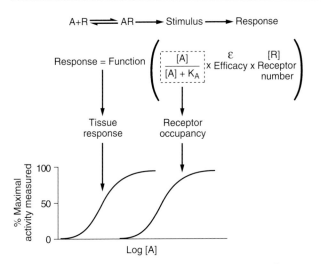

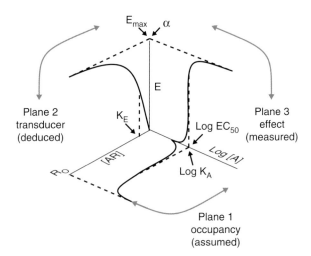

FIGURE 3.5 Major components of classical receptor theory. Stimulus is the product of intrinsic efficacy (ε), receptor number [R], and fractional occupancy as given by the Langmuir adsorption isotherm. A stimulus-response transduction function f translates this stimulus into tissue response. The curves defining receptor occupancy and response are translocated from each other by the stimulus-response function and intrinsic efficacy.

FIGURE 3.7 Principal components of the operational model. The 3D array defines processes of receptor occupation (plane 1), the transduction of the agonist occupancy into response (plane 2) in defining the relationship between agonist concentration, and tissue response (plane 3). The term α refers to the intrinsic activity of the agonist.

receptor and tissue parameters (see Section 3.13.3):

$$\text{Response} = \frac{[A] \cdot \tau \cdot E_{max}}{[A](\tau + 1) + K_A},\qquad(3.11)$$

where the maximal response of the system is E_{max}, the equilibrium dissociation constant of the agonist-receptor complex is K_A, and τ is the term that quantifies the power of the agonist to produce response (efficacy) and the ability of the system to process receptor stimulus into response. Specifically, τ is the ratio $[R_t]/K_E$, which is the receptor density divided by a transducer function expressing the ability of the system to convert agonist-receptor complex to response and the efficacy of the agonist. In this sense, K_E

resembles Stephenson's efficacy term except that it emanates from an experimental and pharmacological rationale (see Section 3.13.3). The essential elements of the operational model can be summarized graphically. In Figure 3.7, the relationship between agonist concentration and receptor binding (plane 1), the amount of agonist-receptor complex and response (plane 2), and agonist concentration and response (plane 3) can be seen. The operational model furnishes a unified view of receptor occupancy, stimulation, and production of response through cellular processing. Figure 3.8a shows the effects of changing τ on dose-response curves. It can be seen that the effects are identical to changes in efficacy in the classical model; namely, an increased maximal response of partial agonism until the

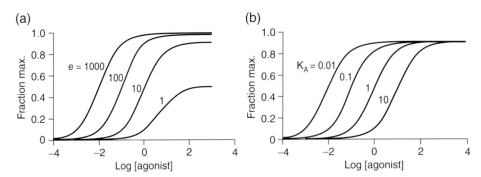

FIGURE 3.6 Classical model of agonism. Ordinates: response as a fraction of the system maximal response. Abscissae: logarithms of molar concentrations of agonist. (a) Effect of changing efficacy as defined by Stephenson [24]. Stimulus-response coupling defined by hyperbolic function Response = stimulus/(stimulus + 0.1). (b) Dose-response curves for agonist of e = 1 and various values for K_A.

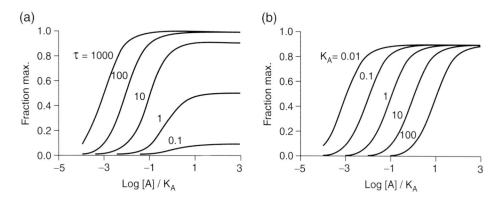

FIGURE 3.8 Operational model of agonism. Ordinates: response as a fraction of the system maximal response. Abscissae: logarithms of molar concentrations of agonist. (a) Effect of changing τ values. (b) Effect of changing K_A.

system maximal response is attained followed by sinistral displacements of the curves. As with the classical model, changes in K_A cause only changes in the location parameter of the curve along the concentration axis (Figure 3.8b).

The operational model, as presented, shows dose-response curves with slopes of unity. This pertains specifically only to stimulus-response cascades where there is no cooperativity and the relationship between stimulus ([AR] complex) and overall response is controlled by a hyperbolic function with slope = 1. In practice, it is known that there are experimental dose-response curves with slopes that are not equal to unity and there is no a priori reason for there not to be cooperativity in the stimulus-response process. To accommodate the fitting of real data (with slopes not equal to unity) and the occurrence of stimulus-response cooperativity, a form of the operational model equation can be used with a variable slope (see Section 3.13.4):

$$E = \frac{E_{max}\tau^n[A]^n}{([A] + K_A)^n + \tau^n[A]^n}.$$ (3.12)

The operational model is used throughout this book for the determination of drug parameters in functional systems.

3.7 Two-state Theory

Two-state theory was originally formulated for ion channels. The earliest form, proposed by Del Castillo and Katz [15], was comprised of a channel that when bound to an agonist changed from a closed to an open state. In the absence of agonist, all of the channels are closed:

$$A + R \rightleftarrows AR_{closed} \rightleftarrows AR_{open}.$$ (3.13)

From theories on cooperative enzymes proposed by Monod and co-workers [16] came the idea that channels could coexist in both open and closed states:

$$
\begin{array}{ccc}
AR_{closed} & \xrightarrow{\alpha L} & AR_{open} \\
K \Big\updownarrow & & \Big\updownarrow \alpha K \\
R_{closed} & \xrightarrow{L} & R_{open} \\
+ & & + \\
A & & A
\end{array}
$$ (3.14)

The number of channels open, as a fraction of the total number of channels, in the presence of a ligand [A] is given as (see Section 3.13.5).

$$\rho_{open} = \frac{\alpha L[A]/K_A + L}{[A]/K_A(1 + \alpha L) + L + 1}.$$ (3.15)

There are some features of this type of system of note. First, it can be seen that there can be a fraction of the channels open in the absence of agonist. Specifically, Equation 3.15 predicts that in the absence of agonist ([A] = 0) the fraction of channels open is equal to $\rho_{open} = L/(1+L)$. For non-zero values of L this indicates that ρ_{open} will be >1. Second, ligands with preferred affinity for the open channel ($\alpha > 1$) cause opening of the channel (will be agonists). This can be seen from the ratio of channels open in the absence and presence of a saturating concentration of ligand $[\rho_\infty/\rho_0 = \alpha(1+L)/(1+\alpha L)]$. This equation reduces to

$$\frac{\rho_\infty}{\rho_0} = \frac{1 + L}{(1/\alpha) + L}.$$ (3.16)

It can be seen that for values $\alpha > 1$, the value $(1/\alpha) < 1$, and the denominator in Equation 3.16 will be less than the numerator. The ratio with the result that ρ_∞/ρ_0 will be >1 (increased channel opening; i.e., agonism). Also, the potency of the agonist will be greater as the spontaneous

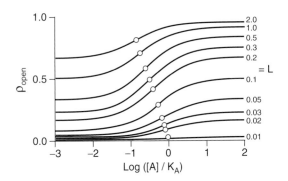

FIGURE 3.9 Dose-response curves to an agonist in a two-state ion-channel system. Ordinates: fraction of open channels. Abscissae: logarithms of molar concentrations of agonist. Numbers next to the curves refer to values of L (ratio of spontaneously open channels to closed channels). Curve calculated for an agonist with a tenfold higher affinity for the open channel ($\alpha = 10$). Open circles show EC_{50} values for the dose-response curves showing the increased potency to the agonist with increasing spontaneously open channels (increasing values of L).

channel opening is greater. This is because the observed EC_{50} of the agonist is

$$EC_{50} = \frac{K_A(1 + L)}{(1 + \alpha L)}. \qquad (3.17)$$

This equation shows that the numerator will always be less than the denominator for $\alpha > 1$ (therefore, the $EC_{50} < K_A$, indicating increased potency over affinity) and that this differential gets larger with increasing values of L (increased spontaneous channel opening). The effects of an agonist, with a tenfold greater affinity for the open channel, in systems of different ratios of spontaneously open channels are shown in Figure 3.9. It can be seen that the maximal agonist activity, the elevated basal activity, and the agonist potency are increased with increasing values of L. Two-state theory has been applied to receptors [17–19] and was required to explain the experimental findings relating to constitutive activity in the late 1980s. Specifically, the ability of channels to spontaneously open with no ligand present was adapted for the model of receptors that could spontaneously form an activated state (in the absence of an agonist *vide infra*).

3.8 The Ternary Complex Model

Numerous lines of evidence in the study of G-protein-coupled receptors indicate that these receptors become activated, translocate in the cell membrane, and subsequently bind with other membrane bound proteins. It was first realized that guanine nucleotides could affect the affinity of agonists but not antagonists, suggesting two-stage binding of ligand to receptor and subsequently the complex to a G-protein [20–22]. The model describing such

a system, first described by De Lean and colleagues [13], is termed the ternary complex model. Schematically, the process is

$$A + R \rightleftharpoons AR + G \rightleftharpoons ARG, \qquad (3.18)$$

where the ligand is A, the receptor R, and the G-protein G. For a number of years this model was used to describe pharmacological receptor effects until new experimental evidence forced modification of the original concept. Specifically, the fact that recombinant G-protein-coupled receptor systems demonstrate constitutive activity shows that receptors spontaneously form activated states capable of producing response through G-proteins in the absence of agonists. This necessitated modification of the ternary complex model.

3.9 The Extended Ternary Model

The resulting modification is called the extended ternary complex model [3], which describes the spontaneous formation of active state receptor ($[R_a]$) from an inactive state receptor ($[R_i]$) according to an allosteric constant ($L = [R_a]/[R_i]$). The active state receptor can form a complex with G-protein ($[G]$) spontaneously to form R_aG, or agonist activation can induce formation of a ternary complex AR_aG:

$$
\begin{array}{ccccc}
 & & G & & \\
AR_i & \xrightleftharpoons{\alpha L} & AR_a & \xrightleftharpoons{\gamma K_g} & AR_aG \\
K_a \updownarrow & & \alpha K_a \updownarrow & & \updownarrow \alpha\gamma K_a \\
R_i & \xrightleftharpoons{L} & R_a & \xrightleftharpoons{K_g} & R_aG \\
 & & G & &
\end{array} \qquad (3.19)
$$

As described in Section 3.13.6, the fraction ρ of G-protein-activating species (producing response)—namely, $[R_aG]$ and $[AR_aG]$—as a fraction of the total number of receptor species $[R_{tot}]$ is given by

$$\rho = \frac{L[G]/K_G(1 + \alpha\gamma[A]/K_A)}{[A]/K_A(1 + \alpha L(1 + \gamma[G]/K_G)) + L(1 + [G]/K_G) + 1}, \qquad (3.20)$$

where the ligand is $[A]$ and K_A and K_G are the equilibrium dissociation constants of the ligand-receptor and G-protein/receptor complexes, respectively. The term α refers to the multiple differences in affinity of the ligand for R_a over R_i (i.e., for $\alpha = 10$ the ligand has a tenfold greater affinity for R_a over R_i). Similarly, the term γ defines the multiple difference in affinity of the receptor for G-protein when the receptor is bound to the ligand. Thus, $\gamma = 10$ means that the

This forms eight vertices of a cube (see Figure 3.12). The equilibrium equations for the various species are

$$[AR_i] = [AR_aG]/\alpha\gamma\delta\beta L[G]K_g, \qquad (3.71)$$

$$[AR_a] = [AR_aG]/\gamma\beta\delta[G]K_g, \qquad (3.72)$$

$$[R_a] = [AR_aG]/\alpha\gamma\delta\beta[G]K_g[A]K_a, \qquad (3.73)$$

$$[R_i] = [AR_aG]/\alpha\gamma\delta\beta L[G]K_g[A]K_a, \qquad (3.74)$$

$$[R_aG] = [AR_aG]/\alpha\gamma\delta[A]K_a, \qquad (3.75)$$

$$[R_iG] = [AR_aG]/\alpha\gamma\delta\beta L[A]K_a, \quad \text{and} \qquad (3.76)$$

$$[AR_iG] = [AR_aG]/\alpha\delta\beta L. \qquad (3.77)$$

The conservation equation for receptor species is

$$[R_{tot}] = [AR_aG] + [AR_iG] + [R_iG] + [R_aG] + [AR_a] \\ + [AR_i] + [R_a] + [R_i]. \qquad (3.78)$$

It is assumed that the receptor species leading to G-protein activation (and therefore physiological response) are complexes between the activated receptor ([Ra]) and the G-protein; namely, $[AR_aG] + [R_aG]$. The fraction of the response-producing species of the total receptor species—$([AR_aG] + [R_aG])/R_{tot}$—is denoted ρ and is given by

$$\rho = \frac{\beta L[G]/K_G(1 + \alpha\gamma\delta[A]/K_A)}{[A]/K_A(1 + \alpha L + \gamma[G]/K_G(1 + \alpha\gamma\beta L)) + [G]/K_G(1 + \beta L) + L + 1}. \qquad (3.79)$$

References

1. Clark, A. J. (1933). The mode of action of drugs on cells. Edward Arnold, London.
2. Costa, T., and Herz, A. (1989). Antagonists with negative intrinsic activity at δ-opioid receptors coupled to GTP-binding proteins. *Proc. Natl. Acad. Sci. U.S.A.* **86:**7321–7325.
3. Samama, P., Cotecchia, S., Costa, T., and Lefkowitz, R. J. (1993). A mutation-induced activated state of the β2-adrenergic receptor: Extending the ternary complex model. *J. Biol. Chem.* **268:**4625–4636.
4. Clark, A. J. (1937). *General pharmacology: Heffter's handbuch d. exp. pharmacology.* Ergband 4, Springer, Berlin.
5. Stephenson, R. P. (1956). A modification of receptor theory. *Br. J. Pharmacol.* **11:**379–393.
6. Ariens, E. J. (1954). Affinity and intrinsic activity in the theory of competitive inhibition. *Arch. Int. Pharmacodyn. Ther.* **99:**32–49.
7. Ariens, E. J. (1964). *Molecular pharmacology,* Vol. 1. Academic Press, New York.
8. MacKay, D. (1977). A critical survey of receptor theories of drug action. In: *Kinetics of drug action,* edited by J. M. Van Rossum, pp. 255–322. Springer-Verlag, Berlin.
9. Furchgott, R. F. (1966). The use of β-haloalkylamines in the differentiation of receptors and in the determination of dissociation constants of receptor-agonist complexes. In: *Advances in drug research, Vol. 3,* edited by N. J. Harper and A. B. Simmonds, pp. 21–55. Academic Press, New York.
10. Furchgott, R. F. (1972). The classification of adrenoreceptors (adrenergic receptors): An evaluation from the standpoint of receptor theory. In: *Handbook of experimental pharmacology, catecholamines, Vol. 33,* edited by H. Blaschko and E. Muscholl, pp. 283–335. Springer-Verlag, Berlin.
11. Black, J. W., and Leff, P. (1983). Operational models of pharmacological agonist. *Proc. R. Soc. Lond. [Biol.]* **220:**141.
12. Cuatrecasas, P. (1974). Membrane receptors. *Ann. Rev. Biochem.* **43:**169–214.
13. DeLean, A., Stadel, J. M. Lefkowitz, R. J. (1980). A ternary complex model explains the agonist-specific binding properties of adenylate cyclase coupled β-adrenergic receptor. *J. Biol. Chem.* **255:**7108–7117.
14. Kenakin, T. P. (2000). The pharmacologic consequences of modeling synoptic receptor systems. In: *Biomedical applications of computer modeling,* edited by A. Christopoulos, pp. 1–20. CRC Press, Boca Raton.
15. Del Castillo, J., and Katz, B. (1957). Interaction at end-plate receptors between different choline derivatives. *Proc. R. Soc. London, B.* **146:**369–381.
16. Monod, J., Wyman, J., and Changeux, J. P. (1965). On the nature of allosteric transition. *J. Mol. Biol.* **12:**306–329.
17. Colquhoun, D. (1973). The relationship between classical and cooperative models for drug action. In: *A symposium on drug receptors,* edited by H. P. Rang, pp. 149–182. University Park Press, Baltimore.
18. Karlin, A. (1967). On the application of "a plausible model" of allosteric proteins to the receptor for acetylcholine. *J. Theoret. Biol.* **16:**306–320.
19. Thron, C. D. (1973). On the analysis of pharmacological experiments in terms of an allosteric receptor model. *Mol. Pharmacol.* **9:**1–9.
20. Hulme, E. C., Birdsall, N. J. M., Burgen, A. S. V., and Metha, P. (1978). The binding of antagonists to brain muscarinic receptors. *Mol. Pharmacol.* **14:**737–750.
21. Lefkowitz, R. J., Mullikin, D., and Caron, M. G. (1976). Regulation of β-adrenergic receptors by guanyl-5′-yl imidodiphosphate and other purine nucleotides. *J. Biol. Chem.* **251:**4686–4692.
22. MaGuire, M. E., Van Arsdale, P. M., and Gilman, A. G. (1976). An agonist-specific effect of guanine nucleotides on the binding of the beta adrenergic receptor. *Mol. Pharmacol.* **12:**335–339.

23. Weiss, J. M., Morgan, P. H., Lutz, M. W., and Kenakin, T. P. (1996a). The cubic ternary complex receptor-occupancy model. I. Model description. *J. Theroet. Biol.* **178:**151–167.

24. Weiss, J. M., Morgan, P. H., Lutz, M. W., and Kenakin, T. P. (1996b). The cubic ternary complex receptor-occupancy model. II. Understanding apparent affinity. *J. Theroet. Biol.* **178:**169–182.

25. Weiss, J. M., Morgan, P. H., Lutz, M. W., and Kenakin, T. P. (1996c). The cubic ternary complex receptor-occupancy model. III. Resurrecting efficacy. *J. Theoret. Biol.* **181:**381–397.

26. Bouaboula, M., Perrachon, S., Milligan, L., Canatt, X., Rinaldi-Carmona, M., Portier, M., Barth, F., Calandra, B., Pecceu, F., Lupker, J., Maffrand, J.-P., Le Fur, G., and Casellas, P. (1997). A selective inverse agonist for central cannabinoid receptor inhibits mitogen-activated protein kinase activation stimulated by insulin or insulin-like growth factor. *J. Biol. Chem.* **272:**22330–22339.

27. Chen, G., Way, J., Armour, S., Watson, C., Queen. K., Jayawrickreme, C., Chen, W.-J., and Kenakin, T. P. (1999). Use of constitutive G-protein-coupled receptor activity for drug discovery. *Mol. Pharmacol.* **57:**125–134.

4

Pharmacological Assay Formats: Binding

the yeoman work in any science ... is done by the experimentalist who must keep the theoreticians honest.

— MICHIO KAKU (1995)

4.1 The Structure of This Chapter

This chapter discusses the application of binding techniques to the study of drug-receptor interaction. It will be seen that the theory of binding and the methods used to quantify drug effect are discussed before the experimental prerequisites for good binding experiments are given. This may appear to be placing the cart before the horse in concept. However, the methods used to detect and rectify nonequilibrium experimental conditions utilize the very methods used to quantify drug effect. Therefore, they must be understood before their application to optimize experimental conditions can be discussed. This chapter first presents what the experiments strive to achieve, and then explores the possible pitfalls of experimental design that may cause the execution to fall short of the intent.

4.2 Binding Theory and Experiment

A direct measure of the binding of a molecule to a protein target can be made if there is some means to distinguish bound molecule from unbound and a means to quantify the amount bound. Historically, the first widely used technique to do this was radioligand binding. Radioactive molecules can be detected by observation of radioactive decay and the amount of quantified through calibration curves relating the amount of molecule to the amount of radioactivity detected. An essential part of this process is the ability to separate the bound from the unbound molecule. This can be done by taking advantage of the size of the protein versus the soluble small molecule.

The protein can be separated by centrifugation, equilibrium dialysis, or filtration. Alternatively, the physical proximity of the molecule to the protein can be used. For example, in scintillation proximity assays the receptor protein adheres to a bead containing scintillant, a chemical that produces light when close to radioactivity. Thus, when radioactive molecules are bound to the receptor (and therefore are near the scintillant) a light signal is produced heralding the binding of the molecule. Other methods of detecting molecules such as fluorescence are increasingly being utilized in binding experiments. For example, molecules that produce different qualities of fluorescence, depending on their proximity to protein, can be used to quantify binding. Similarly, in fluorescence polarization experiments, fluorescent ligands (when not bound to protein) reduce the degree of light polarization of light passing through the medium through free rotation. When these same ligands are bound, their rotation is reduced, thereby concomitantly reducing the effect on polarization. Thus, binding can be quantified in terms of the degree of light polarization in the medium.

In general, there are emerging technologies available to discern bound from unbound molecules and many of these can be applied to receptor studies. It will be assumed from this point that the technological problems associated with determining bound species are not an experimental factor and subsequent discussions will focus on the interpretation of the resulting binding data. Several excellent sources of information on the technology and practical aspects of binding are available [1–3].

Binding experiments can be done in three modes: saturation, displacement, and kinetic. Saturation binding directly observes the binding of a tracer ligand (radioactive, fluorescent, or otherwise detectable) to the receptor. The method quantifies the maximal number of binding sites and the affinity of the ligand for the site (equilibrium dissociation constant of the ligand-receptor complex). This is a direct measure of binding using the Langmuir adsorption isotherm model. A major limitation of this technique is the obvious need for the ligand to be traceable (i.e., it can only done for radioactive or fluorescent molecules). Displacement studies overcome this limitation by allowing measurement of the affinity of nontraceable ligands through their interference with the binding of tracer ligands. Thus, molecules are used to displace or otherwise prevent the binding of tracer ligands and the reduction in signal is used to quantify the affinity of the displacing ligands. Finally, kinetic studies follow the binding of a tracer ligand with time. This can yield first-order rate constants for the onset and offset of binding, which can be used to calculate equilibrium binding constants to assess the temporal approach to equilibrium or to determine binding reversibility or to detect allosteric interactions. Each of these is considered separately. The first step is to discuss some methodological points common to all of these types of binding experiments.

The aim of binding experiments is to define and quantify the relationship between the concentration of ligand in the receptor compartment and the portion of the concentration that is bound to the receptor at any one instant. A first prerequisite is to know that the amount of bound ligand that is measured is bound only to the receptor and not to other sites in the sample tube or well (i.e., cell membrane, wall of the vessel containing the experimental solution, and so on). The amount of ligand bound to these auxiliary sites but not specifically to the target is referred to as *nonspecific binding* (denoted *nsb*). The amount bound only to the pharmacological target of interest is termed the *specific binding*. The amount of specific binding is defined operationally as the bound ligand that can be displaced by an excess concentration of a specific antagonist for the receptor that is not radioactive (or otherwise does not interfere with the signals). Therefore, another prerequisite of binding experiments is the availability of a nontracer ligand (for the specific target defined as one that does not interfere with the signal whether it be radioactivity, fluorescence, or polarized light). Optimally, the chemical structure of the ligand used to define nsb should be different from the binding tracer ligand. This is because the tracer may bind to nonreceptor sites (i.e., adsorption sites, other nonspecific proteins), and if a nonradioactive version of the same molecular structure is used to define specific binding it may protect those very same nonspecific sites (which erroneously define specific binding). A ligand with different chemical structure may not bind to the same nonspecific sites and thus lessen the potential of defining nsb sites as biologically relevant receptors.

The nonspecific binding of low concentrations of biologically active ligands is essentially linear and nonsaturable within the ranges used in pharmacological binding experiments. For a traceable ligand (radioactive, fluorescent, and so on), nonspecific binding is given as

$$nsb = k \cdot [A^*], \qquad (4.1)$$

where k is a constant defining the concentration relationship for nonspecific binding and $[A^*]$ is the concentration of the traceable molecule. The specific binding is saturable and defined by the Langmuir adsorption isotherm

$$\text{Specific binding} = \frac{[A^*]}{[A^*] + K_d}, \qquad (4.2)$$

where K_d is the equilibrium dissociation constant of the ligand-receptor complex. The total binding is the sum of

these and is given as

$$\text{Total binding} = \frac{[A^*] \cdot B_{max}}{[A^*] + K_d} + k \cdot [A^*] \qquad (4.3)$$

The two experimentally derived variables are nsb and total binding. These can be obtained by measuring the relationship between the ligand concentration and the amount of ligand bound (total binding) and the amount bound in the presence of a protecting concentration of receptor specific antagonist. This latter procedure defines the nsb. Theoretically, specific binding can be obtained by subtracting these values for each concentration of ligand, but a more powerful method is to fit the two data sets (total binding and nsb) to Equations 4.1 and 4.3 simultaneously. One reason this is preferable is that more data points are used to define specific binding. A second reason is that a better estimate of the maximal binding (B_{max}) can be made by simultaneously fitting of two functions. Since B_{max} is defined at theoretically infinite ligand concentrations, it is difficult to obtain data in this concentration region. When there is a paucity of data points, nonlinear fitting procedures tend to overestimate the maximal asymptote. The additional experimental data (total plus nonspecific binding) reduces this effect and yields more accurate B_{max} estimates.

In binding, a good first experiment is to determine the time required for the binding reaction to come to equilibrium with the receptor. This is essential to know since most binding reactions are made in stop-time mode and real-time observation of the approach to equilibrium is not possible (this is not true of more recent fluorescent techniques where visualization of binding in real time can be achieved). A useful experiment is to observe the approach to equilibrium of a given concentration of tracer ligand and then to observe reversal of binding by addition of a competitive antagonist of the receptor. An example of this experiment is shown in Figure 4.1. Valuable data is obtained with this approach since it indicates the time needed to reach equilibrium and confirms the fact that the binding is reversible. Reversibility is essential to the attainment of steady states and equilibria (i.e., irreversible binding reactions do not come to equilibrium).

4.2.1 Saturation Binding

A saturation binding experiment consists of the equilibration of the receptor with a range of concentrations of traceable ligand in the absence (total binding) and presence of a high concentration (approximately $100 \times K_d$) of antagonist to protect the receptors (and thus determine the nsb). Simultaneous fitting of the total binding curve (Equation 4.3) and nsb line (Equation 4.1) yields the specific binding with parameters of maximal number of binding sites (B_{max}) and equilibrium dissociation constant of the ligand-receptor complex (K_d). (See Equation 4.2.) An example of this procedure for the human calcitonin

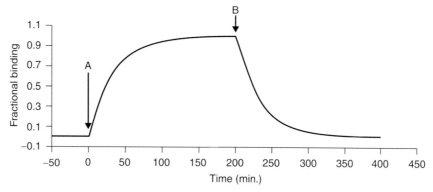

FIGURE 4.1 Time course for the onset of a radioligand onto the receptor and the reversal of radioligand binding upon addition of a high concentration of a nonradioactive antagonist ligand. The object of the experiment is to determine the times required for steady-state receptor occupation by the radioligand and confirmation of reversibility of binding. The radioligand is added at point A and an excess competitive antagonist of the receptor at point B.

receptor is shown in Figure 4.2. Before the widespread use of nonlinear fitting programs, the Langmuir equation was linearized for ease of fitting graphically. Thus, specific binding ([A*R]) according to mass action, represented as

$$\frac{[A^*R]}{B_{max}} = \frac{[A^*]}{[A^*] + K_d}, \qquad (4.4)$$

yields a straight line with the transforms

$$\frac{[A^*R]}{[A^*]} = \frac{B_{max}}{K_d} - \frac{[A^*R]}{K_d}, \qquad (4.5)$$

referred to alternatively as a Scatchard, Eadie, or Eadie-Hofstee plot. From this linear plot, $K_d = -1/\text{slope}$ and the x intercept equals B_{max}.

Alternatively, another method of linearizing the data points is with

$$\frac{1}{[A^*R]} = \frac{1}{[A^*]} \cdot \frac{K_d}{B_{max}} + \frac{1}{B_{max}}. \qquad (4.6)$$

This is referred to as a double reciprocal or lineweaver Burk plot. From this linear plot, $K_d = \text{slope/intercept}$ and the $1/\text{intercept} = B_{max}$. Finally, a linear plot can be achieved with

$$\frac{[A^*]}{[A^*R]} = \frac{[A^*]}{B_{max}} + \frac{K_d}{B_{max}}. \qquad (4.7)$$

This is referred to as a Hanes, Hildebrand-Benesi, or Scott plot. From this linear plot, $K_d = \text{intercept/slope}$ and $1/\text{slope} = B_{max}$.

Examples of these are shown for the saturation data in Figure 4.2. At first glance, these transformations may seem like ideal methods to analyze saturation data. However, transformation of binding data is not generally recommended. This is because transformed plots can distort experimental uncertainty, produce compression of data,

and cause large differences in data placement. Also, these transformations violate the assumptions of linear regression and can be curvilinear simply because of statistical factors (for example, Scatchard plots combine dependent and independent variables). These transformations are valid only for ideal data and are extremely sensitive to different types of experimental errors. They should not be used for estimation of binding parameters. Scatchard plots compress data to the point where a linear plot can be obtained. Figure 4.3 shows a curve with an estimate of B_{max} that falls far short of being able to furnish an experimental estimate of the B_{max}, yet the Scatchard plot is linear with an apparently valid estimate from the abscissal intercept.

In general, nonlinear fitting of the data is essential for parameter estimation. Linear transformations, however, are useful for visualization of trends in data. Variances from a straight edge are more discernible to the human eye than are differences from curvilinear shapes. Therefore, linear transforms can be a useful diagnostic tool. An example of where the Scatchard transformation shows significant deviation from a rectangular hyperbola is shown in Figure 4.4. The direct presentation of the data show little deviation from the saturation binding curve as defined by the Langmuir adsorption isotherm. The data at 10 and 30 nM yield slightly underestimated levels of binding, a common finding if slightly too much protein is used in the binding assay (see Section 4.4.1). While this difference is nearly undetectable when the data is presented as a direct binding curve, it does produce a deviation from linearity in the Scatchard curve (see Figure 4.4b).

Estimating the B_{max} value is technically difficult since it basically is an exercise in estimating an effect at infinite drug concentration. Therefore, the accuracy of the estimate of B_{max} is proportional to the maximal levels of radioligand that can be used in the experiment. The attainment of saturation binding can be deceiving when the ordinates are plotted on a linear scale, as they are in Figure 4.2.

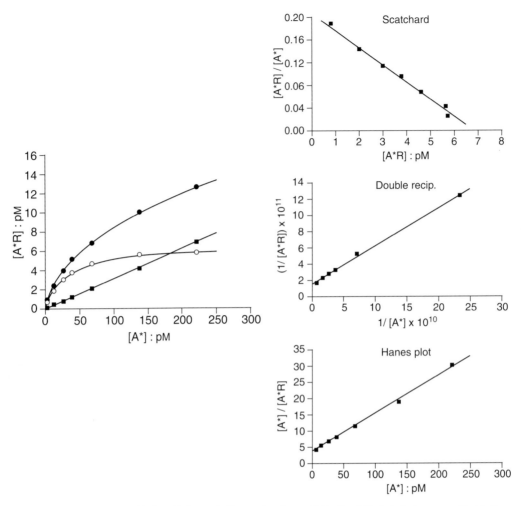

FIGURE 4.2 Saturation binding. Left panel: Curves showing total binding (filled circles), nonspecific binding (filled squares), and specific binding (open circles) of the calcitonin receptor antagonist radiolabel ^{125}I AC512 ($B_{max} = 6.63$ pM; $K_d = 26.8$ pM). Data redrawn from [1]. Panels to the right show linear variants of the specific binding curve: Scatchard (Equation 4.5), double reciprocal (Equation 4.6), and Hanes plots (Equation 4.7) cause distortion and compression of data. Nonlinear curve fitting techniques are preferred.

Figure 4.5 shows a saturation curve for calcitonin binding that appears to reach a maximal asymptote on a linear scale. However, replotting the graph on a semilogarithmic scale illustrates the illusion of maximal binding on the linear scale and, in this case, how far short of true maxima a linear scale can present a saturation binding curve. An example of how to measure the affinity of a radioligand and obtain an estimate of B_{max} (maximal number of binding sites for that radioligand) is given in Section 12.1.1.

4.2.2 Displacement Binding

In practice, there will be a limited number of ligands available that are chemically traceable (i.e., radioactive, fluorescent). Therefore, the bulk of radioligand experiments designed to quantify ligand affinity are done in a displacement mode whereby a ligand is used to displace or otherwise affect the binding of a traceable ligand.

In general, an inverse sigmoidal curve is obtained with reduction in radioligand binding upon addition of non-radioactive antagonist. An example of how to measure the affinity of a displacing ligand is given in Section 12.1.2.

The equations describing the amount of bound radioligand observed in the presence of a range of concentrations of nontraceable ligand vary with the model used for the molecular antagonism. These are provided in material following, with brief descriptions. More detailed discussions of these mechanisms can be found in Chapter 6. If the binding is competitive (both ligands compete for the same binding domain on the receptor), the amount of tracer ligand-receptor complex (ρ^*) is given as (see Section 4.6.1)

$$\rho^* = \frac{[A^*]/K_d}{[A^*]/K_d + [B]/K_B + 1}, \tag{4.8}$$

where the concentration of tracer ligand is $[A^*]$, the nontraceable displacing ligand is $[B]$, and K_d and K_B are

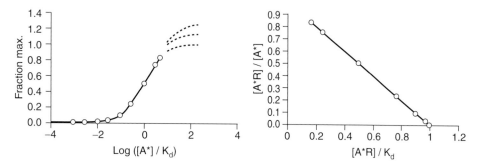

FIGURE 4.3 Erroneous estimation of maximal binding with Scatchard plots. The saturation binding curve shown to the left has no data points available to estimate the true B_{max}. The Scatchard transformation to the right linearizes the existing points, allowing an estimate of the maximum to be made from the x-axis intercept. However, this intercept in no way estimates the true B_{max} since there are no data to define this parameter.

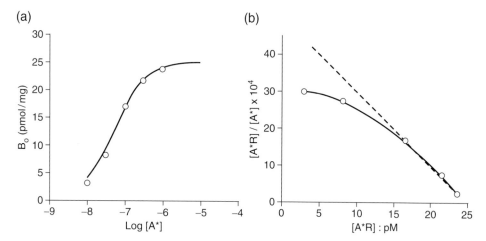

FIGURE 4.4 Saturation binding expressed directly and with a Scatchard plot. (a) Direct representation of a saturation binding plot ($B_{max} = 25$ pmole/mg, $K_d = 50$ nM). Data points are slightly deviated from ideal behavior (lower two concentrations yield slightly lower values for binding as is common when slightly too much receptor protein is used in the assay, *vide infra*). (b) Scatchard plot of the data shown in panel a. It can be seen that the slight deviations in the data lead to considerable deviations from linearity on the Scatchard plot.

respective equilibrium dissociation constants. If the binding is noncompetitive (binding of the antagonist precludes the binding of the tracer ligand), the signal is given by (see Section 4.6.2)

$$\rho^* = \frac{[A^*]/K_d}{[A^*]/K_d([B]/K_B + 1) + [B]/K_B + 1}. \quad (4.9)$$

If the ligand allosterically affects the affinity of the receptor (antagonist binds to a site separate from that for the tracer ligand) to produce a change in receptor conformation to affect the affinity of the tracer (*vide infra*) for the tracer ligand (see Chapter 6 for more detail), the displacement curve is given by (see Section 4.6.3)

$$\rho^* = \frac{[A^*]/K_d(1 + \alpha[B]/K_B)}{[A^*]/K_d(1 + \alpha[B]/K_B) + [B]/K_B + 1}, \quad (4.10)$$

where α is the multiple factor by which the nontracer ligand affects the affinity of the tracer ligand (i.e., $\alpha = 0.1$ indicates that the allosteric displacing ligand produces a tenfold decrease in the affinity of the receptor for the tracer ligand).

As noted previously, in all cases these various functions describe an inverse sigmoidal curve between the displacing ligand and the signal. Therefore, the mechanism of interaction cannot be determined from a single displacement curve. However, observation of a *pattern* of such curves obtained at different tracer ligand concentrations (range of [A*] values) may indicate whether the displacements are due to a competitive, noncompetitive, or allosteric mechanism.

Competitive displacement for a range of [A*] values (Equation 4.8) yields the pattern of curves shown in Figure 4.6a. A useful way to quantify the displacement is to determine the concentration of displacing ligand that produces a diminution of the signal to 50% of the

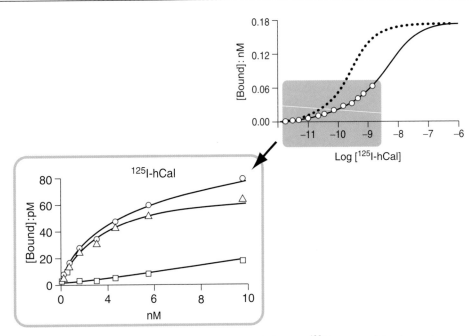

FIGURE 4.5 Saturation binding of the radioligand human ^{125}I-human calcitonin to human calcitonin receptors in a recombinant cell system in human embryonic kidney cells. Left-hand panel shows total binding (open circles), nonspecific binding (open squares), and specific receptor binding (open triangles). The specific binding appears to reach a maximal asymptotic value. The specific binding is plotted on a semi-logarithmic scale (shown in the right-hand panel). The solid line on this curve indicates an estimate of the maximal receptor binding. The data points (open circles) on this curve show that the data defines less than half the computer-estimated total saturation curve. Data redrawn from [4].

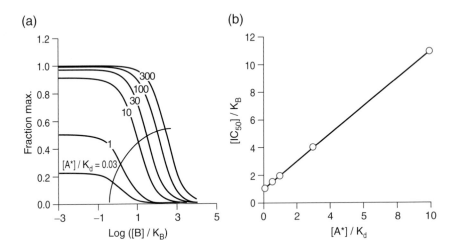

FIGURE 4.6 Displacement of a radioligand by a competitive nonradioactive ligand. (a) Displacement of radioactivity (ordinate scale) as curves shown for a range of concentrations of displacing ligand (abscissae as log scale). Curves shown for a range of radioligand concentrations denoted on the graph in units of $[A^*]/K_d$. Curved line shows the path of the IC_{50} for the displacement curves along the antagonist concentration axis. (b) Multiple values of the K_i for the competitive displacing ligand (ordinate scale) as a function of the concentration of radioligand being displaced (abscissae as linear scale). Linear relationship shows the increase in observed IC_{50} of the antagonist with increasing concentrations of radioligand to be displaced (according to Equation 4.11).

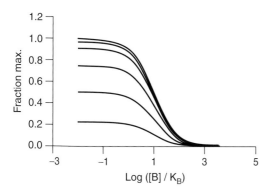

FIGURE 4.7 Displacement curves for a noncompetitive antagonist. Displacement curve according to Equation 4.9 for values of radioligand $[A^*]/K_d = 0.3$ (curve with lowest ordinate scale beginning at 0.25), 1, 3, 10, 30, and 100. While the ordinate scale on these curves increases with increasing $[A^*]/K_d$ values, the location parameter along the x axis does not change.

original value. This concentration of displacing ligand will be referred to as the IC_{50} (inhibitory concentration for 50% decrease). For competitive antagonists, it can be shown that the IC_{50} is related to the concentration of tracer ligand $[A^*]$ by (see Section 4.6.4)

$$IC_{50} = K_B \cdot ([A^*]/K_d + 1). \qquad (4.11)$$

This is a linear relation often referred to as the Cheng-Prusoff relationship [5]. It is characteristic of competitive ligand-receptor interactions. An example is shown in Figure 4.6b.

The displacement of a tracer ligand, for a range of tracer ligand concentrations, by a noncompetitive antagonist is shown in Figure 4.7. In contrast to the pattern shown for competitive antagonists, the IC_{50} for inhibition of tracer binding does not change with increasing tracer ligand concentrations. In fact, it can be shown that the IC_{50} for inhibition is equal to the equilibrium dissociation constant

of the noncompetitive antagonist-receptor complex (see Section 4.6.2).

Allosteric antagonist effects can be an amalgam of competitive and noncompetitive profiles in terms of the relationship between IC_{50} and $[A^*]$. This relates to the magnitude of the term α, specifically the multiple ratio of the affinity of the receptor for $[A^*]$ imposed by the binding of the allosteric antagonist. A hallmark of allosteric inhibition is that it is saturable (i.e., the antagonism maximizes upon saturation of the allosteric binding site). Therefore, if a given antagonist has a value of α of 0.1 this means that the saturation binding curve will shift to the right by a factor of tenfold in the presence of an infinite concentration of allosteric antagonist. Depending on the initial concentration of radioligand, this may cause the displacement binding curve to not reach nsb levels. This effect is illustrated in Figure 4.8. Therefore, in contrast to competitive antagonists, where displacement curves all take binding of the radioligand to nsb values, an allosteric ligand will only displace to a maximum value determined by the initial concentration of radioligand and the value of α for the allosteric antagonist. In fact, if a displacement curve is observed where the radioligand binding is not displaced to nsb values, this is presumptive evidence that the antagonist is operating through an allosteric mechanism. The maximum displacement of a given concentration of radioligand $[A^*]$ by an allosteric antagonist with given values of α is (see Section 4.6.5)

$$\text{Maximal Fractional Inhibition} = \frac{[A^*]K_d + 1}{[A^*]/K_d + 1/\alpha}, \qquad (4.12)$$

where K_d is the equilibrium dissociation constant of the radioligand-receptor complex (obtained from saturation binding studies). The observed displacement for a range of allosteric antagonists for two concentrations of radioligands is shown in Figure 4.9. The effects shown in Figure 4.9 indicate a practical test for the detection of allosteric versus competitive antagonism in displacement

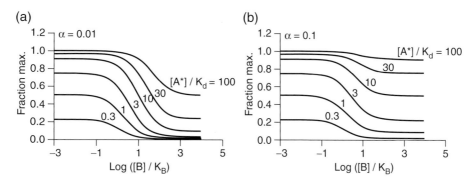

FIGURE 4.8 Displacement curves according to Equation 4.10 for an allosteric antagonist with different cooperativity factors (panel a $\alpha = 0.01$, panel b $\alpha = 0.1$). Curves shown for varying values of radioligand ($[A^*]/K_d$). It can be seen that the curves do not reach nsb values for high values of radioligand and that this effect occurs at lower concentrations of radioligand for antagonists of higher values of α.

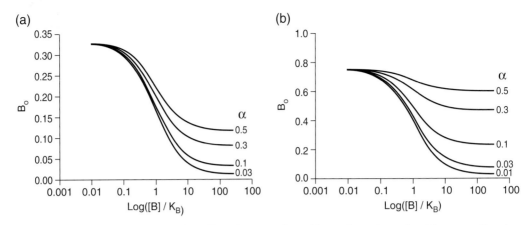

FIGURE 4.9 Displacement curves for allosteric antagonists with varying values of α (shown on figure). Ordinates: bound radioligand. (a) Concentration of radioligand $[A^*]/K_d = 0.1$. (b) Displacement of higher concentration of radioligand $[A^*]/K_A = 3$.

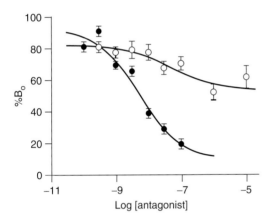

FIGURE 4.10 Displacement of bound ^{125}I-MIP-1α from chemokine C receptors type 1 (CCR1) by MIP-1α (filled circles) and the allosteric ligand UCB35625 (open circles). Note how the displacement by the allosteric ligand is incomplete. Data redrawn from [6].

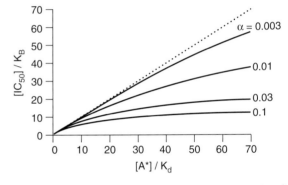

FIGURE 4.11 Relationship between the observed IC$_{50}$ for allosteric antagonists and the amount of radioligand present in the assay according to Equation 4.13. Dotted line shows relationship for a competitive antagonist.

binding studies. If the value of the maximal displacement varies with different concentrations of radioligand, this would suggest that an allosteric mechanism is operative. Figure 4.10 shows the displacement of the radioactive peptide ligand ^{125}I-MIP-1α from chemokine CCR1 receptors by nonradioactive peptide MIP-1α and by the allosteric small molecule modulator UCB35625. Clearly, the nonpeptide ligand does not reduce binding to nsb levels, indicating an allosteric mechanism for this effect [6].

Another, more rigorous, method to detect allosteric mechanisms (and one that may furnish a value of α for the antagonist) is to formally observe the relationship between the concentration of radioligand and the observed antagonism by displacement with the IC$_{50}$ of the antagonist. As shown with Equation 4.11, for a competitive antagonist this relationship is linear (Cheng-Prusoff correction). For

an allosteric antagonist, the relationship is hyperbolic and given by (see Section 4.6.6)

$$IC_{50} = \frac{K_B([A^*]/K_d)}{\alpha[A^*]/K_d + 1}. \tag{4.13}$$

It can be seen from this equation that the maximum of the hyperbola defined by a given antagonist (with ordinate values expressed as the ratio of IC$_{50}$ to K$_B$) will have a maximum asymptote of $1/\alpha$. Therefore, observation of a range of IC$_{50}$ values needed to block a range of radioligand concentrations can be used to estimate the value of α for a given allosteric antagonist. Figure 4.11 shows the relationship between the IC$_{50}$ for allosteric antagonism and the concentration of radioligand used in the assay, as a function of α. It can be seen that unlike the linear relationship predicted by Equation 4.11 (see Figure 4.6b)

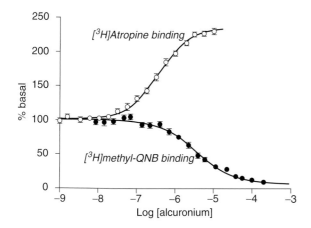

FIGURE 4.12 Effect of alcuronium on the binding of $[^3H]$ methyl-QNB (filled circles) and $[^3H]$ atropine (open circles) on muscarinic receptors. Ordinates are percentage of initial radioligand binding. Alcuronium decreases the binding of $[^3H]$ methyl-QNB and increases the binding of $[^3H]$ atropine. Data redrawn from [7].

TABLE 4.1

Differential effects of the allosteric modulator alcuronium on various probes for the m2 muscarinic receptor.

Agonists[a]	$(1/\alpha)$
Arecoline	1.7
Acetylcholine	10
Bethanechol	10
Carbachol	9.5
Furmethide	8.4
Methylfurmethide	7.3

Antagonists	
Atropine[b]	0.26
Methyl-N-piperidinyl benzilate[b]	0.54
Methyl-N-quinuclidinyl benzilate[c]	63
Methyl-N-scopolamine	0.24

[a]From [9].

[b]From [7].

[c]From [10].

the curves are hyperbolic in nature. This is another hallmark of allosteric versus simple competitive antagonist behavior.

An allosteric ligand changes the shape of the receptor, and in so doing will necessarily alter the rate of association and dissociation of some trace ligands. This means that allosterism is tracer dependent (i.e., an allosteric change detected by one radioligand may not be detected in the same way, or even detected at all, by another). For example, Figure 4.12 shows the displacement binding of two radioligand antagonists, $[^3H]$-methyl-QNB and

$[^3H]$-atropine, on muscarinic receptors by the allosteric ligand alcuronium. It can be seen that quite different effects are observed. In the case of $[^3H]$-methyl-QNB, the allosteric ligand displaces the radioligand and reduces binding to the nsb level. In the case of $[^3H]$-atropine, the allosteric ligand actually enhances binding of the radioligand [7]. There are numerous cases of probe dependence for allosteric effects. For example, the allosteric ligand strychnine has little effect on the affinity of the agonist methylfurmethide (twofold enhanced binding) but a much greater effect on the agonist bethanechol (49-fold enhancement of binding [8]). An example of the striking variation of allosteric effects on different probes by the allosteric modulator alcuronium is shown in Table 4.1 [7, 9, 10].

4.2.3 Kinetic Binding Studies

A more sensitive and rigorous method of detecting and quantifying allosteric effects is through observation of the kinetics of binding.

In general, the kinetics of most allosteric modulators have been shown to be faster than the kinetics of binding of the tracer ligand. This is an initial assumption for this experimental approach. Under these circumstances, the rate of dissociation of the tracer ligand (ρ_{A^*t}) in the presence of the allosteric ligand is given by [11, 12]

$$\rho_{A^*t} = \rho_{A^*} \cdot e^{-k_{\text{off-obs}} \cdot t}, \tag{4.14}$$

where ρ_{A^*} is the tracer-ligand receptor occupancy at equilibrium and $k_{\text{off-obs}}$ is given by

$$k_{\text{off-obs}} = \frac{\alpha[B]k_{\text{off-A*B}}/K_B + k_{\text{off-A*}}}{1 + \alpha[B]/K_B}. \tag{4.15}$$

Therefore, the rate of offset of the tracer ligand in the presence of various concentrations of allosteric ligand can be used to detect allosterism (change in rates with allosteric ligand presence) and to quantify both the affinity ($1/K_B$) and α value for the allosteric ligand. Allosteric modulators (antagonists) will generally decrease the rate of association and/or increase the rate of dissociation of the tracer ligand. Figure 4.13 shows the effect of the allosteric ligand 5-(N-ethyl-N-isopropyl)-amyloride (EPA) on the kinetics of binding (rate of offset) of the tracer ligand $[^3H]$-yohimbine to α_2-adrenoceptors. It can be seen from this figure that EPA produces a concentration-dependent increase in the rate of offset of the tracer ligand, thereby indicating an allosteric effect on the receptor.

4.3 Complex Binding Phenomena: Agonist Affinity from Binding Curves

The foregoing discussion has been restricted to the simple Langmuirian system of the binding of a ligand

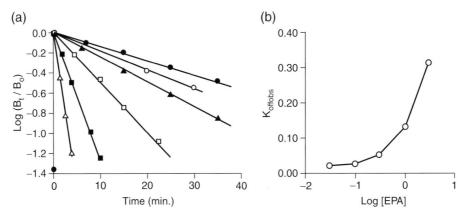

FIGURE 4.13 Effect of the allosteric modulator 5-(N-ethyl-N-isopropyl)-amyloride (EPA) on the kinetics dissociation of [^{3}H] yohimbine from α_2-adrenoceptors. (a) Receptor occupancy of [^{3}H] yohimbine with time in the absence (filled circles) and presence (open circles) of EPA 0.03 mM, 0.1 mM (filled triangles), 0.3 mM (open squares), 1 mM (filled squares), and 3 mM (open triangles). (b) Regression of observed rate constant for offset of concentration of [^{3}H] yohimbine in the presence of various concentrations of EPA on concentrations of EPA (abscissae in mM on a logarithmic scale). Data redrawn from [13].

to a receptor. The assumption is that this process produces no change in the receptor (i.e., analogous to Langmuir's binding of molecules to an inert surface). The conclusions drawn from a system where the binding of the ligand changes the receptor are different. One such process is agonist binding in which due to the molecular property of efficacy the agonist produces a change in the receptor upon binding to elicit response. Under these circumstances, the simple schemes for binding discussed for antagonists may not apply. Specifically, if the binding of the ligand changes the receptor (produces an isomerization to another form) the system can be described as

$$A + R \underset{}{\overset{K_a}{\rightleftarrows}} AR \underset{\sigma}{\overset{\chi}{\rightleftarrows}} AR^*. \qquad (4.16)$$

Under these circumstances, the observed affinity of the ligand for the receptor will not be described by K_A (where $K_A = 1/Ka$) but rather by that microaffinity modified by a term describing the avidity of the isomerization reaction. The observed affinity will be given by (see Section 4.6.7)

$$K_{obs} = \frac{K_A \cdot \chi/\sigma}{1 + \chi/\sigma}. \qquad (4.17)$$

One target type for which the molecular mechanism of efficacy has been partly elucidated is the G-protein-coupled receptor (GPCR). It is known that activation of GPCRs leads to an interaction of the receptor with separate membrane G-proteins to cause dissociation of the G-protein subunits and subsequent activation of effectors (see Chapter 2). For the purposes of binding, this process can lead to an aberration in the binding reaction as perceived in experimental binding studies. Specifically, the activation of the receptor with subsequent binding of that

receptor to another protein (to form a ternary complex of receptor, ligand, and G-protein) can lead to the apparent observation of a "high-affinity" site—a ghost site that has no physical counterpart but appears to be a separate binding site on the receptor. This is caused by two-stage binding reactions, represented as

$$A + R \overset{K_a}{\rightleftarrows} AR + [G] \overset{K_g}{\rightleftarrows} ARG. \qquad (4.18)$$

In the absence of two-stage binding, the relative quantities of [AR] and [R] are controlled by the magnitude of K_a in the presence of ligand [A]. This, in turn, defines the affinity of the ligand for R (affinity = [AR]/([A] [R])). Therefore, if an outside influence alters the quantity of [AR], the observed affinity of the ligand for the receptor R will change. If a ligand predisposes the receptor to bind to G-protein, then the presence of G-protein will drive the binding reaction to the right (i.e., [AR] complex will be removed from the equilibrium defined by K_a). Under these circumstances, more [AR] complex will be produced than that governed by K_a. The observed affinity will be higher than it would be in the absence of G-protein. Therefore, the property of the ligand that causes the formation of the ternary ligand/receptor/G-protein complex (in this case, efficacy) will cause the ligand to have a higher affinity than it would have if the receptor were present in isolation (no G-protein present). Figure 4.14 shows the effect of adding a G-protein to a receptor system on the affinity of an agonist. As shown in this figure, the muscarinic agonist oxotremorine has a receptor equilibrium dissociation constant of 6 μM in a reconstituted phospholipid vesicle devoid of G-proteins. However, upon addition of G_0 protein the affinity increases by a factor of 600 (10 nM).

This effect can actually be used to estimate the efficacy of an agonist (i.e., the propensity of a ligand to demonstrate

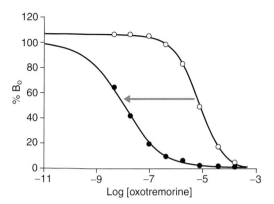

FIGURE 4.14 Effects of G-protein on the displacement of the muscarinic antagonist radioligand [3H]-L-quinuclidinyl benzylate by the agonist oxotremorine. Displacement in reconstituted phospholipid vesicles (devoid of G-protein sububits) shown in filled circles. Addition of G-protein (G$_0$ 5.9 nmol βγ-subunit/3.4 nmol α$_0$-IDP subunit) shifts the displacement curve to the left (higher affinity, see open circles) by a factor of 600. Data redrawn from [14].

high affinity in the presence of G-protein, *vide infra*). The observed affinity of such a ligand is given by (see Section 4.6.8)

$$K_{obs} = \frac{K_A}{1 + [G]/K_G}, \qquad (4.19)$$

where K_G is the equilibrium dissociation constant of the receptor/G-protein complex. A low value for K_G indicates tight binding between receptors and G-proteins (i.e., high efficacy). It can be seen that the observed affinity of the ligand will be increased (decrease in the equilibrium dissociation constant of the ligand-receptor complex) with increasing quantities of G-protein [G] and/or very efficient binding of the ligand-bound receptor to the G-protein (low

value of K_G, the equilibrium dissociation constant for the ternary complex of ligand/receptor/G-protein). The effects of various concentrations of G-protein on the binding saturation curve to an agonist ligand are shown in Figure 4.15a. It can be seen from this figure that increasing concentrations of G-protein in this system cause a progressive shift to the left of the saturation dose-response curve. Similarly, the same effect is observed in displacement experiments. Figure 4.15b shows the effect of different concentrations of G-protein on the displacement of a radioligand by a nonradioactive agonist.

The previous discussion assumes that there is no limitation in the stoichiometry relating receptors and G-proteins. In recombinant systems, where receptors are expressed in surrogate cells (often in large quantities), it is possible that there may be limiting quantities of G-protein available for complexation with receptors. Under these circumstances, complex saturation and/or displacement curves can be observed in binding studies. Figure 4.16a shows the effect of different submaximal effects of G-protein on the saturation binding curve to an agonist radioligand. It can be seen that clear two-phase curves can be obtained. Similarly, two-phase displacement curves also can be seen with agonist ligands displacing a radioligand in binding experiments with subsaturating quantities of G-protein. (Figure 4.16b). Figure 4.17 shows an experimental displacement curve of the antagonist radioligand for human calcitonin receptors [125I]-AC512 by the agonist amylin in a recombinant system where the number of receptors exceeds the amount of G-protein available for complexation to the ternary complex state. It can be seen that the displacement curve has two distinct phases: a high-affinity (presumably due to coupling to G-protein) binding process followed by a lower-affinity binding (no benefit of G-protein coupling).

While high-affinity binding due to ternary complex formation (ligand binding to the receptor followed by

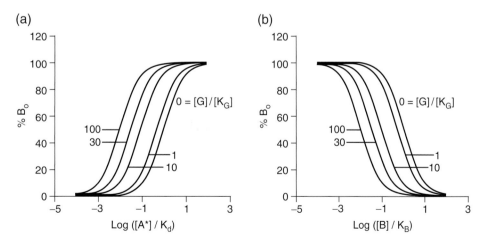

FIGURE 4.15 Complex binding curves for agonists in G-protein unlimited receptor systems. (a) Saturation binding curves for an agonist where there is high-affinity binding due to G-protein complexation. Numbers next to curves refer to the amount of G-protein in the system. (b) Displacement of antagonist radioligand by same agonist in G-protein unlimited system.

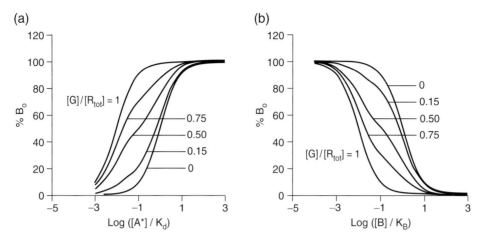

FIGURE 4.16   Complex binding curves for agonists in G-protein limited receptor systems. (a) Saturation binding curves for an agonist where the high-affinity binding due to G-protein complexation $= 100 \times K_d$ (i.e., $K_{obs} = K_d/100$). Numbers next to curves refer to ratio of G-protein to receptor. (b) Displacement of antagonist radioligand by same agonist in G-protein limited system.

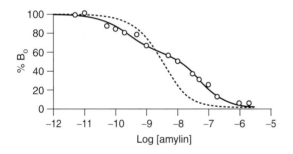

FIGURE 4.17   Displacement of antagonist radioligand ^{125}I-AC512 by the agonist amylin. Ordinates: percentage of initial binding value for AC512. Abscissae: logarithms of molar concentrations of rat amylin. Open circles are data points, solid line fit to two-site model for binding. Dotted line indicates a single phase displacement binding curve with a slope of unity. Data redrawn from [4].

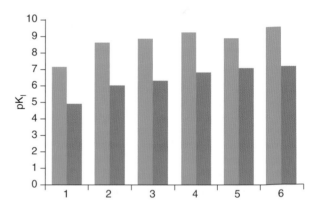

FIGURE 4.18   Affinity of adenosine receptor agonists in whole cells (dark bars) and membranes (cross-hatched bars, high-affinity binding site). Data shown for (1) 2-phenylaminoadenosine, (2) 2-chloroadenosine, (3) 5′-N-ethylcarboxamidoadenosine, (4) N^6-cyclohexyladenosine, (5) (-)-(R)-N^6-phenylisopropyladenosine, and (6) N^6-cyclopentyladenosine. Data redrawn from [15].

binding to a G-protein) can be observed in isolated systems where the ternary complex can accumulate and be quantified, this effect is cancelled in systems where the ternary complex is not allowed to accumulate. Specifically, in the presence of high concentrations of GTP (or a chemically stable analogue of GTP such as GTPγS) the formation of the ternary complex [ARG] is followed immediately by hydrolysis of GTP and the G-protein and dissociation of the G-protein into α- and γβ-subunits (see Chapter 2 for further details). This causes subsequent dissolution of the ternary complex. Under these conditions, the G-protein complex does not accumulate and the coupling reaction promoted by agonists is essentially nullified (with respect to the observable radioactive species in the binding reaction). When this occurs, the high-affinity state is not observed by the binding experiment. This has a

practical consequence in binding experiments. In broken-cell preparations for binding, the concentration of GTP can be depleted and thus the two-stage binding reaction is observed (i.e., the ternary complex accumulates). However, in whole-cell experiments the intracellular concentration of GTP is high and the ternary complex [ARG] species does not accumulate. Under these circumstances, the high-affinity binding of agonists is not observed, only the so-called "low-affinity" state of agonist binding to the receptor. Figure 4.18 shows the binding (by displacement experiments) of a series of adenosine receptor agonists to a broken-cell membrane preparation (where high-affinity binding can be observed) and the same agonists to a whole-cell preparation where the results of G-protein

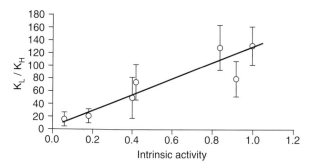

FIGURE 4.19 Correlation of the GTP shift for β-adrenoceptor agonists in turkey erythocytes (ordinates) and intrinsic activity of the agonists in functional studies (abscissae). Data redrawn from [16].

coupling are not observed. It can be seen from this figure that a phase shift for the affinity of the agonists under these two binding experiment conditions is observed. The broken-cell preparation reveals the effects of the ability of the agonists to promote G-protein coupling of the receptor. This latter property, in effect, is the efficacy of the agonist. Thus, ligands that have a high observed affinity in broken cell systems often have a high efficacy. A measure of this efficacy can be obtained by observing the magnitude of the phase shift of the affinities measured in broken-cell and whole-cell systems.

A more controlled experiment to measure the ability of agonists to induce the high-affinity state, in effect a measure of efficacy, can be done in broken-cell preparations in the presence and absence of saturating concentrations of GTP (or GTPγS). Thus, the ratio of the affinity in the absence and presence of GTP (ratio of the high-affinity and low-affinity states) yields an estimate of the efficacy of the agonist. This type of experiment is termed the "GTP shift" after the shift to the right of the displacement curve for agonist ligands after cancellation of G-protein coupling. Figure 4.19 shows the effects of saturating concentrations of GTPγS on the affinity of β-adrenoceptor agonists in turkey erythrocytes. As can be seen from this figure, a correlation of the magnitude of GTP shifts for a series of agonists and their intrinsic activities as measured in functional studies (a more direct measure of agonist efficacy; see Chapter 5). The GTP-shift experiment is a method to estimate the efficacy of an agonist in binding studies.

The previous discussions indicate how binding experiments can be useful in characterizing and quantifying the activity of drugs (provided the effects are detectable as changes in ligand affinity). As for any experimental procedure, there are certain prerequisite conditions that must be attained for the correct application of this technique to the study of drugs and receptors. A short list of required and optimal experimental conditions for successful binding experiments is given in Table 4.2. Some special experimental procedures for determining equilibrium conditions involve the adjustment of biological

TABLE 4.2
Criteria for Binding Experiments.

Minimal criteria and optimal conditions for binding experiments:

- The means of making the ligand chemically detectable (i.e., addition of radioisotope label, fluorescent probe) does not significantly alter the receptor biology of the molecule.
- The binding is saturable.
- The binding is reversible and able to be displaced by other ligands.
- There is a ligand available to determine nonspecific binding.
- There is sufficient biological binding material to yield a good signal-to-noise ratio but not too much so as to cause depletion of the tracer ligand.

For optimum binding experiments, the following conditions should be met:

- There is a high degree of specific binding and a concomitantly low degree of nonspecific binding.
- Agonist and antagonist tracer ligands are available.
- The kinetics of binding are rapid.
- The ligand used for determination of nonspecific binding has a different molecular structure from the tracer ligand.

material (i.e., membrane or cells) for maximal signal-to-noise ratios and/or temporal approach to equilibrium. These are outlined in material following.

4.4 Experimental Prerequisites for Correct Application of Binding Techniques

4.4.1 The Effect of Protein Concentration on Binding Curves

In the quest for optimal conditions for binding experiments, there are two mutually exclusive factors with regard to the amount of receptor used for the binding reaction. On one hand, increasing receptor (B_{max}) also increases the signal strength and usually the signal-to-noise ratio. This is a useful variable to manipulate. On the other hand, a very important prerequisite to the use of the Langmuirian type kinetics for binding curves is that the binding reaction does not change the concentration of tracer ligand being bound. If this is violated (i.e., if the binding is high enough to deplete the ligand), then distortion of the binding curves will result. The amount of tracer ligand-receptor complex as a function of the amount of receptor protein present is given as (see Section 4.6.9)

$$[A^*R] = \frac{1}{2}\left\{[A_T^*] + K_d + B_{max} - \sqrt{([A_T^*] + K_d + B_{max})^2 - 4[A_T^*]B_{max}}\right\}, \quad (4.20)$$

where the radioligand-receptor complex is $[A^*R]$ and $[A_T^*]$ is the total concentration of radioligand. Ideally, the amount of receptor (magnitude of B_{max}) should not limit the amount of $[A^*R]$ complex formed and there should be a linear relationship between $[A^*R]$ and B_{max}. However, Equation 4.20 indicates that the amount of $[A^*R]$ complex formed for a given $[A^*]$ indeed can be limited by the

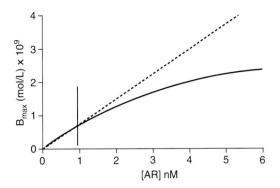

FIGURE 4.20 Effect of increasing protein concentration on the binding of a tracer ligand present at a concentration of $3 \times K_d$. Ordinates: [A*R] in moles/L calculated with Equation 4.20. Abscissae: B_{max} in moles/L $\times 10^9$. Values of B_{max} greater than the vertical solid line indicate region where the relationship between B_{max} and [A*R] begins to be nonlinear and where aberrations in the binding curves will be expected to occur.

amount of receptor present (magnitude of B_{max}) as B_{max} values exceed K_d. A graph of [A*R] for a concentration of [A*] $= 3 \times K_d$ as a function of B_{max} is shown in Figure 4.20. It can be seen that as B_{max} increases the relationship changes from linear to curvilinear as the receptor begins to deplete the tracer ligand. The degree of curvature varies with the initial amount of [A*] present. Lower concentrations are affected at lower B_{max} values than are higher concentrations. The relationship between [AR] and B_{max} for a range of concentrations of [A*] is shown in Figure 4.21a. When B_{max} levels are exceeded (beyond the linear range), saturation curves shift to the right and do not come to an observable maximal asymptotic value. The effect of excess receptor concentrations on a saturation curve is shown in Figure 4.21b.

For displacement curves, a similar error occurs with excess protein concentrations. The concentration of [A*R] in the presence of a nontracer-displacing ligand [B] as a

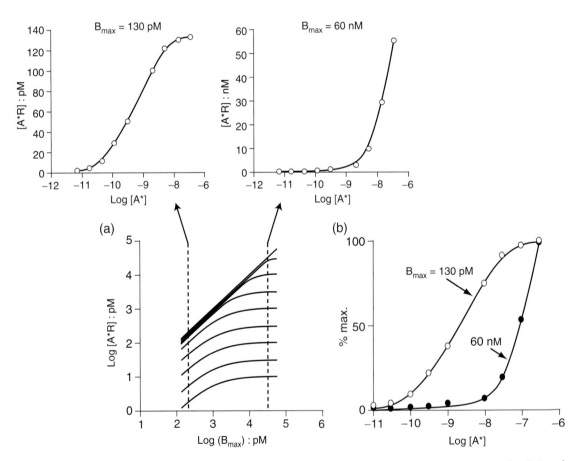

FIGURE 4.21 Effects of excess protein on saturation curves. (a) Bound ligand for a range of concentrations of radioligand, as a function of pM of receptor (Figure 4.20 is one example of these types of curves). The binding of the range of concentrations of radioligands are taken at two values of B_{max} (shown by the dotted lines; namely, 130 pM and 60 nM) and plotted as saturation curves for both B_{max} values on the top panels (note the difference in the ordinate scales). (b) The saturation curves shown on the top panels are replotted as a percentage of the maximal binding for each level of B_{max}. These comparable scales allow comparison of the saturation curves and show the dextral displacement of the curves with increasing protein concentration.

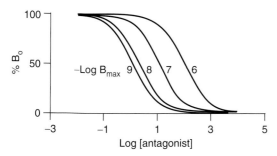

FIGURE 4.22 Effect of excess protein concentration on displacement curves (as predicted by Equation 4.21). As the B_{max} increases ($-\log B_{max}$ values shown next to curves) the displacement curves shift to the right.

function of B_{max} is given by (see Section 4.6.10)

$$[A^*R] = \frac{1}{2}\left\{ [A_T^*] + K_d(1 + [B]/K_B) + B_{max} \right.$$
$$\left. - \sqrt{\left([A_T^*] + K_d(1 + [B]/K_B) + B_{max}\right)^2 - 4[A_T^*]B_{max}} \right\},$$

$$(4.21)$$

where the concentration of the displacing ligand is [B] and K_B is the equilibrium dissociation constant of the displacing ligand-receptor complex. A shift to the right of displacement curves, with a resulting error in the IC_{50} values, occurs with excess protein concentration (see Figure 4.22).

4.4.2 The Importance of Equilibration Time for Equilibrium Between Two Ligands

In terms of ensuring that adequate time is allowed for the attainment of equilibrium between a single ligand and receptors, the experiment shown in Figure 4.1 is useful. However, in displacement experiments there are two ligands (tracer and nontraceable ligand) present and they must compete for the receptor. This competition can take considerably longer than the time required for just a single ligand. This is because the free ligands can only bind to free unbound receptors (except in the case of allosteric mechanisms, *vide infra*). Therefore, the likelihood of a receptor being free to accept a ligand depends on the reversibility of the other ligand, and vice versa. The fractional occupancy at time t for a ligand [A*] bound to a receptor (denoted [A*R_t]) in the presence of another ligand [B] has been derived by [17]

$$[A^*R]_t = \frac{k_1[R_t][A_T^*]}{K_F - K_S}$$
$$\times \left(\frac{k_4(K_F + K_S)}{K_F K_S} + \frac{(k_4 - K_F)e^{-K_F t}}{K_F} - \frac{(k_4 - K_S)e^{-K_S t}}{K_S} \right),$$

$$(4.22)$$

where

$$K_A = k_1[A^*] + k_2$$

$$K_B = k_3[B] + k_4$$

$$K_F = 0.5(K_A + K_B) + \sqrt{(K_A + K_B)^2 + 4k_1k_3[A^*][B]}$$

$$K_F = 0.5(K_A + K_B) - \sqrt{(K_A + K_B)^2 + 4k_1k_3[A^*][B]}.$$

Radioligand binding experiments are usually initiated by addition of the membrane to a premade mixture of radioactive and nonradioactive ligand. After a period of time thought adequate to achieve equilibrium (guided by experiments like that shown in Figure 4.1), the binding reaction is halted and the amount of bound radioligand quantified. Figure 4.23 shows the potential hazard of using kinetics observed for a single ligand (i.e., the radioligand) as being indicative of a two-ligand system. In the absence of another ligand, Figure 4.23a shows that the radioligand comes to equilibrium binding within 30 minutes. However, in the presence of a receptor antagonist (at two concentrations $[B]/K_B = 10$ and 30) a clearly biphasic receptor occupancy pattern by the radioligand can be observed where the radioligand binds to free receptors quickly (before occupancy by the slower acting antagonist) and then a reequilibration occurs as the radioligand and antagonist redistribute according to the rate constants for receptor occupancy of each. The equilibrium for the two ligands does not occur until > 240 minutes. Figure 4.23b shows the difference in the measured affinity of the antagonist at times = 30 and 240 minutes. It can be seen from this figure that the times thought adequate from observation of a single ligand to the receptor (as that shown in Figure 4.1) may be quite inadequate compared to the time needed for two ligands to come to temporal equilibrium with the receptor. Therefore, in the case of displacement experiments utilizing more than one ligand temporal experiments should be carried out to ensure that adequate times are allowed for complete equilibrium to be achieved for two ligands.

4.5 Chapter Summary and Conclusions

- If there is a means to detect (i.e., radioactivity, fluorescence) and differentiate between protein-bound and free ligand in solution, then binding can directly quantify the interaction between ligands and receptors.
- Binding experiments are done in three general modes: saturation, displacement, and kinetic binding.
- Saturation binding requires a traceable ligand but directly measures the interaction between a ligand and a receptor.
- Displacement binding can be done with any molecule and measures the interference of the molecule with a bound tracer.
- Displacement experiments yield an inverse sigmoidal curve for nearly all modes of antagonism.

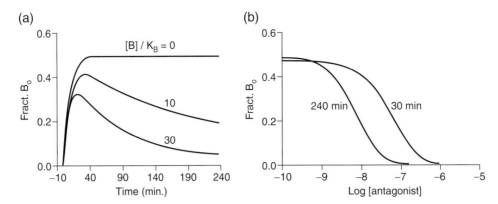

FIGURE 4.23 Time course for equilibration of two ligands for a single receptor. (a) Time course for displacement of a radioligand present at a concentration of $[A^*]/K_d = 1$. Kinetic parameter for the radioligand $k_1 = 10^5\,s^{-1}\,mol^{-1}$, $k_2 = 0.05\,s^{-1}$. Equilibrium is attained within 30 minutes in the absence of a second ligand ($[B]/K_B = 0$). Addition of an antagonist (kinetic parameters $= k_1 = 10^6\,s^{-1}\,mol^{-1}$, $k_2 = 0.001\,s^{-1}$) at concentrations of $[B]/K_B = 10$ and 30, as shown in panel A. (b) Displacement of radioligand $[A^*]$ by the antagonist B measured at 30 minutes and at 240 minutes. It can be seen that a tenfold error in the potency of the displacing ligand [B] is introduced into the experiment by inadequate equilibration time.

Competitive, noncompetitive, and allosteric antagonism can be discerned from the pattern of multiple displacement curves.

- Allosteric antagonism is characterized by the fact that it attains a maximal value. A sensitive method for the detection of allosteric effects is through studying the kinetics of binding.
- Kinetic experiments are also useful to determine the time needed for attainment of equilibria and to confirm reversibility of binding.
- Agonists can produce complex binding profiles due to the formation of different protein species (i.e., ternary complexes with G-proteins). The extent of this phenomenon is related to the magnitude of agonist efficacy and can be used to quantify efficacy.
- While the signal-to-noise ratio can be improved with increasing the amount of membrane used in binding studies, too much membrane can lead to depletion of radioligand with a concomitant introduction of errors in the estimates of ligand affinity.
- The time to reach equilibrium for two ligands and a receptor can be much greater than that required for a single receptor and a single ligand.

4.6 Derivations

- Displacement binding: competitive interaction (4.6.1)
- Displacement binding: noncompetitive interaction (4.6.2)
- Displacement of a radioligand by an allosteric antagonist (4.6.3)
- Relationship between IC_{50} and K_I for competitive antagonists (4.6.4)
- Maximal inhibition of binding by an allosteric antagonist (4.6.5)

- Relationship between IC_{50} and K_I for allosteric antagonists (4.6.6)
- Two-stage binding reactions (4.6.7)
- The effect of G-protein coupling on observed agonist affinity (4.6.8)
- Effect of excess receptor in binding experiments: saturation binding curve (4.6.9)
- Effect of excess receptor in binding experiments: displacement experiments (4.6.10)

4.6.1 Displacement Binding: Competitive Interactions

The effect of a nonradioactive ligand [B] displacing a radioligand $[A^*]$ by a competitive interaction is shown schematically as

$$
\begin{array}{c}
B \\
+ \\
A + R \xrightleftharpoons{\;K_a\;} AR \\
K_b \big\downarrow\big\uparrow \\
BR
\end{array}
\qquad (4.23)
$$

where K_a and K_b are the respective ligand-receptor association constants for radioligand and nonradioactive ligand. The following equilibrium constants are defined

$$[R] = \frac{[A^*R]}{[A^*]K_a} \qquad (4.24)$$

$$[BR] = K_b[B][R] = \frac{K_b[B][A^*R]}{[A^*]K_a} \qquad (4.25)$$

Total receptor concentration $[R_{tot}] = [R] + [A^*R] + [BR]$.

$$(4.26)$$

This leads to the expression for the radioactive species $[A^*R]/[R_{tot}]$ (denoted as ρ^*):

$$\rho^* = \frac{[A^*]K_a}{[A^*]K_a + [B]K_b + 1}. \qquad (4.27)$$

Converting to equilibrium dissociation constants (i.e., $K_d = 1/K_a$) leads to the equation

$$\rho^* = \frac{[A^*]/K_d}{[A^*]/K_d + [B]/K_B + 1}. \qquad (4.28)$$

4.6.2 Displacement Binding: Noncompetitive Interaction

It is assumed that mass action defines the binding of the radioligand to the receptor and that the nonradioactive ligand precludes binding of the radioligand $[A^*]$ to receptor. There is no interaction between the radioligand and displacing ligand. Therefore, the receptor occupancy by the radioligand is defined by mass action times the fraction q of receptor not occupied by noncompetitive antagonist:

$$\rho^* = \frac{[A^*]/K_d}{[A^*]/K_d + 1} \cdot q, \qquad (4.29)$$

where K_d is the equilibrium dissociation constant of the radioligand-receptor complex. The fraction of receptor bound by the noncompetitive antagonist is given as $(1 - q)$. This yields the following expression for q:

$$q = (1 + [B]/K_B)^{-1}. \qquad (4.30)$$

Combining Equations 4.29 and 4.30 and rearranging yield the following expression for radioligand bound in the presence of a noncompetititve antagonist:

$$\rho^* = \frac{[A^*]/K_d}{[A^*]/K_d([B]/K_B + 1) + [B]/K_B + 1}. \qquad (4.31)$$

The concentration that reduces binding by 50% is denoted as the IC_{50}. The following relation can be defined:

$$\frac{[A^*]/K_d}{[A^*]/K_d(IC_{50}/K_B + 1) + IC_{50}/K_B + 1} = \frac{0.5[A^*]/K_d}{[A^*]/K_d + 1}. \qquad (4.32)$$

It can be seen that the equality defined in Equation 4.32 is true only when $IC_{50} = K_B$ (i.e., the concentration of a non-competitive antagonist that reduces the binding of a tracer ligand by 50% is equal to the equilibrium dissociation constant of the antagonist-receptor complex).

4.6.3 Displacement of a Radioligand by an Allosteric Antagonist

It is assumed that the radioligand $[A^*]$ binds to a site separate from one binding an allosteric antagonist $[B]$.

Both ligands have equilibrium association constants for receptor complexes of K_a and K_b, respectively. The binding of either ligand to the receptor modifies the affinity of the receptor for the other ligand by a factor α. There can be three ligand-bound receptor species; namely $[A^*R]$, $[BR]$, and $[BA^*R]$:

$$
\begin{array}{ccc}
B & & B \\
+ & & + \\
A^* + R & \xrightarrow{K_a} A^*R \\
\Big\downarrow\Big\uparrow K_b & & \Big\downarrow\Big\uparrow \alpha K_b \\
A^* + BR & \xrightarrow{\alpha K_a} A^*RB
\end{array}
\qquad (4.33)
$$

The resulting equilibrium equations are:

$$K_a = [A^*R]/[A^*][R] \qquad (4.34)$$

$$K_b = [BR]/[B][R] \qquad (4.35)$$

$$\alpha K_a = [A^*RB]/[BR][A^*] \qquad (4.36)$$

$$\alpha K_b = [A^*RB]/[A^*R][B]. \qquad (4.37)$$

Solving for the radioligand-bound receptor species $[A^*R]$ and $[A^*RB]$ as a function of the total receptor species $([R_{tot}] = [R] + [A^*R] + [BR] + [A^*RB])$ yields

$$
\frac{[A^*R] + [A^*RB]}{[R_{tot}]}
$$

$$
= \frac{((1/\alpha[B]K_b) + 1)}{((1/\alpha[B]K_b) + (1/\alpha K_a) + (1/\alpha[A^*]K_aK_b) + 1)}. \qquad (4.38)
$$

Simplifying and changing association to dissociation constants (i.e., $K_d = 1/K_a$) yield (as defined by Ehlert, [18])

$$\rho^* = \frac{[A^*]/K_d(1 + \alpha[B]/K_B)}{[A^*]/K_d(1 + \alpha[B]/K_B) + [B]/K_B + 1}. \qquad (4.39)$$

4.6.4 Relationship Between IC_{50} and K_I for Competitive Antagonists

A concentration of displacing ligand that produces a 50% decrease in ρ^* is defined as the IC_{50}. The following relation can be defined:

$$\frac{[A^*]/K_d}{[A^*]/K_d + 1} = \frac{0.5[A^*]/K_d}{[A^*]/K_d + IC_{50}/K_B + 1}. \qquad (4.40)$$

From this, the relationship between the IC_{50} and the amount of tracer ligand $[A^*]$ is defined as [2]

$$IC_{50} = K_B \cdot ([A^*]/K_d + 1).$$

4.6.5 Maximal Inhibition of Binding by an Allosteric Antagonist

From Equation 4.39, the ratio of bound radioligand [A*] in the absence and presence of an allosteric antagonist [B], denoted by ρ_{A*}/ρ_{A*B}, is given by

$$\frac{\rho_{A*B}}{\rho_{A*}} = \frac{[A^*]/K_d(1 + \alpha[B]/K_B) + [B]/K_B + 1}{([A^*]/K_d + 1) \cdot (1 + \alpha[B]/K_B)}. \quad (4.41)$$

The fractional inhibition is the reciprocal; namely, ρ_{A*}/ρ_{A*B}. The maximal fractional inhibition occurs as $[B]/K_B \to \infty$. Under these circumstances, maximal inhibition is given by

$$\text{Maximal Inhibition} = \frac{[A^*]/K_d + 1}{[A^*]/K_d + 1/\alpha}. \quad (4.42)$$

4.6.6 Relationship Between IC$_{50}$ and K$_I$ for Allosteric Antagonists

The concentration of allosteric antagonist [B] that reduces a signal from a bound amount [A*] of radioligand by 50% is defined as the IC$_{50}$:

$$\frac{(1 + [A^*]/K_d)}{([A^*]/K_d(1 + \alpha IC_{50}/K_B) + IC_{50}/K_B + 1)} = 0.5. \quad (4.43)$$

This equation reduces to:

$$IC_{50} = K_B \frac{(1 + ([A^*]/K_d))}{(1 + \alpha([A^*]/K_d))}. \quad (4.44)$$

4.6.7 Two-stage Binding Reactions

Assume that the ligand [A] binds to receptor [R] to produce a complex [AR], and by that reaction changes the receptor from [R] to [R*]:

$$A + R \underset{}{\overset{K_a}{\rightleftarrows}} AR \underset{\sigma}{\overset{\chi}{\rightleftarrows}} AR^*. \quad (4.45)$$

The equilibrium equations are:

$$K_a = [A][R]/[AR] \ldots \quad (4.46)$$

$$\chi/\sigma = [AR]/[AR^*]. \quad (4.47)$$

The receptor conservation equation is

$$[R_{tot}] = [R] + [AR] + [AR^*]. \quad (4.48)$$

Therefore, the quantity of end product [AR*] formed for various concentrations of [A] is given as

$$\frac{[AR^*]}{[R_{tot}]} = \frac{[A]/K_A}{[A]/K_A(1 + \chi/\sigma) + \chi/\sigma}, \quad (4.49)$$

where $K_A = 1/K_a$. The observed equilibrium dissociation constant (K_{obs}) of the complete two-stage process is given as

$$K_{obs} = \frac{K_A \cdot \chi/\sigma}{1 + \chi/\sigma}. \quad (4.50)$$

It can be seen that for non-zero positive values of χ/σ (binding promotes formation of R*), $K_{obs} < K_A$.

4.6.8 The Effect of G-protein Coupling on Observed Agonist Affinity

Receptor [R] binds to agonist [A] and goes on to form a ternary complex with G-protein [G]:

$$A + R \overset{K_a}{\rightleftarrows} AR + [G] \overset{K_g}{\rightleftarrows} ARG. \quad (4.51)$$

The equilibrium equations are:

$$K_a = [A][R]/[AR] \ldots \quad (4.52)$$

$$K_g = [AR][G]/[ARG]. \quad (4.53)$$

The receptor conservation equation is

$$[R_{tot}] = [R] + [AR] + [ARG]. \quad (4.54)$$

Converting association to dissociation constants (i.e., $1/K_a = K_A$):

$$\frac{[ARG]}{[R_{tot}]} = \frac{([A]/K_A)([G]/K_G)}{[A]/K_A(1 + [G]/K_G) + 1}. \quad (4.55)$$

The observed affinity according to Equation 4.5.5 is

$$K_{obs} = \frac{K_A}{1 + ([G]/K_G)}. \quad (4.56)$$

4.6.9 Effect of Excess Receptor in Binding Experiments: Saturation Binding Curve

The Langmuir adsorption isotherm for radioligand binding [A*] to a receptor to form a radioligand-receptor complex [A*R] can be rewritten in terms of one where it is not assumed that receptor binding produces a negligible effect on the free concentration of ligand ([A*$_{free}$]):

$$[A^*R] = \frac{([A_T^*] - [A^*R])B_{max}}{[A_T^*] - [A^*R] + K_d}, \quad (4.57)$$

where B_{max} reflects the maximal binding (in this case, the maximal amount of radioligand-receptor complex). Under these circumstances, analogous to the derivation shown in Section 2.11.4, the concentration of radioligand bound is

$$[A^*R]^2 - [A^*R](B_{max} + [A_T^*] + K_d) + [A_T^*]B_{max} = 0. \quad (4.58)$$

One solution to Equation 4.58 is

$$[A^*R] = \frac{1}{2}\left\{[A_T^*] + K_d + B_{max} \right.$$
$$\left. - \sqrt{([A_T^*] + K_d + B_{max})^2 - 4[A_T^*]B_{max}} \right\}. \qquad (4.59)$$

4.6.10 Effect of Excess Receptor in Binding Experiments: Displacement Experiments

The equation for displacement of a radioligand [A*] by a nonradioactive ligand [B] can be rewritten in terms of one where it is not assumed that receptor binding does not deplete the amount of radioligand in the medium (no change in [A*$_{free}$]):

$$[A^*R] = \frac{([A_T^*] - [A^*R])B_{max}}{[A_T^*] - [A^*R] + K_d + [B]/K_B}, \qquad (4.60)$$

where B_{max} reflects the maximal binding the maximal amount of radioligand-receptor complex. Under these circumstances, the concentration of radioligand bound in the presence of a nonradioactive ligand displacement is

$$[A^*R]^2 - [A^*R](B_{max} + [A_T^*] + K_d(1 + [B]/K_B)$$
$$+ [A_T^*]B_{max} = 0. \qquad (4.61)$$

One solution to Equation 4.61 is

$$[A^*R] = \frac{1}{2}\left\{[A_T^*] + K_d(1 + [B]/K_B) + B_{max} \right.$$
$$\left. - \sqrt{([A_T^*] + K_d(1 + [B]/K_B) + B_{max})^2 - 4[A_T^*]B_{max}} \right\} \qquad (4.62)$$

References

1. Hulme, E. C. (1990). *Receptor biochemistry: A practical approach*. Oxford University Press, Oxford.
2. Klotz, I. M. (1997). *Ligand-receptor energetics: A guide for the perplexed*. John Wiley and Sons, New York.
3. Limbird, L. E. (1995). *Cell surface receptors: A short course on theory and methods*. Martinus Nihjoff, Boston.
4. Chen, W.-J., Armour, S., Way, J., Chen, G., Watson, C., Irving, P., Cobb, J., Kadwell, S., Beaumont, K., Rimele, T., and Kenakin, T. P. (1997). Expression cloning and receptor pharmacology of human calcitonin receptors from MCF-7 cells and their relationship to amylin receptors. *Mol. Pharmacol.* **52:**1164–1175.
5. Cheng, Y. C., and Prusoff, W. H. (1973). Relationship between the inhibition constant (Ki) and the concentration of inhibitor which causes 50 percent inhibition (I50) of an enzymatic reaction. *Biochem. Pharmacol.* **22:**3099–3108.
6. Sabroe, I., Peck, M. J., Van Keulen, B. J., Jorritsma, A., Simmons, G., Clapham, P. R., Williams, T. J., and Pease, J. E. (2000). A small molecule antagonist of chemokine receptors CCR1 and CCR3. *J. Biol. Chem.* **275:**25985–25992.
7. Hejnova, L., Tucek, S., and El-Fakahany, E. E. (1995). Positive and negative allosteric interactions on muscarinic receptors. *Eur. J. Pharmacol.* **291:**427–430.
8. Jakubic, J., and El-Fakahany, E. E. (1997). Positive cooperativity of acetylcholine and other agonists with allosteric ligands on muscarinic acetylcholine receptors. *Mol. Pharmacol.* **52:**172–177.
9. Jakubic, J., Bacakova, L., El-Fakahany, E. E., and Tucek, S. (1997). Positive cooperativity of acetylcholine and other agonists with allosteric ligands on muscarinic acetylcholine receptors. *Mol. Pharmacol.* **52:**172–179.
10. Proska, J., and Tucek, S. (1994). Mechanisms of steric and cooperative interactions of alcuronium on cardiac muscarinic acetylcholine receptors. *Mol. Pharmacol.* **45:**709–717.
11. Christopoulos, A. (2000). Quantification of allosteric interactions at G-protein coupled receptors using radioligand assays. In: *Current protocol in pharmacology*, edited by Enna, S. J., pp. 1.22.21–1.22.40. Wiley and Sons, New York.
12. Lazareno, S., and Birdsall, N. J. M. (1995). Detection, quantitation, and verification of allosteric interactions of agents with labeled and unlabeled ligands at G protein-coupled receptors: Interactions of strychnine and acetylcholine at muscarinic receptors. *Mol. Pharmacol.* **48:**362–378.
13. Leppick, R. A., Lazareno, S., Mynett, A., and Birdsall, N. J. (1998). Characterization of the allosteric interactions between antagonists and amiloride at the human α_{2A}-adrenergic receptor. *Mol. Pharmacol.* **53:**916–925.
14. Florio, V. A., and Sternweis, P. C. (1989). Mechanism of muscarinic receptor action on G$_o$ in reconstituted phospholipid vesicles. *J. Biol. Chem.* **264:**3909–3915.
15. Gerwins, P., Nordstedt, C., and Fredholm, B. B. (1990). Characterization of adenosine A$_1$ receptors in intact DDT$_1$ MF-2 smooth muscle cells. *Mol. Pharmacol.* **38:**660–666.
16. Lefkowitz, R. J., Caron, M. G., Michel, T., and Stadel, J. M. (1982). Mechanisms of hormone-effector coupling: The β-adrenergic receptor and adenylate cyclase. *Fed. Proc.* **41:**2664–2670.
17. Motulsky, H. J., and Mahan, L. C. (1984). The kinetics of competitive radioligand binding predicted by the law of mass action. *Mol. Pharmacol.* **25:**1–9.
18. Ehlert, F. J. (1985). The relationship between muscarinic receptor occupancy and adenylate cyclase inhibition in the rabbit myocardium. *Mol. Pharmacol.* **28:**410–421.

5

Agonists: The Measurement of Affinity and Efficacy in Functional Assays

Cells let us walk, talk, think, make love, and realize the bath water is cold.
— **LORRAINE LEE CUDMORE**

5.1 Functional Pharmacological Experiments

Another major approach to the testing of drug activity is with functional assays. These are comprised of any biological system that yields a biochemical product or physiological response to drug stimulation. Such assays detect molecules that produce biological response or those that block the production of physiological response. These can be whole tissues, cells in culture, or membrane preparations. Like biochemical binding studies, the pharmacological output can be tailored by using selective stimulation. Whereas the output can be selected by the choice of radioligand or other traceable probe with binding studies, in functional studies the output can be selected by choice of agonist. When necessary, selective antagonists can be used to obviate unwanted functional responses and isolate the receptor of interest. This practice was more prevalent in isolated tissue studies where the tissue was chosen for the presence of the target receptor, and in some cases this came with concomitant presence of other related and obfuscating receptor responses. In recombinant systems, a surrogate host cell line with a blank cellular background can often be chosen. This results in much more selective systems and less need for selective agonist probes.

There are two main differences between binding and functional experiments. The first is that functional responses are usually highly amplified translations of receptor stimulus (see Chapter 2). Therefore, while binding signals emanate from complete receptor populations functional readouts often utilize only a small fraction of the receptor population in the preparation. This can lead to a greatly increased sensitivity to drugs that possess efficacy. No differences should be seen for antagonists. This amplification can be especially important for the detection of agonism since potency may be more a function of ligand efficacy than affinity. Thus, a highly efficacious agonist may produce detectable responses at 100 to 1,000 times lower concentrations than those that produce measurable amounts of displacement of a tracer in binding studies. The complex interplay between affinity and efficacy can be misleading in structure activity studies for agonists. For example, Figure 5.1 shows the lack of correlation of relative agonist potency for two dopamine receptor subtypes and the binding affinity on those receptor subtypes for a series of dopamine agonists. This data show, that for these molecules changes in chemical structure lead to changes in relative efficacy not reflected in the affinity measurement. The relevant activity is relative agonist potency. Therefore, the affinity data is misleading. In this case, a functional assay is the correct approach for optimization of these molecules.

Functional assays give flexibility in terms of what biochemical functional response can be monitored for drug activity. Figure 5.2 shows some of the possibilities. In some cases, the immediate receptor stimulus can be observed, such as the activation of G-proteins by agonist-activated receptor. Specifically, this is in the observation of an increased rate of exchange of GDP to GTP on the G-protein α-subunit. Following G-protein activation comes initiation of effector mechanisms. For example, this can include activation of the enzyme adenylyl cyclase to produce the second messenger cyclic AMP. This and other second messengers go on to activate enzymatic biochemical cascades within the cell. A second layer of response observation is the measurement of the quantity of these second messengers. Yet another layer of response is the observation of the effects of the second messengers. Thus, activation of enzymes such as MAPKinase can be used to monitor drug activity.

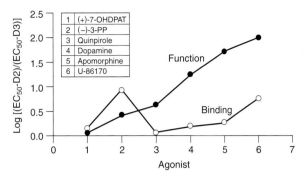

FIGURE 5.1 Ratio of affinity (open circles) and agonist potency (filled circles) for dopamine agonists on dopamine D2 vs D3 receptors. Abscissae: numbers referring to agonist key on right. Data calculated from [1].

A second difference between binding and function is the quality of drug effect that can be observed. Specifically, functional studies reveal interactions between receptors and cellular components that may not be observed in binding studies such as some allosteric effects or other responses in a receptor's pharmacological repertoire (i.e., receptor internalization). For example, the cholecystokinin (CCK) receptor antagonist D-Tyr-Gly-[(Nle28,31,D-Trp30)cholecystokinin-26-32]-phenethyl ester is a receptor antagonist and does not produce receptor stimulation. While ostensibly this may appear to indicate a lack of

efficacy, this ligand does produce profound receptor internalization [2]. Therefore, a different kind of efficacy is revealed in functional studies that would not have been evident in binding.

A practical consideration is the need for a radioactive ligand in binding studies. There are instances where there is no such traceable probe or it is too expensive to be a viable approach. Functional studies require only that an endogenous agonist be available. As with binding studies, dissimulations in the value of the independent variable (namely, drug concentration) lead to corresponding errors in the observed value of the dependent variable (in the case of functional experiments, cellular response). The factors involved (namely, drug solubility and adsorption, see Chapter 2) are equally important in functional experiments. However, there are some additional factors unique to functional studies that should be considered. These are dealt with in Section 5.4.

5.2 The Choice of Functional Assays

There are a number of assay formats available to test drugs in a functional mode. As discussed in Chapter 2, a main theme throughout the various stimulus-response cascades found in cells is the amplification of receptor stimulus occurring as a function of the distance, in biochemical steps and reactions, away from the initial receptor event. Specifically, the further down the stimulus-

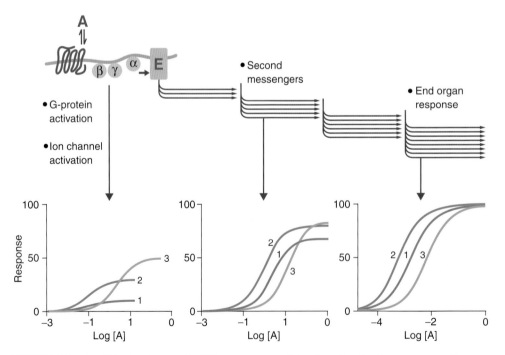

FIGURE 5.2 Amplification inherent in different vantage points along the stimulus-response pathway in cells. Agonists have a rank order of efficacy of 3 > 2 > 1 and a rank order of potency of 2 > 1 > 3. Assays proximal to the agonist-receptor interaction have the least amplification. The product of the initial interaction goes on to activate other processes in the cell. The signal is generally amplified. As this continues, texture with respect to differences in efficacy is lost and the agonists all demonstrate full agonism.

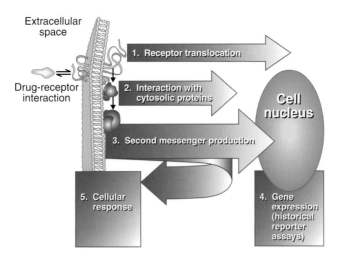

FIGURE 5.3 Different types of functional readouts of agonism. Receptors need not mediate cellular response but may demonstrate behaviors such as internalization into the cytoplasm of the cell (mechanism 1). Receptors can also interact with membrane proteins such as G-proteins (mechanism 2) and produce cytosolic messenger molecules (mechanism 3), which can go on to mediate gene expression (mechanism 4). Receptors can also mediate changes in cellular metabolism (mechanism 5).

response pathway the agonism is observed the more amplified the signal. Figure 5.3 illustrates the effects of three agonists at different points along the stimulus-response cascade of a hypothetical cell. At the initial step (i.e., G-protein activation, ion channel opening), all are partial agonists and it can be seen that the order of potency is $2 > 1 > 3$ and the order of efficacy is $3 > 2 > 1$. If the effects of these agonists were to be observed at a step further in the stimulus-response cascade (i.e., production of second messenger), it can be seen that agonists 2 and 3 are full agonists while agonist 1 is a partial agonist. Their rank order of potency does not change but now there is no distinction between the relative efficacies of agonists 2 and 3. At yet another step in the cascade (namely, end organ response), all are full agonists with the same rank order of potency. The point of this simulation is to note the differences, in terms of the characterization of the agonists (full versus partial agonists, relative orders of efficacy), that occur by simply viewing their effects at different points along the stimulus-response pathway.

Historically, isolated tissues have been used as the primary form of functional assay, but since these usually come from animals the species differences coupled with the fact that human recombinant systems now can be used have made this approach obsolete. Functional assays in whole-cell formats, where end organ response is observed (these will be referred to as group I assays), can be found as specialized cells such as melanophores, yeast cells, or microphysiometry assays. Group II assays record the product of a pharmacological stimulation (for example, an induction of a gene that goes on to produce a traceable product such as light sensitive protein). Second messengers

(such as cyclic AMP, calcium, and inositol triphosphate) can also be monitored directly either in whole-cell or broken-cell formats (group III assays). Finally, membrane assays such as the observation of binding of GTPγS to G-proteins can be used. While this is an assay done in binding mode, it measures the ability of agonists to induce response and thus may also be considered a functional assay. It is worth considering the strengths and shortcomings of all of these approaches:

Group I assays (end organ response) are the most highly amplified and therefore most sensitive assays. This is an advantage in screening for weakly efficacious agonists but has the disadvantage of showing all agonists above a given level of efficacy to be full agonists. Under these circumstances, information about efficacy cannot be discerned from the assay since at least for all the agonists that produce maximal system response no information regarding relative efficacy can be obtained. There are cell culture group I assays. One such approach uses microphysiometry. All cells respond to changes in metabolism by adjustment of internal hydrogen ion concentration. This process is tightly controlled by hydrogen ion pumps that extrude hydrogen ions into the medium surrounding the cell. Therefore, with extremely sensitive monitoring of the pH surrounding cells in culture a sensitive indicator of cellular function can be obtained. Microphysiometry measures the hydrogen ion extrusion of cells to yield a generic readout of cellular function. Agonists can perturb this control of hydrogen ion output. One of the major advantages of this format is that it is generic (i.e., the observed does not depend on the nature of the biochemical coupling mechanisms in the cytosol of the cell). For example, the success of cell transfection

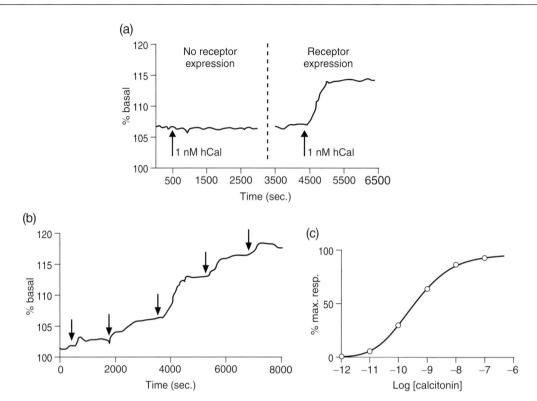

FIGURE 5.4 Microphysiometry responses of HEK 293 cells transfected with human calcitonin receptor. (a) Use of microphysiometry to detect receptor expression. Before transfection with human calcitonin receptor cDNA, HEK cells do not respond to human calcitonin. After transfection, calcitonin produces a metabolic response, thereby indicating successful membrane expression of receptors. (b) Cumulative concentration-response curve to human calcitonin shown in real time. Calcitonin added at the arrows in concentrations of 0.01, 0.1, 1.10, and 100 nM. Dose-response curve for the effects seen in panel B.

experiments can be monitored with microphysiometry. Unless receptors are biochemically tagged, it may be difficult to determine whether the transfection of cDNA for a receptor into a cell actually results in membrane expression of the receptor. On occasion, the cell is unable to process the cDNA to form the complete receptor and it is not expressed on the cell surface. Figure 5.4a shows microphysiometry responses to calcitonin (an agonist for the human calcitonin receptor) before and after transfection of the cells with cDNA for the human calcitonin receptor. The appearance of the calcitonin response indicates that successful membrane expression of the receptor occurred. Another positive feature of this format is the fact that responses can be observed in real time. This allows the observation of steady states and the possibility of obtaining cumulative dose-response curves to agonists (see Figure 5.4b and c).

A specialized cell type that is extremely valuable in drug discovery is the *Xenopus laevis* melanophore. This is a cell derived from the skin of frogs that controls the dispersion of pigment in response to receptor stimulation. Thus, activation of Gi protein causes the formation of small granules of pigment in the cell rendering them transparent to visible light. In contrast, activation of Gs and Gq protein causes dispersion of the melanin resulting in an opaque cell (loss of transmittance of visible light). Therefore, the activation of receptors can be observed in real time through changes in the transmittance of visible light through a cell monolayer. Figure 5.5 shows the activation of human β-adrenoceptors in melanophores by β-adrenoceptor agonists. It can be seen that activation of Gs protein by the activated β-adrenoceptor leads to an increase in pigmentation of the melanophore. This, in turn, is quantified as a reduced transmittance of visible light to yield graded responses to the agonists. One of the key features of this format is that the responses can be observed in real time. Figure 5.6a shows the reduced transmittance to visible light of melanophores transfected with human calcitonin receptor acitvated with the agonist human calcitonin. Another feature of this format is that the transfected receptors are very efficiently coupled (i.e., agonists are extremely potent in these systems). Figure 5.6b shows the dose-response curve to human calcitonin in transfected melanophores compared to the less efficiently coupled calcium fluorescence assay in human embryonic kidney cells for this same receptor.

Another specialized cell line that has been utilized for functional drug screening are yeast cells. A major

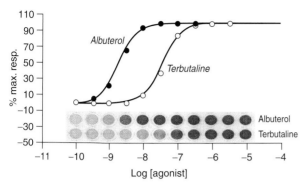

FIGURE 5.5 Melanophores, transfected with human β-adreno-ceptors, disperse melanin to become opaque when stimulated with β-adrenoceptors agonists such as albuterol and terbutaline. Inset shows light transmission through a melanophore cell monolayer with increasing concentration of agonist. Light transmission is quantified and can be used to calculate graded responses to the agonists.

advantage of this format is that there are few endogenous receptors and G-proteins, leading to a very low background signal (i.e., the major signal is the transfected receptor of interest). Yeast can be genetically altered to not grow in medium not containing histidine unless a previously transfected receptor is present. Coupled with the low maintenance and high growth rate, yeast cells are a viable system of high-throughput screening and secondary testing of drugs.

Group II assays consist of those monitoring *cellular second messengers*. Thus, activation of receptors to cause Gs-protein activation of adenylate cyclase will lead to elevation of cytosolic or extracellularly secreted cyclic AMP. This second messenger phosphorylates numerous cyclic AMP-dependent protein kinases, which go on to phosphorylate metabolic enzymes and transport and regulatory proteins (see Chapter 2). Cyclic AMP can be detected either radiometrically or with fluorescent probe technology.

Another major second messenger in cells is calcium ion. Virtually any mammalian cell line can be used to measure transient calcium currents in fluorescence assays when cells are preloaded with an indicator dye that allows monitoring of changes in cytosolic calcium concentration. These responses can be observed in real time, but a characteristic of these responses is that they are transient. This may lead to problems with hemi-equilibria in antagonist studies whereby the maximal responses to agonists may be depressed in the presence of antagonists. These effects are discussed more fully in Chapter 6.

Another approach to the measurement of functional cellular responses is through reporter assays (group III). Reporter assays yield an amount of cellular product made in response to stimulation of the cell. For example, elevation of cyclic AMP causes activation of protein kinase A. The activated subunits resulting from protein kinase A activation bind to cyclic AMP response element binding (CREB) protein, which then binds to a promoter region of cyclic-AMP-inducible genes. If the cell is previously stably transfected with genes for the transcription of luciferase in the nucleus of the cell, elevation of cyclic AMP will induce the transcription of this protein. Luciferase produces visible light when brought into contact with the substrate LucLite, and the amount of light produced is proportional to the amount of cyclic AMP produced. Therefore, the cyclic AMP produced through receptor stimulation leads to a measurable increase in the observed light produced upon lysis of the cell. There are numerous other reporter systems for cyclic AMP and inositol triphosphate, two prevalent second messengers in cells (see Chapter 2). It can be seen that such a transcription system has the potential for great sensitivity, since the time of exposure can be somewhat tailored to amplify the observed response. However, this very advantage can also be a disadvantage, since the time of exposure to possible toxic effects of drugs is also increased. One advantage of real-time assays such as melanophores and microphysio-metry is the ability to obtain responses in a short period of time and thereby possibly reducing toxic effects that require

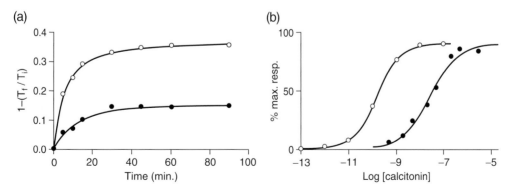

FIGURE 5.6 Calcitonin receptor responses. (a) Real-time melanin dispersion (reduced light transmittance) caused by agonist activation (with human calcitonin) of transfected human calcitonin receptors type II in melanophores. Responses to 0.1 nM (filled circles) and 10 nM (open circles) human calcitonin. (c) Dose-response curves to calcitonin in melanophores (open circles) and HEK 293 cells, indicating calcium transient responses (filled circles).

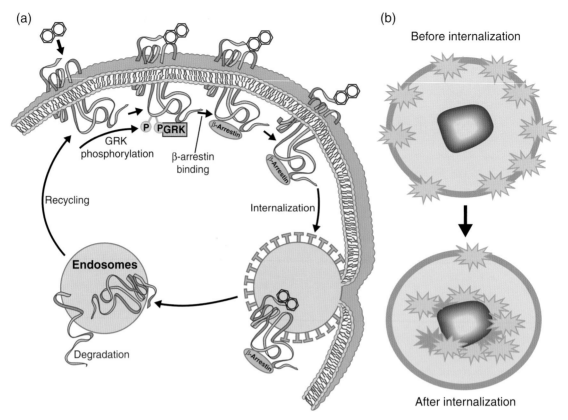

FIGURE 5.7   Internalization of GPCRs. (a) Receptors adopt an active conformation either spontaneously or through interaction with a ligand and become phosphorylated. This promotes β-arrestin binding, which precedes internalization of the receptor into clatherin pits. Receptors then are either degraded in endosomes or recycled to the cell surface. (b) A fluorescent analog of β-arrestin can be visualized and tracked according to location either at the cell membrane (receptors not internalized) or near the cell nucleus (internalized receptors). This enables detection of changes in GPCRs.

longer periods of time to become manifest. Reporter responses are routinely measured after a 24-hour incubation (to give sufficient time for gene transcription). Therefore, the exposure time to drug is increased with a concomitant possible increase in toxic effects.

Finally, receptor stimulus can be measured through membrane assays directly monitoring G-protein activation (group IV assays). In these assays, radiolabeled GTP (in a stable form; for example, GTPγS) is present in the medium. As receptor activation takes place, the GDP previously bound to the inactive state of the G-protein is released and the radiolabeled GTPγS binds to the G-protein. This is quantified to yield a measure of the rate of GDP/GTPγS exchange and hence receptor stimulus.

The majority of functional assays involve primary signaling. In the case of GPCRs, this involves activation of G-proteins. However, receptors have other behaviors—some of which can be monitored to detect ligand activity. For example, upon stimulation many receptors are desensitized through phosphorylation and subsequently taken into the cell and either recycled back to the cell surface or digested. This process can be monitored by observing ligand-mediated receptor internalization. For

many receptors this involves the migration of a cytosolic protein called β-arrestin. Therefore, the transfection of fluorescent β-arrestin to cells furnishes a method to track the movement of the fluorescent β-arrestin from the cytosol to the inner membrane surface as receptors are activated (Figure 5.7). Alternative approaches to detecting internalization of GPCRs involve pH-sensitive cyanine dyes such as CypHer-5 that fluoresce when irradiated with red laser light, but only in an acidic environment. Therefore, epitope tagging of GPCRs allows binding of antibodies labeled with CypHer-5 to allow detection of internalized receptors (those that are in the acidic internal environment of the cell and thus fluoresce to laser light) [3]. A general list of minimal and optimal conditions for functional assays is given in Table 5.1.

5.3 Recombinant Functional Systems

With the advent of molecular biology and the ability to express transfected genes (through transfection with cDNA) into surrogate cells to create functional recombinant systems has come a revolution in pharmacology.

TABLE 5.1
Minimal and optimal criteria for experiments utilizing cellular function.

Minimal

- An agonist and antagonist to define the response on the target are available.
- The agonist is reversible (after washing with drug-free medium).

Optimal

- The response should be sustained and not transient. No significant desensitization of the response occurs within the time span of the experiment.
- The response production should be rapid.
- The responses can be visualized in real time.
- There are independent methods to either modulate or potentiate functional responses.
- There is a capability to alter the receptor density (or cells available with a range of receptor densities).

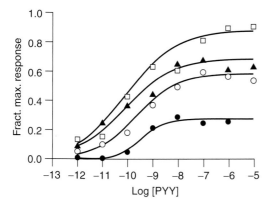

FIGURE 5.8 Dose-response curves to peptide PYY (YPAKPEAPGEDASPEELSRYYASLRHYLNLVTRQRY$_{NH2}$) in melanophores. Ordinates: minus values for $1 - T_f/T_i$ reflecting increases in light transmission. Abscissae: logarithms of molar concentrations of PYY. Cells transiently transfected with cDNA for the human NPY1 receptor. Levels of cDNA = 10 μg (filled circles), 20 μg (open circles), 40 μg (filled triangles), and 80 μg (open squares). Data redrawn from [4].

Previously, pharmacologists were constrained to the pre-wired sensitivity of isolated tissues for agonist study. As discussed in Chapter 2, different tissues possess different densities of receptor, different receptor co-proteins in the membranes, and different efficiencies of stimulus-response mechanisms. Judicious choice of tissue type could yield uniquely useful pharmacologic systems (i.e., sensitive screening tissues). However, before the availability of recombinant systems these choices were limited. With the ability to express different densities of human target proteins such as receptors has come a transformation in drug discovery. Recombinant cellular systems can now

be made with a range of sensitivities to agonists. The techniques involved in the construction of recombinant receptor systems is beyond the scope of this Chapter. However, some general ideas are useful in that they can be used for the creation of optimal systems for drug discovery.

The first idea to consider is the effect of receptor density on sensitivity of a functional system to agonists. Clearly, if quanta of stimulus are delivered to the stimulus-response mechanism of a cell per activated receptor the amount of the total stimulus will be directly proportional to the number of receptors activated. Figure 5.8 shows Gi-protein-mediated responses of melanophores transiently transfected with cDNA for human neuropeptide Y-1 receptors. As can be seen from this figure, increasing receptor expression (transfection with increasing concentrations of receptor cDNA) causes an increased potency and maximal response to the neuropeptide Y agonist PYY.

Receptor density has disparate effects on the potency and maximal responses to agonists. The operational model predicts that the EC$_{50}$ to an agonist will vary with receptor density according to the following relationship (see Section 3.13.3)

$$EC_{50} = \frac{K_A \cdot K_E}{[R_t] + K_E}, \qquad (5.1)$$

where $[R_t]$ is the receptor density, K_A is the equilibrium dissociation constant of the agonist-receptor complex, and K_E is the concentration of activated receptor that produces half maximal response (a measure of the efficiency of the stimulus-response mechanism of the system) (see Section 3.13.3 for further details). Similarly, the agonist maximal response is given by

$$\text{Maximal Response} = \frac{[R_t] \cdot E_{max}}{[R_t] + K_E}, \qquad (5.2)$$

where E_{max} is the maximal response capability of the system. It can be seen that increases in receptor density will cause an increase in agonist maximal response to the limit of the system maximum (i.e., until the agonist is a full agonist). Thereafter, increases in receptor density will have no further effect on the maximal response to the agonist. In contrast, Equation 5.1 predicts that increases in receptor density will have concomitant increases in the potency of full agonist with no limit. These effects are shown in Figure 5.9. It can be seen from this figure that at receptor density levels where the maximal response reaches an asymptote agonist potency increases linearly with increases in receptor density. Figure 5.9b shows the relationship between the pEC$_{50}$ for the β$_2$-adrenoceptor agonist isoproterenol and β$_2$-adrenoceptor density in rat C$_6$ glioma cells. It can be seen that while no further increases in maximal response are obtained the agonist potency increases with increasing receptor density.

Recombinant systems can also be engineered to produce receptor-mediated responses by introducing adjunct proteins. For example, it has been shown that the G$_{\alpha16}$ G-protein subunit couples universally to nearly all

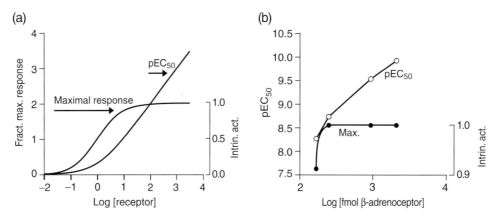

FIGURE 5.9 Effects of receptor density on functional assays. (a) Effect of increasing receptor density on potency (pEC$_{50}$) and maximal response to an agonist. Left ordinal axis is ratio of observed EC$_{50}$ and K$_A$ as −Log Scale; Right ordinal axis fraction of system maximal response (intrinsic activity). (b) Observed pEC$_{50}$ values for isoproterenol for increases in cyclic AMP in rat glioma cells transfected with human β$_2$-adrenoceptors (open circles) and maximal response to isoproterenol (as a fraction of system maxima, filled circles) as a function of β$_2$-adrenoceptor density on a log scale (fmol/mg protein). Data redrawn from [5].

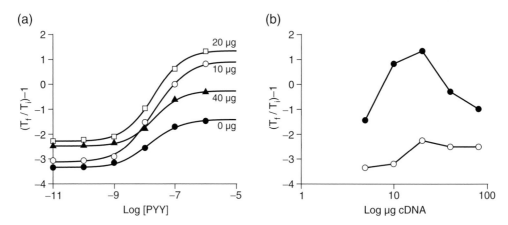

FIGURE 5.10 Effects of co-expressed G-protein (G$_{\alpha16}$) on neuropeptide NPY4 receptor responses (NPY-4). (a) Dose-response curves for NPY-4. Ordinates: *Xenopus laevis* melanophore responses (increases light transmission). Ordinates: logarithms of molar concentrations of neuropeptide Y peptide agonist PYY. Curves obtained after no co-transfection (labeled 0 μg) and co-transfection with cDNA for G$_{\alpha16}$. Numbers next to the curves indicate μg of cDNA of G$_{\alpha16}$ used for co-transfection. (b) Maximal response to neuropeptide Y (filled circles) and constitutive activity (open circles) as a function of μg cDNA of co-transfected G$_{\alpha16}$.

receptors [6]. In recombinant systems, where expression of the receptor does not produce a robust agonist response, co-transfection of the G$_{\alpha16}$ subunit can substantially enhance observed responses. Figure 5.10 shows that both the maximal response and potency of the neuropeptide Y peptide agonist PYY is enhanced when neuropeptide Y-4 receptors are co-transfected with cDNA for receptor and G$_{\alpha16}$. Similarly, other elements may be required for a useful functional assay. For example, expression of the gluatamate transporter EAAT1 (a glutamate aspirate transporter) is required in some cell lines to control extracellular glutamate levels (which lead to receptor desensitization) [7].

While high receptor density may strengthen an agonist signal, it may also reduce its fidelity. In cases where receptors are pleiotropic with respect to the G-proteins with which they interact (receptors interact with more than one G-protein), high receptor numbers may complicate signaling by recruitment of modulating signaling pathways. For example, Figure 5.11 shows a microphysiometry response to human calcitonin produced in human embryonic kidney cells transfected with human calcitonin receptor. It can be seen that the response is sustained. In a transfected cell line with a much higher receptor density, the response is not of higher magnitude and is also transient, presumably because

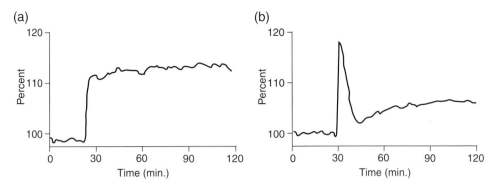

FIGURE 5.11 Microphysiometry responses to 1 nM human calcitonin. (a) Responses obtained from HEK 293 cells stably transfected with low levels of human calcitonin receptor (68 pM/mg protein). Response is sustained. (b) Response from HEK 293 cells stably transfected with high levels of receptor (30,000 pM/mg protein). Data redrawn from [8].

of complications due to the known pleiotropy of this receptor with other G-proteins. The responses in such systems are more difficult to quantify, and cumulative dose-response curves are not possible. These factors make a high-receptor-density system less desirable for pharmacological testing. This factor must be weighed against the possible therapeutic relevance of multiple G-protein coupling to the assay.

5.4 Functional Experiments: Dissimulation in Time

A potential problem when measuring drug activity relates to the temporal ability of systems to come to equilibrium, or at least to a steady state. Specifically, if there are temporal factors that interfere with the ability of the system to return cellular response or if real-time observation of response is not possible the time of exposure to drugs, especially agonists, becomes an important experimental variable. In practice, if responses are observed in real time then steady states can be observed and the experiment designed accordingly. The rate of response production can be described as a first-order process. Thus, the effect of a drug ([E]) expressed as a fraction of the maximal effect of that drug (receptors saturated by the drug, [E_m]) is

$$\frac{[E]}{[E_m]} = 1 - e^{-k_{on}t}, \tag{5.3}$$

where k_{on} is a first-order rate constant for approach of the response to the equilibrium value and t is time. The process of drug binding to a receptor will have a temporal component. Figure 5.12 shows three different rates of response production by an agonist or binding of a ligand in general. The absolute magnitude of the equilibrium binding is the same, but the time taken to achieve the effect is quite different. It can be seen from this figure that if response is measured at $t = 1,000$ s only drug A is at steady state. If comparisons are made at this time point, the effect of the other two drugs will be underestimated. As previously

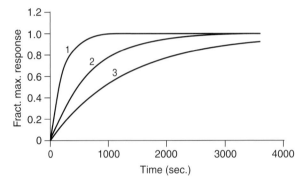

FIGURE 5.12 First-order rate of onset of response for three agonists of equal potency but differing rates of receptor onset. Ordinates: response at time t as a fraction of equilibrium response value. Abscissae: time in seconds. Curve 1: $k_1 = 3 \times 10^6 \, s^{-1} \, mol^{-1}$, $k_2 = 0.003 \, s^{-1}$. Curve 2: $k_1 = 10^6 \, s^{-1} \, mol^{-1}$, $k_2 = 0.001 \, s^{-1}$. Curve 3: $k_1 = 5 \times 10^5 \, s^{-1} \, mol^{-1}$, $k_2 = 0.0005 \, s^{-1}$.

noted, if responses are observed in real time steady states can be observed and temporal inequality ceases to be an issue. However, this can be an issue in stop-time experiments where real-time observation is not possible and the product of a drug response interaction is measured at a given time point. This is further discussed later in the chapter.

Another potential complication can occur if the responsiveness of the receptor system changes temporally. This can happen if the receptor (or host system, or both) demonstrates desensitization (tachyphylaxis) to drug stimulation (see Chapter 2). There are numerous systems where constant stimulation with a drug does not lead to a constant steady-state response. Rather, a "fade" of the response occurs. This can be due to depletion of a cofactor in the system producing the cellular response or a conformational change in the receptor protein. Such phenomena protect against overactive stimulation of

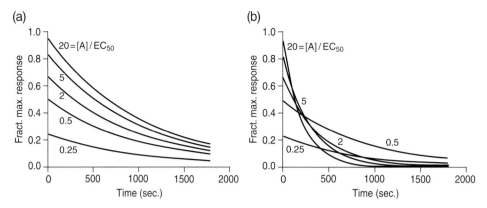

FIGURE 5.13 Fade of agonist-induced responses in systems with a uniform rate constant for desensitization (panel a) or a rate of desensitization proportional to the magnitude of the response (panel b). Abscissae: time in seconds. Ordinates: fractions of maximal response; responses ranging from 0.25 to 0.95x maximum. (a) Temporal response multiplied by an exponential decay of rate constant $10^{-3} \, s^{-1}$. Numbers refer to the concentration of agonist expressed as a fraction of the EC_{50}. (b) Rate constant for exponential decay equals the magnitude of the fractional response multiplied by a uniform rate constant $10^{-3} \, s^{-1}$. For panel b, the rate of desensitization increases with increasing response.

systems to physiological detriment. Whatever the cause, the resulting response to the drug is temporally unstable, leading to a dependence of the magnitude of the response on the time at which the response was recorded. The process of desensitization can be a first-order decay according to an exponential function, the time constant for which is independent of the magnitude of the response. Under these circumstances, the response tracings would resemble those shown in Figure 5.13a. Alternatively, the rate of desensitization may be dependent on the intensity of the stimulation (i.e., the greater the response the more rapid will be the desensitization). Under these circumstances, the fade in response will resemble a pattern shown in Figure 5.13b. These temporal instabilities can lead to underestimation of the response to the agonist. If the wrong time point for measurement of response is taken this can lead to a shift to the right of the agonist dose-response curve (Figure 5.14a) or a diminution of the true maximal response (see Figure 5.14b). Temporal studies must be done to ensure that the response values are not dependent on the time chosen for measurement.

5.5 Experiments in Real Time Versus Stop Time

The observation of dependent variable values (in functional experiments this is cellular response) as they happen (i.e., as the agonist or antagonist binds to the receptor and as the cell responds) is referred to as real time. In contrast, a response chosen at a single point in time is referred to as stop-time experimentation. There are certain experimental formats that must utilize stop-time measurement of responses since the preparation is irreparably altered by the process of measuring response. For example, measurement of gene activation through reporter molecules necessitates lysis of the cell. Therefore, only one

measurement of response can be made. In these instances, the response is a history of the temporal process of response production from the initiation of the experiment to the time of measurement (for example, the production of the second cellular messenger cyclic AMP as a function of time). In specially constructed reporter cells, such as those containing an 8-base-pair palindrome sequence called cyclic AMP response element (CRE), receptor activation causes this element to activate a p-promoter region of cyclic-AMP-inducible genes. This, in turn, causes an increase in transcription of a protein called luciferase. This protein produces light when brought into contact with an appropriate substrate, making it detectable and quantifiable. Therefore, any agonist increasing cyclic AMP will lead to an increase in luciferase. This is one of a general type of functional assays (called reporter assays) where agonism results in the production and accumulation of a detectable product. The amount of product accumulated after agonism can be measured only once. Therefore, an appropriate time must be allowed for assumed equilibrium before reading of the response. The addition of an agonist to such an assay causes the production of the second (reporter) messenger, which then goes on to produce the detectable product. The total amount of product made from the beginning of the process to the point where the reaction is terminated is given by the area under the curve defining cyclic AMP production. This is shown in Figure 5.15. Usually the experimenter is not able to see the approach to equilibrium (real-time response shown in Figure 5.15a) and must choose a time point as the best estimate regarding when equilibrium has been attained. Figure 5.15b shows the area under the curve as a function of time. This area is the stop-time response. This function is not linear in the early stages during approach to equilibrium but is linear when a steady state or true

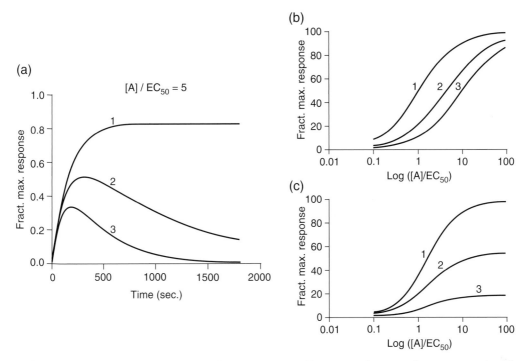

FIGURE 5.14 Temporal desensitization of agonist response. (a) Patterns of response for a concentration of agonist producing 80% maximal response. Curve 1: no desensitization. For concentration of agonist $[A] = 5x$ EC_{50}, first-order rate of onset $k_1 = sec^{-1} mol^{-1}$, $k_2 = 10^{-3} sec^{-1}$. Curve 2 equals constant desensitization rate $= k_{desen} = 10^{-3}$. Curve 3: variable desensitization rate equals ρk_{desen}, where ρ equals fractional receptor occupancy. (b) Complete dose-response curves to the agonist taken at equilibrium with no desensitization (curve 1), and at peak response for constant desensitization rate (curve 2) and variable desensitization rate (curve 3). (c) Curves as per panel B but response measured after 10 minutes equilibration with the agonist.

equilibrium has been attained. Therefore, a useful method to determine whether equilibrium has been achieved in stop-time experiments is to stop the reaction at more than one time point and ensure that the resulting signal (product formed) is linear with time. If the relationship between three stop-time responses obtained at three different time points is linear, then it can be assumed that the responses are being measured at equilibrium.

A potential pitfall with stop-time experiments comes with temporal instability of responses. When a steady-state sustained response is observed with time, then a linear portion of the production of reporter can be found (see Figure 5.15b). However, if there is desensitization or any other process that makes the temporal responsiveness of the system change the area under the curve will not assume the linear character seen with sustained equilibrium reactions. For example, Figure 5.16 shows a case where the production of cyclic AMP with time is transient. Under these circumstances, the area under the curve does not assume linearity. Moreover, if the desensitization is linked to the strength of signal (i.e., becomes more prominent at higher stimulations) the dose-response relationship may be lost. Figure 5.16 shows a stop-time reaction dose-response curve to a temporally stable system and a temporally unstable system where the desensitization is linked to the

strength of signal. It can be seen that the dose-response curve to the agonist is lost in the stop-time temporally unstable system.

5.6 The Measurement of Agonist Affinity in Functional Experiments

Binding experiments can yield direct measurements of ligand affinity (Chapter 4). However, with the use of null techniques these same estimates can also be obtained in functional studies. The concepts and procedures used to do this differ for partial and full agonists.

5.6.1 Partial Agonists

As noted in Chapter 2, the functional EC_{50} for a full agonist may not, and most often will not, correspond to the binding affinity of the agonist. This is due to the fact that the agonist possesses efficacy and the coupling of agonist binding to production of response is nonlinear. Usually, a hyperbolic function links the binding reaction to the observed dose-response curve—leading to a phase shift of the location parameters (midpoint values) of the two curves. The relationship of the EC_{50} for an agonist in any

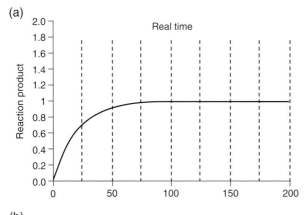

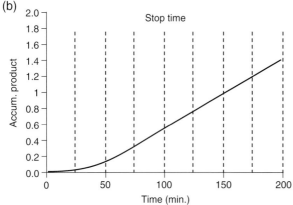

FIGURE 5.15 Different modes of response measurement. (a) Real time shows the time course of the production of response such as the agonist-stimulated formation of a second messenger in the cytosol. (b) The stop-time mode measures the area under the curve shown in panel A. The reaction is stopped at a designated time (indicated by the dotted lines joining the panels) and the amount of reaction product is measured. It can be seen that in the early stages of the reaction, before a steady state has been attained (i.e., a plateau has not yet been reached in panel A), the area under the curve is curvilinear. Once the rate of product formation has attained a steady state, the stop-time mode takes on a linear character.

system to the affinity, as defined by the classical model, is given by (see Section 5.9.1)

$$EC_{50} = \frac{K_A \cdot \beta}{(e + \beta)}, \qquad (5.4)$$

where β refers to the hyperbolic coupling constant relating receptor stimulus to response, affinity is K_A (equilibrium dissociation constant of the agonist-receptor complex), and e is efficacy. The steepness of the hyperbolic relationship between agonist receptor occupancy (and resulting stimulus) and tissue response is given by the magnitude of β (see Figure 5.17). It can be seen that low values of β or high values of efficacy displace the EC_{50} from the K_A along the concentration axis. A similar effect can be seen in terms of the operational model (see Section 5.9.1), where the EC_{50} is

related to the K_A by

$$EC_{50} = \frac{K_A}{(1 + \tau)} = K_A, \qquad (5.5)$$

where τ is the term relating efficacy of the agonist and the efficiency of the receptor system in converting receptor activation to response (high values of τ reflect either high efficacy, highly efficient receptor coupling, or both). High values of τ are associated with full agonism. It can be seen from Equation 5.5 that full agonism produces differences between the observed EC_{50} and the affinity (K_A).

Both Equation 5.4 and Equation 5.5 show that as the efficacy of agonist decreases the $EC_{50} \rightarrow K_A$. Thus, as e $\rightarrow 0$ in Equation 5.4 $EC_{50} \rightarrow K_A$. Similarly, as $\tau \rightarrow 0$ $EC_{50} \rightarrow K_A$ (Equation 5.5). Therefore, in general the EC_{50} of a weak partial agonist can be a reasonable approximation of the K_A (see Section 5.9.1 for further details). The lower the magnitude of the maximal response (lower τ) the closer the EC_{50} will approximate the K_A. Figure 5.18 shows the relationship between agonist receptor occupancy for partial agonists and the response for different levels of maximal response (different values of τ). It can be seen that as the maximal response $\rightarrow 0$ the relationship between agonist receptor occupancy and tissue response becomes linear and the $EC_{50} \rightarrow K_A$.

By utilizing complete dose-response curves, the method devised by Barlow, Scott, and Stephenson [9] can be used to measure the affinity of a partial agonist. Using null procedures, the effects of stimulus-response mechanisms are neutralized and receptor-specific effects of agonists are isolated. This method, based on classical or operational receptor theory, depends on the concept of equiactive concentrations of drug. Under these circumstances, receptor stimuli can be equated since it is assumed that equal responses emanate from equal stimuli in any given system. An example of this procedure is given in Section 12.2.1.

Dose-response curves to a full agonist [A] and a partial agonist [P] are obtained in the same receptor preparation. From these curves, reciprocals of equiactive concentrations of the full and partial agonist are used in the following linear equation (derived for the operational model; see Section 5.9.2)

$$\frac{1}{[A]} = \frac{1}{[P]} \cdot \frac{\tau_a \cdot K_P}{\tau_p \cdot K_A} + \frac{\tau_a - \tau_p}{\tau_p \cdot K_A}, \qquad (5.6)$$

where τ_a and τ_p are efficacy terms for the full and partial agonist, respectively, and K_A and K_P their respective ligand-receptor equilibrium dissociation constants. Thus, a regression of 1/[A] upon 1/[P] yields the K_B modified by an efficacy term with the following parameters from Equation 5.6:

$$K_P = \frac{Slope}{Intercept} \left(1 - \frac{\tau_p}{\tau_a} \right). \qquad (5.7)$$

It can be seen from Equation 5.7 that a more accurate estimate of the affinity will be obtained with partial agonists of low efficacy (i.e., as $\tau_a \gg \tau_p$, $\tau_p/\tau_a \rightarrow 0$). Double reciprocal plots are known to produce

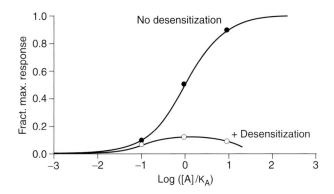

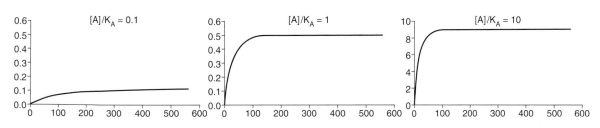

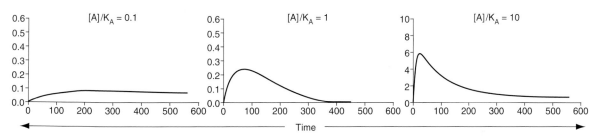

FIGURE 5.16 The effect of desensitization on stop-time mode measurements. Bottom panels show the time course of response production for a system with no desensitization, and one in which the rate of response production fades with time. The top dose response curves indicate the area under the curve for the responses shown. It can be seen that whereas an accurate reflection of response production is observed when there is no desensitization the system with fading response yields an extremely truncated dose-response curve.

over-emphasis of some values, skew the distribution of data points, and be heterogeneously sensitive to error. For these reasons, it may be useful to use a metameter of Equation 5.6 as a linear plot to measure the K_P. Thus, the K_P can be estimated from a plot according to

$$\frac{[P]}{[A]} = \frac{[P]}{K_A}((\tau_A/\tau_p) - 1) + \frac{\tau_a K_P}{\tau_p K_A}, \qquad (5.8)$$

where

$$K_P = \frac{\text{Intercept}}{\text{Slope}}(1 - \tau_p/\tau_a). \qquad (5.9)$$

Another variant is

$$\frac{[A]}{[P]} = \frac{\tau_p K_A}{\tau_a K_P} - [A] \cdot \frac{(1 - \tau_p/\tau_a)}{K_P}, \qquad (5.10)$$

where

$$K_P = \frac{(\tau_p/\tau_a - 1)}{\text{slope}}. \qquad (5.11)$$

An example of the application of this method to the measurement of the affinity of the histamine receptor partial agonist E-2-P (with full agonist histamine) is shown in Figure 5.19. A full example of the application of this method for the measurement of partial agonists is given in Section 12.2.2.

5.6.2 Full Agonists

For full agonists, the approximation of the EC_{50} as affinity is not useful and other methods must be employed to estimate affinity. A method to measure the affinity of

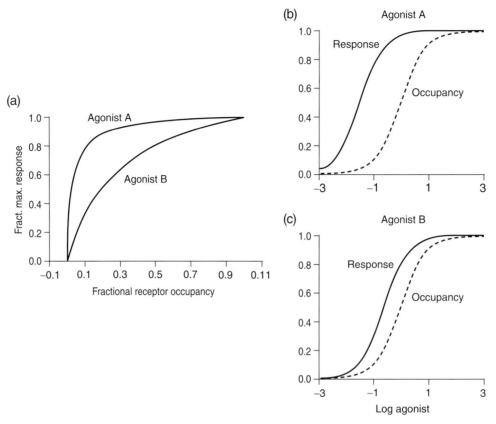

FIGURE 5.17 Relationship between receptor occupancy and tissue response for two agonists. (a) Occupancy-response curves for two agonists that differ in efficacy (agonist A $\tau = 3$ and agonist B $\tau = 30$). (b) Doseresponse (solid line) and receptor occupancyresponse (dotted line) for agonist A. Stimulus-response coupling and efficacy cause a 166-fold phase shift between the curves. (b) Dose-response curve (solid line) and receptor occupancy (dotted line) curves for agonist B. Stimulus-response coupling and efficacy cause a twofold phase shift between the curves.

high-efficacy agonists has been described by Furchgott [11]. This method is based on the comparison of the responses to an agonist in a given receptor system under control conditions and again after a fraction of the receptor population has been irreversibly inactivated. For some receptors—such as α-adrenoceptors, muscarinic, serotonin, and histamine receptors—this can be accomplished through controlled chemical alkylation with site-directed alkylating agents such as β-haloalkylamines. Thus, equiactive responses obtained before and after receptor alkylation are compared in the following double reciprocal relation (see Section 5.9.3):

$$\frac{1}{[A]} = \frac{1}{[A']} \cdot \frac{1}{q} + \frac{1}{K_A} \cdot \frac{1-q}{q}, \qquad (5.12)$$

where [A] and [A'] are equiactive agonist concentrations measured before and after receptor alkylation, respectively, q is the fraction of receptors remaining after alkylation, and K_A is the equilibrium dissociation constant of the agonist-receptor complex. Thus, a regression of $1/[A]$ upon $1/[A']$ yields a straight line with given slope and intercept. From

these, the equilibrium dissociation constant of the agonist-receptor complex can be calculated:

$$K_A = \frac{\text{Slope} - 1}{\text{Intercept}}. \qquad (5.13)$$

An example of the use of this approach is given in Figure 5.20. The method of Furchgott indicates that the affinity of the muscarinic agonist oxotremorine in guinea pig ileal smooth muscle is 8.2 μM. The EC_{50} for half maximal contractile response to this agonist is 25 nM (a 330-fold difference). This underscores the fact that the EC_{50} for full agonists can differ considerably from the K_A. A full example of the use of this method to measure the affinity of a full agonist is given in Section 12.2.3.

This method can also be employed with the operational model. Specifically, the operational model defines receptor response as

$$\text{Response} = \frac{[A] \cdot \tau \cdot E_{\max}}{[A](1 + \tau) + K_A}, \qquad (5.14)$$

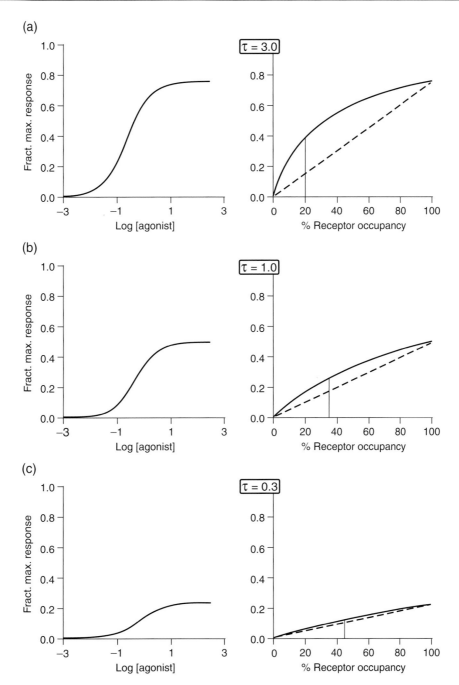

FIGURE 5.18 The relationship between the EC_{50} for partial agonists and the affinity (K_A). For higher-efficacy partial agonists ($\tau = 3$), the relationship between receptor occupancy and response is hyperbolic (note solid versus dotted line in right-hand panel where the dotted line represents a linear and direct relationship between the occupancy of the receptor by the agonist and the production of response). This deviation lessens with lower efficacy values for the partial agonist (note panels for agonist with $\tau = 1$). With weak partial agonists, the EC_{50} and K_A values nearly coincide (see panels with $\tau = 0$).

where E_{max} is the maximal response of the system, K_A is the equilibrium dissociation constant of the agonist-receptor complex, and τ is the ratio of the receptor density divided by the transducer function for the system (defined as K_E).

The transducer function defines the efficiency of the system to translate receptor stimulus into response and defines the efficacy of the agonist. Specifically, it is the fitting parameter of the hyperbolic function linking receptor

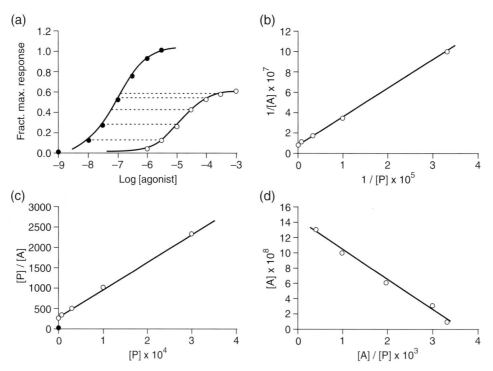

FIGURE 5.19 Method of Barlow for measurement of affinity of a partial agonist. (a) Guinea pig ileal smooth muscle contraction to histamine (filled circles) and partial histamine receptor agonist E-2-P (N,N-diethyl-2-(1-pyridyl)ethylamine (open circles). Dotted lines show equiactive concentrations of each agonist used for the double reciprocal plot shown in panel b. (b) Double reciprocal plot of equiactive concentrations of histamine (ordinates) and E-2-P (abscissae). Linear plot has a slope of 55.47 and an intercept of 1.79×10^6. This yields a $K_B \cdot (1 - \tau_p/\tau_A) = 30.9 \, \mu M$. (c) Variant of double reciprocal plot according to Equation 5.8. (d) Variant of double reciprocal plot according to Equation 5.10. Data redrawn from [10].

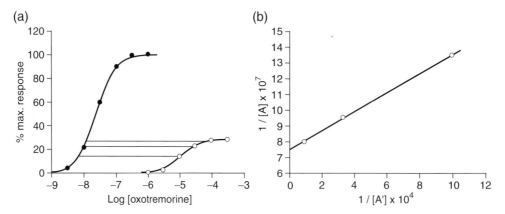

FIGURE 5.20 Measurement of the affinity of a full agonist by the method of Furchgott. (a) Concentration response curves to oxotremorine in guinea pig ileal smooth muscle strips. Ordinates: percent maximal contraction. Abscissae: logarithms of molar concentrations of oxotremorine. Control curve (filled circles) and after partial alkylation of muscarinic receptors with phenoxybenzamine $10 \, \mu M$ for 12 minutes (open circles). Lines represent equiactive concentrations of oxotremorine before and after receptor alkylation. (b) Regression of reciprocals of equiactive concentrations of oxotremorine before (ordinates) and after (abscissae) receptor alkylation. The regression is linear with a slope of 609 and an intercept of 7.4×10^7. Resulting K_A estimate for oxotremorine according to Equation 5.12 is $8.2 \, \mu M$. Data redrawn from [12].

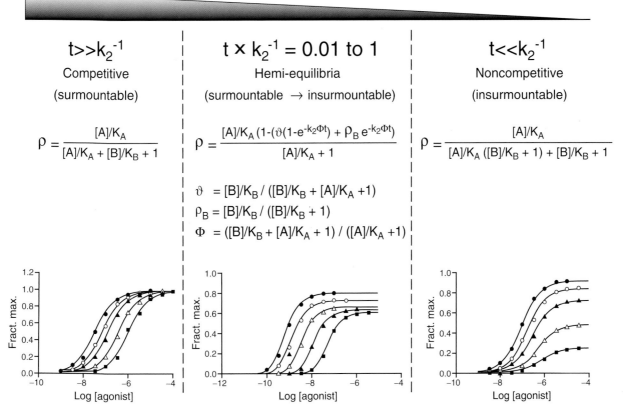

$$\rho_t = \frac{[B]/K_B}{[B]/K_B + [A]/K_A + 1} - \left(\frac{[B]/K_B}{[B]/K_B + [A]/K_A + 1} - \frac{[B]/K_B}{[B]/K_B + 1}\right)e^{-k_2t}\left[\frac{[B]/K_B + [A]/K_A + 1}{[A]/K_A + 1}\right]$$

Equilibration time x rate of offset

$t \gg k_2^{-1}$	$t \times k_2^{-1} = 0.01$ to 1	$t \ll k_2^{-1}$
Competitive	Hemi-equilibria	Noncompetitive
(surmountable)	(surmountable $\rightarrow$ insurmountable)	(insurmountable)

$$\rho = \frac{[A]/K_A}{[A]/K_A + [B]/K_B + 1}$$

$$\rho = \frac{[A]/K_A \left(1 - (\vartheta(1 - e^{-k_2\Phi t}) + \rho_B e^{-k_2\Phi t}\right)}{[A]/K_A + 1}$$

$$\rho = \frac{[A]/K_A}{[A]/K_A ([B]/K_B + 1) + [B]/K_B + 1}$$

$\vartheta = [B]/K_B / ([B]/K_B + [A]/K_A + 1)$
$\rho_B = [B]/K_B / ([B]/K_B + 1)$
$\Phi = ([B]/K_B + [A]/K_A + 1) / ([A]/K_A + 1)$

FIGURE 6.4 The range of antagonist behaviors observed under different kinetic conditions. When there is sufficient time for complete reequilibration ($t \gg k_2^{-1}$), surmountable antagonism is observed (panel furthest to the left). As the time for reequilibration diminishes (relative to the rate of offset of the antagonist from the receptor; $t \times k_2^{-1} = 0.1$ to 0.01), the curves shift according to competitive kinetics (as in the case for surmountable antagonism) but the maxima of the curves are truncated (middle panel). When there is insufficient time for reequilibration, the antagonist essentially irreversibly occludes the fraction of receptors it binds to during the equilibration period ($t \ll k_2^{-1}$) and depression of the maxima occurs with dextral displacement is determined by the extent of receptor reserve for the agonist (panel to the right).

response) are used to calculate dose ratios. An example calculation of a DR is shown in Figure 6.5. Thus, for every concentration of antagonist [B] there will be a corresponding DR value. These are plotted as a regression of Log (DR-1) upon Log [B]. If the antagonism is competitive, there will be a linear relationship between Log (DR-1) and Log [B] according to the Schild equation. Under these circumstances it can be seen that a value of zero for the ordinate will give an intercept of the x axis where Log [B] = Log K_B. Therefore, the concentration of antagonist that produces a Log (DR-1) = 0 value will be equal to the Log K_B, the equilibrium dissociation constant of the antagonist-receptor complex. This is a system-independent and molecular quantification of the antagonist affinity that should be accurate for every cellular system containing the

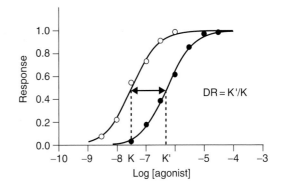

FIGURE 6.5 Calculation of equiactive dose ratios (DR values) from two dose-response curves.

receptor. When the concentration of antagonist in the receptor compartment is equal to the K_B value (the concentration that binds to 50% of the receptors), then the dose-ratio will be 2. Since K_B values are obtained from a logarithmic plot, they are log normally distributed and are therefore conventionally reported as pK_B values. These are the negative logarithm of the K_B used much like pEC_{50} values are used to quantify agonist potency. The negative logarithm of this particular concentration is also referred to empirically as the pA_2, the concentration of antagonist that produces a twofold shift of the agonist dose-response curve. Antagonist potency can be quantified by calculating the pA_2 from a single concentration of antagonist producing a single value for the dose ratio from the equation:

$$pA_2 = Log(DR-1) - Log[B]. \qquad (6.13)$$

It should be noted that this is a single measurement. Therefore, comparison to the model of competitive antagonism cannot be done. The pA_2 only serves as an empirical measure of potency. Only if a series of DR values for a series of antagonist concentrations yields a linear Schild regression with a slope of unity can the pA_2 value (obtained from the intercept of the Schild plot) be considered a molecular measure of the actual affinity of the antagonist for the receptor (pK_B). Therefore, a pK_B value is always equal to the pA_2. However, the converse (namely, that the pA_2 can always be considered an estimate of the pK_B) is not necessarily true. For this to occur, a range of antagonist concentrations must be tested and shown to comply with the requirements of Schild analysis (linear plot with slope equal to unity). A precept of Schild analysis is that the magnitude of DR values must not be dependent on the level of response used to make the measurement. This occurs if the dose-response curves (control plus those obtained in the presence of antagonist) are parallel and all have a common maximal asymptote response (as seen in Figure 6.5).

There are statistical procedures available to determine whether the data can be fit to a model of dose-response curves that are parallel with respect to slope and all share a common maximal response (see Chapter 11). In general, dose-response data can be fit to a three-parameter logistic equation of the form

$$Response = \frac{E_{max}}{1 + 10^{(LogEC_{50} - Log[A])^n}}, \qquad (6.14)$$

where the concentration of the agonist is [A], E_{max} refers to the maximal asymptote response, EC_{50} is the location parameter of the curve along the concentration axis, and n is a fitting parameter defining the slope of the curve. A variant four-parameter logistic curve can be used if the baseline of the curves does not begin at zero response ([i.e., if there is a measurable response in the absence of agonist basal]):

$$Response = Basal + \frac{E_{max} - Basal}{1 + 10^{(LogEC_{50} - Log[A])^n}}. \qquad (6.15)$$

In practice, a sample of data will be subject to random variation, and curve fitting with nonlinear models most likely will produce differences in slope and/or maxima for the various dose-response curves. Therefore, the question to be answered is, does the sample of data come from a population that consists of parallel dose-response curves with common maxima? Hypothesis testing can be used to determine this (see Chapter 11). Specifically, a value for the statistic F is calculated by fitting the data to a complex model (where each curve is fit to its own value of n, EC_{50}, and E_{max}) and to a more simple model (where a common E_{max} and n values are used for all of the curves and the only differences between them are values of EC_{50}). (See Chapter 11 for further details.) If the F statistic indicates that a significantly better fit is not obtained with the complex model (separate parameters for each curve), then this allows fitting of the complete data set to a pattern of curves with common maxima and slope. This latter condition fulfils the theoretical requirements of Schild analysis. An example of this procedure is shown in Chapter 11 (see Figure 11.14).

If the data set can be fit to a family of curves of common slope and maximum asymptote, then the EC_{50}s of each curve can be used to calculate DR values. Specifically, the EC_{50} values for each curve obtained in the presence of antagonist are divided by the EC_{50} for the control curve (obtained in the absence of antagonist). This yields a set of equiactive dose ratios. If hypothesis testing indicates that individually fit curves must be used, then a set of EC_{50} values must be obtained graphically. A common level of response (i.e., 50%) is chosen and EC_{50} values are either calculated from the equation or determined from the graph. With slopes of the dose-response curves near unity, this approximation is not likely to produce substantial error in the calculation of DR values and should still be suitable for Schild analysis. However, this approach is still an approximation and fitting to curves of common slope and maxima is preferred. It should be noted that an inability to fit the curves to a common maximum and slope indicates a departure from the assumptions required for assigning simple competitive antagonism.

The measured dose ratios are then used to calculate Log (DR-1) ordinates for the corresponding abscissal logarithm of the antagonist concentration that produced the shift in the control curve. A linear equation of the form

$$y = mx + b \qquad (6.16)$$

is used to fit the regression of Log (DR-1) upon Log [B]. Usually a statistical software tool can furnish an estimate of the error on the slope.

The model of simple competitive antagonism predicts that the slope of the Schild regression should be unity. However, experimental data is a sample from the complete population of infinite DR values for infinite concentrations of the antagonist. Therefore, random sample variation may produce a slope that is not unity. Under these circumstances, a statistical estimation of the 95% confidence limits of the slope (available in most

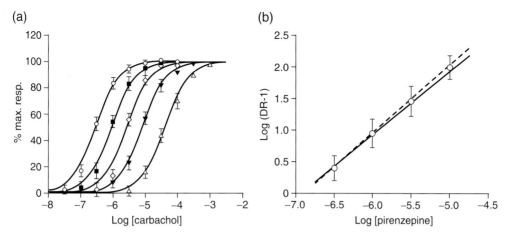

FIGURE 6.6 Schild regression for pirenzepine antagonism of rat tracheal responses to carbachol. (a) Dose-response curves to carbachol in the absence (open circles, n = 20) and presence of pirenzepine 300 nM (filled squares, n = 4), 1 μM (open diamonds, n = 4), 3 μM (filled inverted triangles, n = 6), and 10 μM (open triangles, n = 6). Data fit to functions of constant maximum and slope. (b) Schild plot for antagonism shown in panel A. Ordinates: Log (DR-1) values. Abscissae: logarithms of molar concentrations of pirenzepine. Dotted line shows best line linear plot. Slope = 1.1 + 0.2; 95% confidence limits = 0.9 to 1.15. Solid line is the best fit line with linear slope. $pK_B = 6.92$. Redrawn from [5].

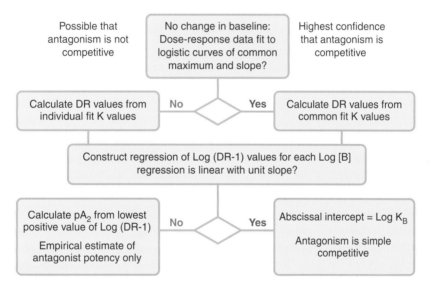

FIGURE 6.7 Schematic diagram of some of the logic used in Schild analysis.

fitting software) is used to determine whether the sample data could have come from the population describing simple competitive antagonism (i.e., have unit slope). If the 95% confidence limits of the experimentally fit slope include unity, then it can be concluded that the antagonism is of the simple competitive type and that random variation caused the deviation from unit slope. The regression is then *refit to an equation where m = 1* and the abscissal intercept taken to be the logarithm of the K_B. An example of Schild analysis for the inhibition of muscarinic-receptor-mediated responses of rat tracheae, to the agonist carbachol by the antagonist pirenzepine, is shown in Figure 6.6.

If the slope of the regression is not unity or if the regression is not linear, then the complete data set cannot be used to estimate the antagonist potency. Under these circumstances, either the antagonism is not competitive or some other factor is obscuring the competitive antagonism. An estimate of the potency of the antagonist can still be obtained by calculating a pA_2 according to Equation 6.13. This should be done using the *lowest positive Log(DR-1) value*. Hypothesis testing can be used to determine the lowest statistically different value for DR from the family of curves (see Figure 11.16).

A schematic diagram of some of the logic used in Schild analysis is shown in Figure 6.7. It should be pointed out

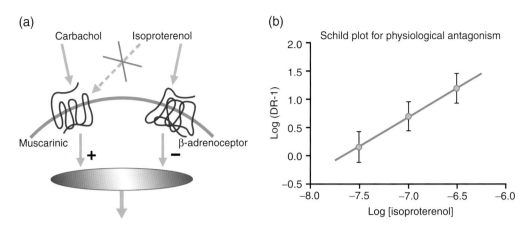

FIGURE 6.8 Apparent simple competitive antagonism of carbachol-induced contraction of guinea pig trachea through physiological antagonism of tracheal contractile mechanisms by β-adrenoceptor relaxation of the muscle. (a) Schematic diagram of the physiological interaction of the muscarinic receptor-induced contraction and β-adrenoceptor-induced relaxation of tracheal tissue. (b) Schild regression for isoproterenol (β-adrenoceptor agonist) antagonism of carbachol-induced contraction. The regression is linear with unit slope (slope = 1.02 + 0.02) apparently, but erroneously indicative of simple competitive antagonism. Redrawn from [6].

that a linear Schild regression with a unit slope are the minimal requirements for Schild analysis, but that it does not necessarily prove that a given inhibition is of the simple competitive type. For example, in guinea pig tracheae relaxant β-adrenoceptors and contractile muscarinic receptors coexist. The former cause the tissue to relax, while the latter counteract this relaxation and cause the tissue to contract. Thus, the β-adrenoceptor agonist isoproterenol, by actively producing relaxation, will physiologically antagonize contractile responses to the muscarinic agonist carbachol. Figure 6.8 shows a Schild plot constructed from the concentration-dependent relaxation of guinea pig trachea of the contractile dose-response curves to carbachol. It can be seen that the plot is linear with a slope of unity, apparently in agreement with a mechanism of simple competitive antagonism. However, these opposing responses occur at totally different cell surface receptors and the interaction is further down the stimulus-response cascade in the cytoplasm. Thus, the apparent agreement with the competitive model for this data is spurious (i.e., the plot cannot be used as evidence of simple competitive antagonism). An example of the use of this method is given in Section 12.2.4.

6.3.2 Patterns of Dose-Response Curves That Preclude Schild Analysis

There are patterns of dose-response curves that preclude Schild analysis. The model of simple competitive antagonism predicts parallel shifts of agonist dose-response curves with no diminution of maxima. If this is not observed it could be because the antagonism is not of the competitive type or some other factor is obscuring the competitive nature of the antagonism. The shapes of dose-response curves can prevent measurement of response-independent

dose ratios. For example, Figure 6.9a shows antagonism where clearly there is a departure from parallelism, and in fact a distinct decrease in slope of the curve for the agonist in the presence of the antagonist is observed. This is indicative of noncompetitive antagonism. Irrespective of the mechanism, this pattern of curves prevents estimation of response-independent DR values and thus Schild analysis would be inappropriate for this system. Figure 6.9b shows a pattern of curves with depressed maximal responses but shifts that are near parallel in nature. This is a pattern indicative of hemi-equilibrium conditions whereby the agonist and antagonist do not have sufficient time (due to the response collection window) to come to temporal equilibrium. If this could be determined, then Schild analysis can estimate antagonist potency from values of response below where depression of responses occurs (i.e., EC_{30}). The differentiation of hemi-equilibria from noncompetitive blockade is discussed in Section 6.5.

The pattern shown in Figure 6.9c is one of parallel shift of the dose-response curves up to a maximal shift. Further increases in antagonist concentration do not produce further shifts of the dose-response curves beyond a limiting value. This is suggestive of an allosteric modification of the agonist affinity by the antagonist, and other models can be used to estimate antagonist affinity under these conditions (see Chapter 7). This is discussed further on in this chapter. Finally, if the agonist has secondary properties that affect the response characteristics of the system (i.e., toxic effects at high concentrations), then dextral displacement of the dose-response curve into these regions of agonist concentration may affect the observed antagonism. Figure 6.9 shows depression of the maximal response at high agonist concentrations. This pattern may preclude full Schild analysis but a pA_2 may be estimated.

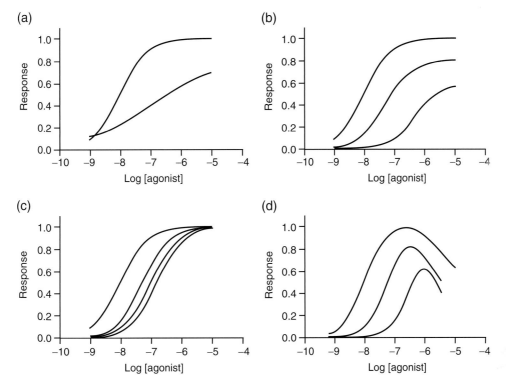

FIGURE 6.9 Patterns of dose-response curves produced by antagonists that may preclude Schild analysis. (a) Depression of maximal response with nonparallelism indicative of noncompetitive blockade. DR values are not response independent. (b) Depressed maxima with apparent parallel displacement indicative of hemi-equilibrium conditions (*vide infra*). (c) Loss of concentration dependence of antagonism as a maximal shift is attained with increasing concentrations of antagonist indicative of saturable allosteric blockade. (d) Depressed maximal responses at high concentration of agonist where the antagonist shifts the agonist response range into this region of depression (indicative of toxic or nonspecific effects of agonist at high concentrations).

6.3.3 Best Practice for the Use of Schild Analysis

There are two ways to make Schild analysis more effective. The first is to obtain Log (DR-1) values as near to zero as possible (i.e., use concentrations of the antagonist that produce a low level of antagonism such as a twofold to fivefold shift in the control dose-response curve). This will ensure that real data is in close proximity to the most important parameter sought by the analysis; namely, the abscissal intercept (pK_B or pA_2 value). If Log (DR-1) values are greater than 1.0, then the pK_B (or pA_2) will need to be extrapolated from the regression. Under these circumstances any secondary effects of the antagonist that influence the slope of the Schild regression will subsequently affect the estimate of antagonist potency. Second, at least a 30-fold (and preferably 100-fold) concentration range of antagonist (concentrations that produce an effect on the control dose-response curve) should be utilized. This will yield a statistically firm estimate of the slope of the regression. If the concentration range is below this, then the linear fit of the Log(DR-1) versus Log [B] will produce large 95% confidence limits for the slope. While unity most likely will reside within this broad range, the fit will be much less useful as an indicator of whether or not unity

actually is a correct slope for the antagonist. That unity is included could simply reflect the fact that the confidence range is so large.

There are Schild regressions that deviate from ideal behavior but can still be useful either to quantify antagonist potency or to indicate the mechanism of antagonism. For example, Figure 6.10a shows a linear Schild regression at low antagonist concentrations that departs from ideal behavior (increased slope) at higher antagonist concentrations. This is frequently encountered experimentally as secondary effects from higher concentrations of either the agonist or the antagonist come into play, leading to toxicity or other depressant effects on the system. The linear portion of the regressions at lower antagonist concentrations can still be used for estimation of the pK_B (if a large enough concentration range of antagonist is used) or for the pA_2 (if not).

Figure 6.10b shows a pattern of antagonism often observed in isolated tissue studies but not so often in cell-based assays. Saturation of uptake systems for the agonist or saturation of an adsorption site for the agonist can account for this effect. The linear portion of the regression can be used to estimate the pK_B or the pA_2. If there is a loss of concentration dependence of antagonism, as seen in

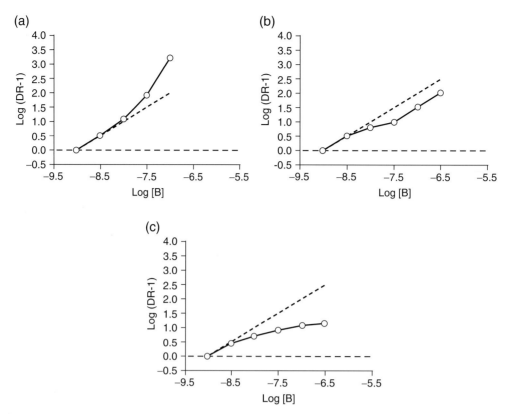

FIGURE 6.10 Some commonly encountered patterns of Schild regressions. (a) Initial linearity with increased slope at higher concentration indicative of toxic effects of either the agonist or antagonist at higher concentrations. (b) Region of decreased slope with reestablishment of linearity often observed for saturation of uptake or other adsorption effects. (c) Hyperbolic loss of antagonism indicative of saturable allosteric antagonism.

Figure 6.10c, this indicates a possible allosteric mechanism whereby a saturation of binding to an allosteric site is operative. This is dealt with further on in this chapter.

One of the strengths of Schild analysis is the capability of unveiling nonequilibrium conditions in experimental preparations such as inadequate time of equilibration or removal of drugs from the receptor compartment. Figure 6.11 shows a range of possible experimentally observed but problematic linear Schild regressions that could be encountered for competitive antagonists.

6.3.4 Analyses for Inverse Agonists in Constitutively Active Receptor Systems

In constitutively active receptor systems (where the baseline is elevated due to spontaneous formation of receptor active states, see Chapter 3 for full discussion), unless the antagonist has identical affinities for the inactive receptor state, the spontaneously formed active state, and the spontaneously G-protein coupled state (three different receptor conformations, see discussion in Chapter 1 on receptor conformation) it will alter the relative concentrations of these species—and in so doing alter the baseline response. If the antagonist has higher affinity for the

receptor active state, it will be a partial agonist in an efficiently coupled receptor system. This is discussed in the next section. If the antagonist has higher affinity for the inactive receptor, then it will demonstrate simple competitive antagonism in a quiescent system and *inverse agonism* in a constitutively active system.

The dose-response curves reflecting inverse agonism do not conform to the strict requirements of Schild analysis (i.e., parallel shift of the dose-response curves with no diminution of maxima). In the case of inverse agonists in a constitutively active receptor system, the dextral displacement of the agonist concentration-response curve is accompanied by a depression of the elevated basal response (due to constitutive activity). (See Figure 6.12a.) This figure shows the nonparallel nature of the curves as the constitutively elevated baseline is reduced by the inverse agonist activity. In quiescent receptor systems (nonconstitutively active), both competitive antagonists and inverse agonists produce parallel shifts to the right of the agonist dose-response curves (see Figure 6.12b).

The effects of high values of constitutive activity can be determined for functional systems where function is defined by the operational model. Thus, it can be assumed in a simplified system that the receptor exists in an active (R*) and inactive (R) form and that agonists stabilize

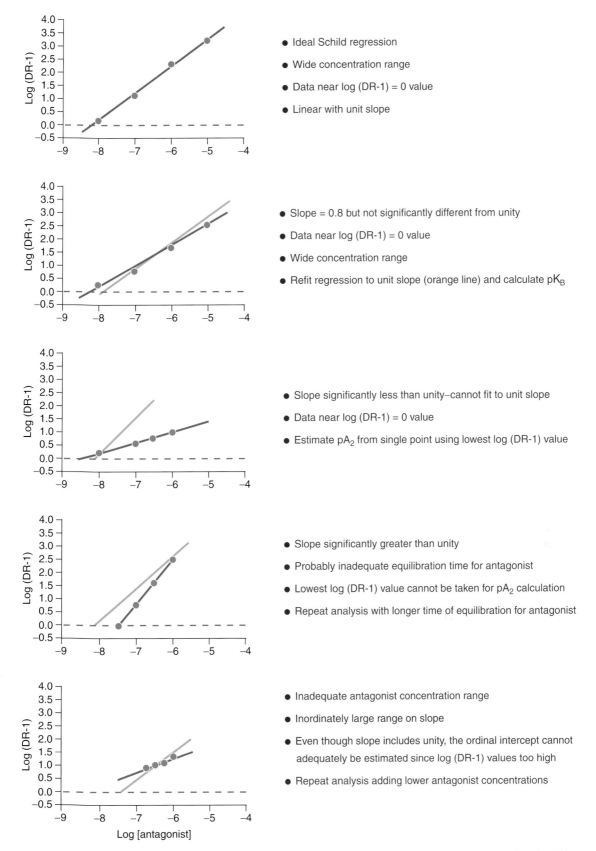

- Ideal Schild regression
- Wide concentration range
- Data near log (DR-1) = 0 value
- Linear with unit slope

- Slope = 0.8 but not significantly different from unity
- Data near log (DR-1) = 0 value
- Wide concentration range
- Refit regression to unit slope (orange line) and calculate pK_B

- Slope significantly less than unity–cannot fit to unit slope
- Data near log (DR-1) = 0 value
- Estimate pA_2 from single point using lowest log (DR-1) value

- Slope significantly greater than unity
- Probably inadequate equilibration time for antagonist
- Lowest log (DR-1) value cannot be taken for pA_2 calculation
- Repeat analysis with longer time of equilibration for antagonist

- Inadequate antagonist concentration range
- Inordinately large range on slope
- Even though slope includes unity, the ordinal intercept cannot adequately be estimated since log (DR-1) values too high
- Repeat analysis adding lower antagonist concentrations

FIGURE 6.11 Some examples of commonly encountered Schild data and some suggestions as to how antagonism should be quantified for these systems.

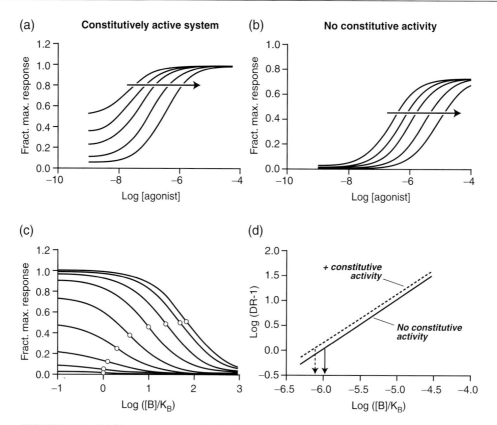

FIGURE 6.12 Schild analysis for constitutively active receptor systems. (a) Competitive antagonism by the inverse agonist in a constitutively active receptor system with DR values calculated at the EC_{80}. (b) Competitive antagonism by the same inverse agonist in a nonconstitutively active receptor system. (c) Direct effects of an inverse agonist in systems of differing levels of constitutive activity. Open circles show midpoints of the concentration-response curves. (d) Schild regression for an inverse agonist in a nonconstitutive assay where the inverse agonist produces no change in baseline (solid line) and in a constitutively active assay where depression of elevated baseline is observed (dotted line). A small shift to the left of the Schild regression is observed, leading to a slight overestimation of inverse agonist potency.

(and therefore enrich the prevalence of) the active form while inverse agonists prefer the inactive form. It also is assumed that response emanates from the active form of the receptor:

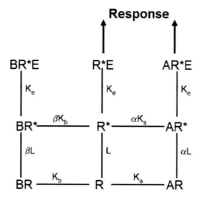

Under these circumstances, the fractional response in a functional system can be derived from the expression

defining the amount of active-state receptor coupled to G-protein. This yields the following expression for response with a Hill coefficient of unity (see Section 6.8.4):

Response

$$= \frac{\alpha L[A]/K_A \tau + \beta L[B]/K_B \tau + L\tau}{[A]/K_A(1+\alpha L(1+\tau)) + [B]/K_B(1+\beta L(1+\tau)) + L(\tau+1)+1},$$
(6.17)

where τ is the efficacy of the full agonist, n is a fitting parameter for the slope of the agonist concentration-response curve, K_A and K_B are the respective equilibrium dissociation constants of the full agonist and inverse agonist for the inactive state of the receptor, α and β are the relative ratios of the affinity of the full and inverse agonist for the active state of the receptor, and L is the allosteric constant for the receptor $(L = [R^*]/[R])$.

There are two ways to estimate the potency of an inverse agonist from the system described by Equation 6.17. The first is to observe the concentration of inverse agonist that reduces the level of constitutive activity by 50%, the IC_{50} of the compound as an active inverse agonist. This is done by observing the level of constitutive response in the absence of full agonist ([A] = 0) with a variant of Equation 6.17:

Constitutive Response

$$= \frac{\beta L[B]/K_B \tau + L\tau}{[B]/K_B(1 + \beta L(1 + \tau)) + L(\tau + 1) + 1}. \quad (6.18)$$

Figure 6.12c shows the effect of increasing levels of constitutive activity on the midpoint of a curve to an inverse agonist. This shows that with increasing levels of inverse agonism—either through increasing intrinsic constitutive activity, increased L, or increasing levels of receptor and/or efficiency of receptor coupling (increasing τ)—the IC_{50} of the inverse agonist will increasingly be larger than the true K_B. This is important to note since it predicts that the value of the pIC_{50} for an inverse agonist will be system dependent and can vary from cell type to cell type (just as observed potency for positive agonists). However, in the case of inverse agonists the effects of increasing receptor density and/or receptor coupling are opposite those observed for positive agonists where increases cause a concomitant increase in observed potency. This trend in the observed potency of inverse agonism on system conditions (L and τ) can be seen from the midpoint of the curve defined by Equation 6.18. This is the IC_{50} for an inverse agonist inhibition of constitutive activity:

$$\text{Observed } IC_{50} = \frac{K_B(L(\tau + 1) + 1)}{(\beta L(1 + \tau) + 1)}. \quad (6.19)$$

Equation 6.19 predicts an increasing IC_{50} with either increases in L or τ. In systems with low-efficacy inverse agonists or in systems with low levels of constitutive activity, the observed location parameter is still a close estimate of the K_B (equilibrium dissociation constant of the ligand-receptor complex, a molecular quantity that transcends test system type). In general, the observed potency of inverse agonists only defines the lower *limit* of affinity.

As observed in Figure 6.12a, inverse agonists produce dextral displacement of concentration-response curves to full agonists and thus produce dose ratios that may be used in Schild analysis. It is worth considering the use of dose ratios from such curves and the error in the calculated pK_B and pA_2 produced by the negative efficacy of the inverse agonist and changes in basal response levels. It can be shown that the pA_2 value for an inverse agonist in a constitutively active receptor system is given by (see section 6.8.5)

$$pA_2 = pK_B - \text{Log}([A](\alpha - 1)/([A](\alpha - 1) + (1 - \beta))). \quad (6.20)$$

This expression predicts that the modifying term will always be <1 for an inverse agonist (β < 1). Therefore, the calculation of the affinity of an inverse agonist from dextral displacement data (pA_2 measurement) will always overestimate the potency of the inverse agonist. However, since

β < 1 and the α value for a full agonist will be $\gg$ 1, the error most likely will be very small. Figure 6.12d shows the effect of utilizing dextral displacements for an inverse agonist in a constitutively active system. The Schild regression is linear but is phase shifted to the right in accordance with the slight overestimation of inverse agonist potency.

6.3.5 *Analyses for Partial Agonists*

Schematically, response is produced by the full agonist ([AR]) complex—which interacts with the stimulus response system with equilibrium association constant K_e—and the partial agonist (lower efficacy), which interacts with an equilibrium association constant K_e':

Therefore, there are two efficacies for the agonism: one for the full agonist (denoted τ) and one for the partial agonist (denoted τ'). In terms of the operational model for functional response, this leads to the following expression for response to a full agonist [A] in the presence of a partial agonist [B] (see Section 6.8.6):

$$\text{Response} = \frac{[A]/K_A \tau + [B]/K_B \tau'}{[A]/K_A(1 + \tau) + [B]/K_B(1 + \tau') + 1}. \quad (6.21)$$

If the partial agonism is sufficiently low so as to allow a full agonist to produce further response, then a pattern of curves of elevated baseline (due to the partial agonism) shifted to the right of the control curve (due to the antagonist properties of the partial agonist) will be obtained. (See Figure 6.13a.) However, low-efficacy agonists can be complete antagonists in poorly coupled receptor systems and partial agonists in systems of higher receptor density and/or coupling efficiency (Figure 6.13b).

The observed EC_{50} for partial agonism can be a good estimate for the affinity (K_B). However, in systems of high receptor density and/or efficient receptor coupling where the responses approach full agonism, the observed EC_{50} will overestimate the true potency of the partial agonist. This can be seen from the location parameter of the partial agonist in Equation 6.22 in the absence of full agonist ([A] = 0):

$$\text{Observed } EC_{50} = \frac{[B]/K_B}{(1 + \tau')}. \quad (6.22)$$

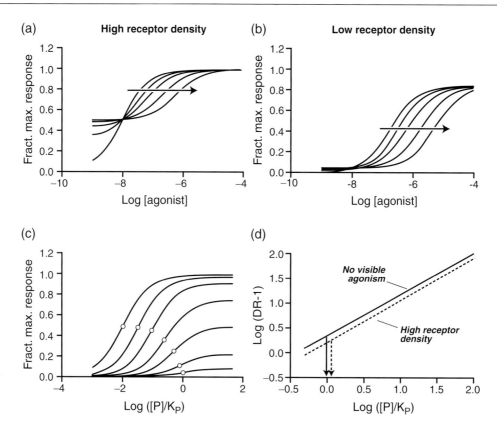

FIGURE 6.13 Schild analysis for a partial agonist. (a) Competitive antagonism by a partial agonist. DR values calculated at EC_{75} for agonist response. (b) Schild regressions for antagonism of same receptor in a low receptor-density/coupling-efficiency receptor where no partial agonism is observed. (c) Dose-response curve for directly observed partial agonism. Under some conditions, the EC_{50} for the partial agonist closely approximates the K_B. (d) Schild regression for a partial agonist in a low receptor/ coupling assay where the partial agonist produces no observed response (solid line) and in a high receptor/coupling assay where agonism is observed (dotted line). A small shift to the right of the Schild regression is observed, leading to a slight underestimation of partial agonist potency.

Figure 6.13C shows the effect of increasing receptor density and/or efficiency of receptor coupling on the magnitude of the EC_{50} of the partial agonist. Equiactive dose ratios still can be estimated from the agonist-dependent region of the dose-response curves. For example, Figure 6.13a shows DR values obtained as ratios of the EC_{75}. The resulting Schild regression slightly underestimates the K_B (see Figure 6.13d). However, the error will be minimal. Underestimation of the true pK_B is also predicted by the operational model (Section 6.8.7):

$$pA_2 = pK_B - Log(\tau/(\tau - \tau')). \qquad (6.23)$$

It can be seen that the modifying term will always be >1, but will also have a relatively low magnitude (especially for low values of partial agonist efficacy τ'). Also, in systems where the partial agonist does not produce response ($t' \rightarrow 0$), the $pA_2 = pK_B$ as required by simple competitive antagonism (as shown in Figure 6.13b). The use of dose ratios for partial agonists where the partial agonist produces response will always slightly underestimate affinity by the Schild method (or calculation of the pA_2). The Schild regression for

a partial agonist reflects this in that it is still linear but slightly shifted to the right of the true regression for simple competitive antagonism (Figure 6.13d).

Another method to measure the affinity of a partial agonist has been presented by Stephenson [7] and modified by Kaumann and Marano [8]. The method of Stephenson compares equiactive concentrations of full agonist in the absence of and the presence of a concentration of partial agonist to estimate the affinity of the partial agonist. The following equation is used (see Section 6.8.8):

$$[A] = \frac{[A']}{1 + \left(1 - (\tau_p/\tau_a)\right) \cdot ([P]/K_p)} + \frac{(\tau_p/\tau_a) \cdot ([P]/K_p) \cdot K_A}{1 + \left(1 - (\tau_p/\tau_a)\right) \cdot ([P]/K_p)}. \qquad (6.24)$$

A regression of [A] upon [A'] yields a straight line. The K_p can be estimated by

$$K_p = \frac{[P]slope}{1 - slope} \cdot \left(1 - (\tau_p/\tau_a)\right). \qquad (6.25)$$

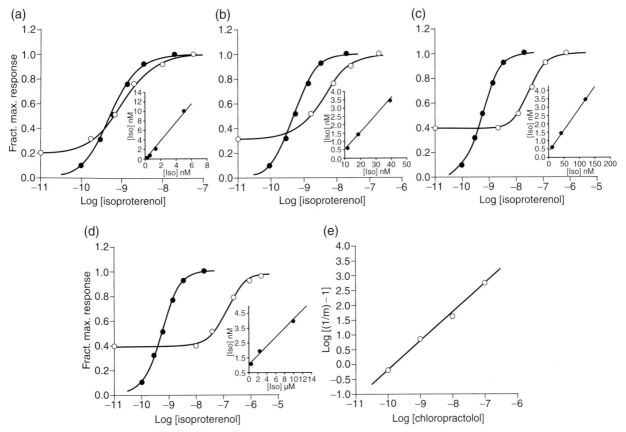

FIGURE 6.14 Method of Stephenson [7] and Kaumann and Marano [8] used to measure the affinity of the partial β-adrenoceptor agonist chloropractolol in rat atria. Panels a to d show responses to isoproterenol in the absence (filled circles) and presence of chloropractolol (open circles). Curves shown in the presence of 10 nM (panel a), 100 nM (panel b), 1 μM (panel c) and 10 μM (panel d) chloropractolol. Note elevated basal responses in response to the partial agonist chloropractolol. Insets to panels a through d show plots of equiactive concentrations of isoproterenol in the absence (ordinates) and presence of chloropractolol according to Equation 6.24. Slopes from these graphs used for plot shown in panel E according to the method of Kaumann and Marano [9] (see Equation 6.26). This plot is linear with a slope of 0.95, yielding a K_P estimate of 16.5 nM. Data redrawn from [9].

A full example of the use of this method is given in Section 12.2.6.

A more rigorous version of this method has been presented by Kauman and Marano [8]. In this method, the slopes from a range of equiactive agonist concentration plots are utilized in another regression (see Section 6.8.8):

$$Log\left(\frac{1}{slope} - 1\right) = Log[P] - LogK_p \qquad (6.26)$$

where m is the slope for a particular regression of equiactive concentrations of an agonist in the absence and presence of a particular concentration of partial agonist [P]. An example of the use of this method for the measurement of the partial agonist chloropractolol is shown in Figure 6.14. The various plots of equiactive concentrations (insets to panels A to D) furnish a series of values of m for a series of concentrations of chloropractolol. These are used in a regression according to Equation 6.26 (see Figure 6.14) to yield an estimate of the K_P for chloropractolol from the

intercept of the regression. Further details on the use of this method is given in Section 12.2.7.

6.3.6 The Method of Lew and Angus: Nonlinear Regressional Analysis

One shortcoming of Schild analysis is an overemphasized use of the control dose-response curve (i.e., the accuracy of every DR value depends on the accuracy of the control EC_{50} value). An alternative method utilizes nonlinear regression of the Gaddum equation (with visualization of the data with a Clark plot [10], named for A. J. Clark). This method, unlike Schild analysis, does not emphasize control pEC_{50}, thereby giving a more balanced estimate of antagonist affinity. This method, first described by Lew and Angus [11], is robust and theoretically more sound than Schild analysis. On the other hand, it is not as visual. Schild analysis is rapid and intuitive, and can be used to detect nonequilibrium steady states in the system that can corrupt estimates of pK_B. Also, nonlinear regression requires matrix algebra to estimate the error of the pK_B. While error estimates are given with many commercially available software packages for curve fitting, they are difficult to obtain without these (from first principles). In contrast, Schild analysis furnishes an estimate of the error for the pK_B from the linear regression using all of the data. If an estimate of the error is required and the means to calculate it are not available in the curve fitting software, manual calculation with Schild analysis is a viable alternative. In general, the method of Lew and Angus still holds definite advantages for the measurement of competitive antagonist potency. One approach to rigorously describe competitive antagonism is to use Schild analysis to visualize the data and the method of Lew and Angus to estimate the pK_B.

To apply this method, the pEC_{50} values of the control and shifted dose-response curves and the corresponding concentrations of antagonist [B] values associated with those pEC_{50}s are used to construct a Clark plot [10] according to the equation

$$pEC_{50} = -Log([B] + 10^{-pK_B}) - Log\,c, \qquad (6.27)$$

where pK_B and c are fitting constants. Note that the control pEC_{50} is used with [B] = 0. The relationship between the pEC_{50} and increments of antagonist concentration can be shown in a Clark plot of pEC_{50} versus $-Log([B] + 10^{-pK_B})$. Constructing such a plot is useful because although it is not used in any calculation of the pK_B it allows visualization of the data to ensure that the plot is linear and has a slope of unity.

Although the Clark plot can be used to visualize the slope relationship between pEC_{50} and $-Log([B] + 10^{-pK_B})$, deviation of the slope from unity is better obtained by refitting the data to a "power departure" version of Equation 6.27:

$$pEC_{50} = -Log([B]^m + 10^{-pK_B}) - Log\,c, \qquad (6.28)$$

where m is allowed to vary as part of the nonlinear fit. A value of F is calculated for comparison of the fits to Equations 6.27 and 6.28, respectively. If the value of F is not significant, then there is no reason to use the power departure equation and the antagonism can be considered to be simple competitive. To test for significant deviation from linearity of the Clark plot (indicating a departure from simple competitive antagonism at some concentration used in the experiment), the data is fit to a "quadratic departure" version of Equation 6.27:

$$pEC_{50} = -Log([B](1 + n[B]10^{-pK_B}) + 10^{-pK_B}) - Log\,c, \qquad (6.29)$$

where n is allowed to vary with the nonlinear fitting procedure. As with the analysis for slope, a value for F is calculated. If the quadratic departure is not statistically supported, then the regression can be considered linear.

The method of Lew and Angus uses nonlinear curve fitting procedures to estimate the pK_B. An estimate of the error calculated with Equation 6.27 is provided by the estimate of the fitting error. This is obtained from most if not all commercially available fitting programs (or can be calculated with matrix algebra). An example of this type of analysis is shown in Figure 6.15a. The pEC_{50} values for the dose-response curves and the concentrations of antagonist were fit to the equation shown in panel in Figure 6.15b to yield the Clark plot shown in panel B. The resulting pK_B value is 8.09 + 0.145. The data was then refit to the power departure version of the equation to yield the Clark plot shown in panel C. The calculated F for comparison of the simple model (slope = unity) to the more complex model (slope fit independently) yielded a value for F which is not greater than that required for 95% confidence of difference. Therefore, the slope can be considered not significantly different from unity. Finally, the data was again refit to the quadratic departure version of the equation to yield the Clark plot shown in panel D to test for nonlinearity. The resulting F indicates that the plot is not significantly nonlinear.

6.4 Noncompetitive Antagonism

From an examination of Equation 6.1, and noted in Figure 6.4, if the rate of offset of the orthosteric antagonist is slow such that a correct reequilibration cannot occur between the agonist, antagonist, and receptors during the period of response collection in the presence of antagonist, then essentially a pseudo-irreversible blockade of receptors will occur. Thus, when $t \ll k_2^{-1}$ in Equation 6.1 the agonist will not access antagonist-bound receptors and a noncompetitive antagonism will result. This is the opposite extreme of the case for simple competitive antagonism discussed in Section 6.3.

The term *competitive antagonism* connotes an obvious mechanism of action (i.e., two drugs compete for the same binding site on the receptor to achieve effect). Similarly, the term *noncompetitive* indicates that two drugs bind to the receptor and that these interactions are mutually exclusive (i.e., when one drug occupies the binding site then another cannot exert its influence on the receptor). However, this should not necessarily be related to binding loci on the receptor. Two drugs may interact noncompetitively but still require occupancy of the same receptor binding site. Alternatively, the sites may be separate as in allosteric effects (see next chapter).

In an operational sense, noncompetitive antagonism is defined as the case where the antagonist binds to the receptor and makes it functionally inoperative. This can occur through preclusion of agonist binding or through some other biochemical mechanism that obviates agonist effect on the receptor and thereby blocks response due to agonist. Under these circumstances, no amount of increase in the agonist concentration can reverse the effect of a noncompetitive antagonist. A distinctive feature of non-competitive antagonists is the effect they may have on the

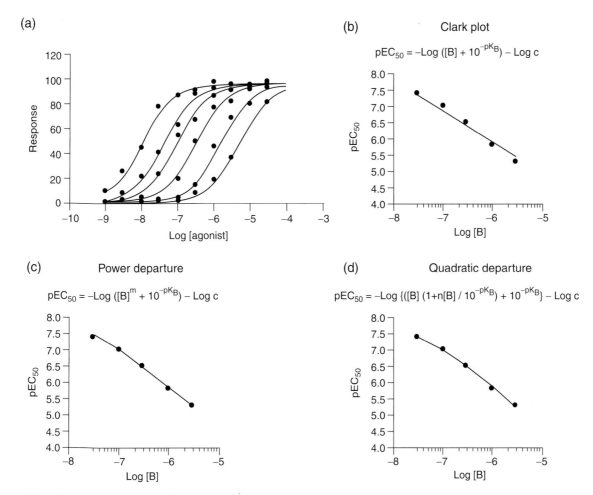

FIGURE 6.15 Example of application of method of Lew and Angus [10]. (a) Dose-response data. (b) Clark plot according to Equation 6.27 shown. (c) Data refit to "power departure" version of Equation 6.27 to detect slopes different from unity (Equation 6.28). (d) Data refit to "quadratic departure" version of Equation 6.27 to detect deviation from linearity (Equation 6.29).

maximal agonist response. In situations where 100% of the receptors need be occupied to achieve the maximal response to the agonist (i.e., partial agonists), any amount of noncompetitive antagonism will lead to a diminution of the maximal response. However, in systems where there is a receptor reserve there will not be a depression of the maximal response until such a point where there is sufficient antagonism to block a fraction of receptor larger than that required to achieve maximal response. As discussed in Chapter 2, the magnitude of the receptor reserve is both system dependent (dependent on receptor number *and* the efficiency of stimulus-response coupling) and agonist dependent (intrinsic efficacy). Therefore, noncompetitive antagonists will have differing capabilities to depress the maximal response to the same agonist in different systems. The same will be true for different agonists in the same system.

The equation describing agonist receptor occupancy under conditions of noncompetitive antagonism is given by Equation 6.8. The effect of antagonist on the maximal agonist receptor occupancy (i.e., as $[A] \rightarrow \infty$) and comparison to the control maximal stimulus from Equation 6.8 is

$$\text{Maximal agonist occupancy} = \frac{1}{1 + [B]/K_B}. \qquad (6.30)$$

It can be seen that at non-zero values of $[B]/K_B$ the maximal agonist *receptor* occupancy will be depressed. However, as discussed in Chapter 2, some high efficacy agonists and/or some highly coupled receptor systems (high receptor density) yield maximal tissue response by activation of only a fraction of the receptor population ("spare receptors"). Thus, a noncompetitive antagonist may preclude binding of the agonist to all of the receptors, but this may or may not result in a depression of the maximal response to the agonist. To discuss this further requires conversion of the agonist receptor occupancy curve

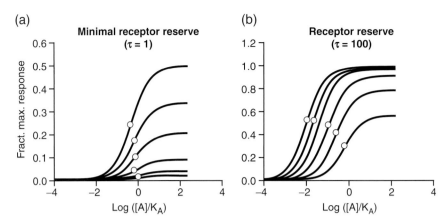

FIGURE 6.16 Effects of a slow offset orthosteric antagonist that essentially does not reequilibrate with agonist and receptors upon addition of agonist to the system (pseudo-irreversible receptor blockade). (a) In this system a low value of τ is operative (i.e., the efficacy of the agonist is low) if there is a low receptor density and/or poor coupling of receptors. Under these circumstances, little to no dextral displacement is observed for the concentration-response curves upon antagonism (insurmountable blockade). (b) If the τ value is high (high efficacy, high receptor density, highly efficient receptor coupling, high receptor reserve), then the same antagonist may produce dextral displacement of the concentration-response curves with no depression of maximal response until relatively large portions of the receptor population are blocked.

(Equation 6.8) into tissue response through the operational model:

$$B$$
$$+$$
$$A \; + \; R \; \underset{K_A}{\rightleftharpoons} \; AR \; \underset{K_E}{\rightleftharpoons} \; Response,$$
$$\Big\Updownarrow K_B$$
$$BR$$

whereby the antagonist precludes agonist activation and response is produced through interaction of the [AR] complex with the tissue stimulus-response cascade through the constant K_E according to the operational model. Under these circumstances, the response to an agonist obtained in the presence of a noncompetitive antagonist is given by:

$$Response = \frac{[A]/K_A \tau \, E_{max}}{[A]/K_A(1 + \tau + [B]/K_B) + [B]/K_B + 1}.$$
$$(6.31)$$

Now it can be seen that the maximal response (as a fraction of the control maximal response) to the agonist (as [A] → ∞) is given by

$$Maximal \; Response = \frac{(1 + \tau)}{(1 + \tau + [B]/K_B)}. \quad (6.32)$$

Here it can be seen that for very efficacious agonists, or in systems of high receptor density or very efficient receptor coupling (all leading to high values of τ), the maximal

response to the agonist may not be depressed in the presence of the noncompetitive antagonist. In Figure 6.16, the effect of a noncompetitive antagonist on the receptor response to an agonist in a system with no receptor reserve (τ = 1) is shown. It can be seen that the maximal response to the agonist is depressed at all non-zero values of [B]/K_B. In Figure 6.16b, the same antagonist is used to block responses to a highly efficacious agonist in a system with high receptor reserve (τ = 100). From these simulations it can be seen that observation of insurmountable antagonism is not necessarily a prerequisite for a noncompetitive receptor mechanism.

In terms of measuring the potency of insurmountable antagonists, the data can be fit to an explicit model. As shown in Figure 6.17a, responses to an agonist in the absence and presence of various concentrations of an insurmountable antagonist are fit to Equation 6.31 (Figure 6.17b) and an estimate of the K_B for the antagonist obtained. One shortcoming of this approach is the complexity of the model itself. It will be seen in the next chapter that allosteric models of receptor antagonism can also yield patterns of agonist concentration response curves like those shown in Figure 6.17, and that these can be fit equally well with allosteric models. Thus, model fitting can be ambiguous if the molecular mechanism of the antagonism is not known beforehand.

Historically, Gaddum and colleagues [3] devised a method to measure the affinity of insurmountable antagonists based on a double reciprocal linear transformation. With this method, equiactive concentrations of agonist in the absence ([A]) and presence ([A′]) of a noncompetitive antagonist ([B]) are compared in a double reciprocal plot

7

Allosteric Drug Antagonism

When one tugs at a single thing in nature, he finds it attached to the rest of the world.
— **JOHN MUIR**

Whatever affects one directly, affects all indirectly... This is the interrelated structure of reality.
— **MARTIN LUTHER KING JR.**

7.1 Introduction

A major molecular mechanism of receptor antagonism involves the binding of the antagonist to its own site on the receptor separate from the binding site of the endogenous agonist. When this occurs, the interaction between the agonist and antagonist takes place through the receptor protein. This is referred to as an *allosteric* interaction (for schematic diagram see Figure 6.2). Thus, an allosteric antagonist produces a conformational change of shape of the receptor, which in turn changes the affinity of the receptor for the agonist and/or changes the receptor function.

Some operational classifications of antagonism relate solely to certain molecular mechanisms. For example, allosteric antagonists produce saturable effects (i.e., a maximum antagonism is produced after which further increases in antagonist concentration have no further effect). However, operational effects on dose-response curves do not always unambiguously indicate a molecular mechanism in that experiments can reveal combinations of compatible operational and mechanistic classifications (i.e., an allosteric molecular mechanism can produce either surmountable and insurmountable effects on dose-response curves depending on the system). Finally, since allosteric effects produce a change in shape of the receptor it cannot be assumed a priori that a uniform antagonistic effect on agonism will result. In fact, it will be seen that some

allosteric ligands produce an increase in the affinity of the receptor for ligands (note the stimulation of the binding of [^{3}H]-atropine by alcuronium in Figure 4.12). In addition, the effect of an allosteric ligand on a receptor probe (this can be an agonist or radioligand) is totally dependent on the nature of the probe (i.e., a conformational change that increases the affinity of the receptor for one agonist may decrease it for another). For example, while the allosteric ligand alcuronium produces a tenfold change in the affinity of the muscarinic m2 receptor for acetylcholine it produces only a 1.7-fold change in the affinity for arecoline [1]. These effects make consistent nomenclature for allosteric ligands difficult. For this reason, allosteric ligands will be referred to as allosteric modulators with the understanding that modulation in this sense means modification, either in a positive or negative direction.

7.2 The Nature of Receptor Allosterism

The word allosteric comes from the Greek *allos*, meaning different, and *steric*, which refers to arrangement of atoms in space. As a word allostery literally means a change in shape. Specifically in the case of allosterism of proteins, the change in shape is detected by its interaction with a probe. Therefore, there can be no steric interference at this probe site. In fact, allosteric effects are defined by the interaction of an allosteric modulator at a so-called allosteric binding site on the protein to affect the conformation at the probe site of the protein. Since the probe and modulator molecules do not interact directly, their influence on each other must take place through a change in shape of the protein. Historically, allosteric effects have been studied and described for enzymes. Early discussions of allosteric enzyme effects centered on the geography of substrate and modulator binding. Koshland [2], a pioneer of allosteric enzyme research, classified binding geography of enzymes in terms of "contact amino acids" and intimate parts of the active site for substrate binding and "contributing amino acids," those important for preservation of the tertiary structure of the active site but not playing a role in substrate binding. Finally, he defined "noncontributing amino acids" as those not essential for enzyme catalysis but perhaps serving a structural role in the enzyme. Within Koshland's hypothesis, binding to these latter two categories of amino acids constituted a mechanism of

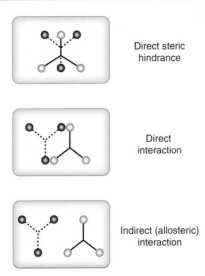

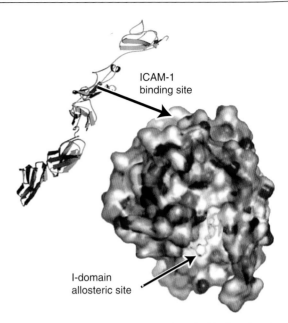

FIGURE 7.1 Enzyme ortho- and allosterism as presented by Koshland [2]. Steric hindrance whereby the competing molecules physically interfered with each other as they bound to the substrate site was differentiated from a direct interaction where only portions of the competing molecules interfered with each other. If no direct physical interaction between the molecules occurred, then the effects were solely due to effects transmitted through the protein structure (allosteric).

FIGURE 7.2 Model of LFA-1 showing the binding domain of ICAM-1 (the endogenous ligand for this protein) and the binding site for lovastatin, an allosteric modulator for this protein. Redrawn from [7].

allosterism rather than pure endogenous ligand competition. Within this context, pharmacological antagonists can bind to sites distinct from those utilized by the endogenous agonist (i.e., hormone, neurotransmitter) to alter binding and subsequent tissue response (Figure 7.1). Some of these differences in binding loci can be discerned through point mutation of receptors. For example differences in amino acids required for competitive antagonist binding and allosteric effector binding can be seen in mutant muscarinic m1 receptors where substitution of an aspartate residue at position 71, but not at positions 99 and 122, affects the affinity of the allosteric modulator gallamine but not the affinity of the competitive antagonist radiolabeled [³H]-N-methylscopolamine [3].

Allosteric sites can be remote from an enzyme active site. For example, the binding site for nevirapine, an allosteric modulator of HIV-reverse transcriptase, is 10 angstroms away from the enzyme active site [4]. Similarly, allosteric inhibition of β-lactamase occurs 16 angstroms away from the active site [5]. The binding site for CP320626 for glycogen phosphorylase b is 33 angstroms from the catalytic site and 15 angstroms from the site for cyclic AMP [6]. A visual demonstration of the relative geography of allosteric binding and receptor active sites can be seen in Figure 7.2. Here, the integrin LFA1, which binds to molecules on other cell membranes to mediate cell adhesion, has a receptor probe active site binding intercellular adhesion molecule-1 (ICAM1) and an allosteric binding site for the drug lovastatin, in a deep hydrophobic cleft next to the α7 helix (see Figure 7.2) [7].

While visualization of the relative binding sites for receptor probes and allosteric modulators is conceptually helpful, preoccupation with the geography of ligand binding is needlessly confining since the actual binding sites involved are secondary to the mechanism of allosterism. As shown by the above examples, the modulator and probe binding sites need not be near each other for allosteric effects to occur (i.e., the binding of the modulator does not necessarily need to produce a deformation near the receptor probe site). In fact, there is data to suggest that the relative geometry of binding is immaterial except for the fact that the receptor probe and modulator must bind to exclusively different sites.

Just as the location of allosteric sites is secondary to the consequences of allosteric effect, there is evidence to suggest that the structural requirements of allosteric sites may be somewhat more permissive with respect to the chemical structures bound to them (i.e., the structure activity relationships for allosteric sites may be more relaxed due to the fact that allosteric proteins are more flexible than other proteins). For example, as shown in Figure 7.3a, structurally diverse molecules such as efavirenz, nevirapine, UC-781, and Cl-TIBO all bind to HIV reverse transcriptase [8]. Similarly, HIV-entry inhibitors Sch-C, Sch-D, UK427,857, aplaviroc, and TAK779 all demonstrate prohibitive binding (consistent with binding at the same site) for the CCR5 receptor of Figure 7.3b [9].

It is useful to think of the allosteric binding not in terms of deformation of the receptor active site but rather as a lever to lock the receptor into a given conformation.

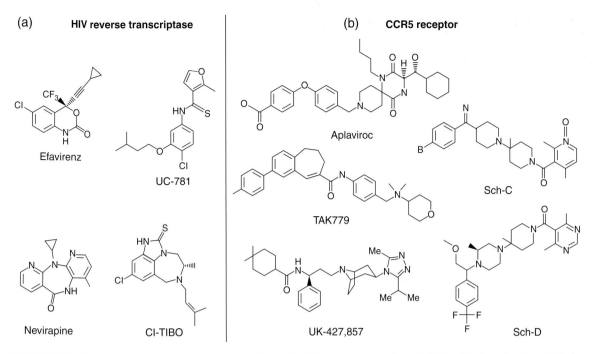

FIGURE 7.3 Diversity of structures that interact with the (a) HIV reverse transcriptase inhibitor binding site [8] and (b) the CCR5 receptor mediating HIV-1 fusion [9].

As discussed in Chapter 1, receptors and other biologically relevant proteins are a dynamic system of interchanging conformations referred to as an ensemble. These various conformations are sampled according to the thermal energy of the system, in essence the protein roams on a conceptual "energy landscape." While there are preferred low energy conformations, the protein has the capacity to form a large number of conformations. An allosteric modulator may have a high affinity for some of these and thus bind to them preferentially when they are formed. Thus, by selectively binding to these conformations the allosteric modulators stabilize them at the expense of other conformations. This creates a bias and a shift in the number of conformations toward the ligand-bound conformation (Figure 7.4; see Section 1.10 for further details).

The fact that the allosterically preferred conformation may be relatively rare in the library of conformations available to the receptor may have kinetic implications. Specifically, if the binding site for the modulator appears only when the preferred conformation is formed spontaneously, then complete conversion to allosterically modified receptor may require a relatively long period of equilibration. For example, the allosteric p38 MAP kinase inhibitor BIRB 796 binds to a conformation of MAP kinase requiring movement of a Phe residue by 10 angstroms (so-called "out" conformation). The association rate for this modulator is 8.5×10^5 M^{-1} s^{-1}, 50 times slower than that required for other inhibitors (4.3×10^7 M^{-1} s^{-1}). The result is that while other inhibitors reach equilibrium within 30 minutes, BIRB 376 requires 2 full hours of equilibration time [8].

7.3 Properties of Allosteric Modulators

The fact that global conformations of the receptor are stabilized by allosteric modulators has implications for their effects. Specifically, this opens the possibility of changes in multiple regions of the receptor instead of a single point change in conformation and with this, comes the possibility of changing multiple points of contact between the receptor and other proteins (see Figure 7.5). An example of the global nature of the conformational changes due to allosteric interaction is made evident in the interaction of CP320626 on glycogen phosphorylase b. In this case, the binding of this allosteric modulator causes the release of 9 of 30 water molecules from a cavity capped by α-helices of the enzyme subunits [6]. Such global conformational effects make possible the influence of the interaction of large proteins by small allosteric molecules. For example, HIV-1 entry is mediated by the interaction of the chemokine receptor CCR5 and the HIV viral coat protein gp120, both large (70- to 100-K Daltons) proteins. Analysis by point mutation indicates that all four extracellular loops of the receptor and multiple regions of gp120 associate for HIV fusion [10–13] yet small allosteric molecules such as aplaviroc and Sch-D (0.6% of their size) are able to block this interaction at nanomolar concentrations (see Figure 7.6). In general, the stabilization of receptor conformations by allosteric ligands makes possible the alteration of large protein-protein interaction making this a potentially very powerful molecular mechanism of action.

Another particularly unique aspect of allosteric mechanisms is that they can be very probe specific (i.e., a conformational change that is catastrophic for one receptor

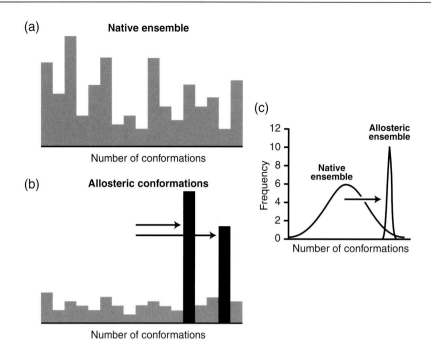

FIGURE 7.4 Histograms depicting the frequency of occurrence of various receptor conformations. (a) The natural "native" ensemble of receptor structures consists of various conformations in varying numbers at any one instant. (b) The addition of an allosteric modulator that preferentially binds to two receptor conformations causes these to become stabilized and thus enriched in the native milieu of conformations. (c) Ordering the conformations by frequency of occurrence forms Gaussian distributions for the ensemble. The addition of the selective ligand enriches certain conformations, reducing the frequency of sampling of other conformations. The mean conformation is shifted to an allosterically altered one.

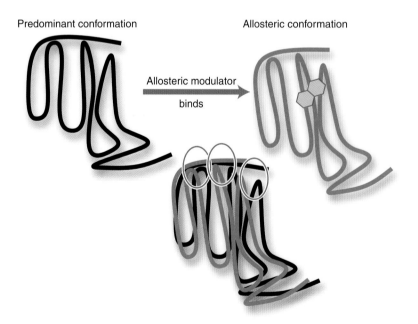

FIGURE 7.5 Schematic diagram of a GPCR in a native conformation (black) and allosterically altered conformation (red). When these are superimposed upon each other it can be seen that more than one region of the receptor is altered upon allosteric modulation (see circled areas).

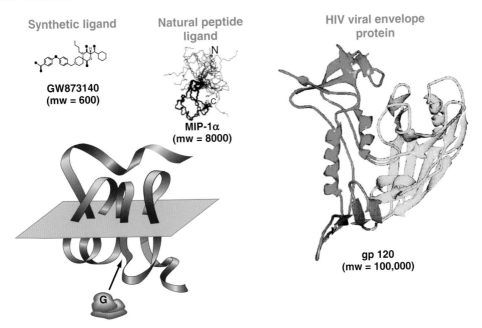

FIGURE 7.6 Cartoons showing the relative size of the CCR5 receptor, gp120 HIV viral coat protein, the natural ligand for the CCR5 receptor (the chemokine MIP-1α), and GW873140 (aplaviroc) [9], an allosteric modulator that blocks the interaction of CCR5 with both MIP-1α and gp120.

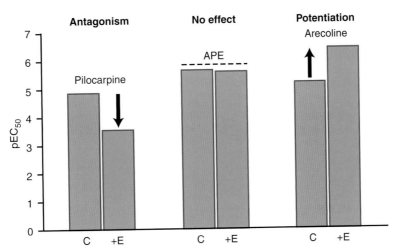

FIGURE 7.7 Effect of the allosteric modulator eburnamonine on the potency of muscarinic agonists on m2 receptors. It can be seen that while no change in potency is observed for APE (arecaidine propargyl ester) pilocarpine is antagonized and arecoline is potentiated, illustrating the probe dependence of allosterism. From [1].

probe may be inconsequential to another). This is illustrated in Figure 7.7. where it can be seen that the allosteric modulator eburnamonine produces a 25-fold antagonism of the muscarinic agonist pilocarpine, no effect on the agonist arecaidine propargyl ester (APE), and a 15-fold *potentiation* of the agonist arecoline [1]. Also, because allosteric modulation involves a change in the receptor conformation there is the potential of texture in antagonism. Orthosteric antagonists which occlude the agonist binding site prevent all agonist signaling equally

(i.e., the end result of all orthosteric antagonist-bound receptors is the same; namely, a receptor uniformly insensitive to all agonists). This may not necessarily be true for allosteric modulators. Just as a given allosteric modulator can produce different effects on different receptor probes, different modulators can produce different effects on the same modulator. For example, Table 7.1 shows the effects of different allosteric modulators on common agonists of muscarinic receptors. It can be seen from these data that different allosteric modulators have

TABLE 7.1
The effects of different allosteric modulators on common agonists of muscarinic receptors.

Receptor	Receptor Probe	Modulator	Effect[1]	Difference[2]
m3	Bethanechol	Strychnine Brucine	49x potentiation 0.67x inhibition	73x
m2	P-TZTP[3]	Alcuronium Brucine	4.7x potentiation 0.13x inhibition	36x
m2	Acetylcholine	Vincamine Eburnamonine	18x potentiation 0.32x inhibition	31x

[1] α value for changes in potency.

[2] ratio of α values for the two modulators.

[3] 3-(3-pentylthio-1,2,5-thiadiazol-4-yl)-1,2,5,6-tetrahydro-1-methylpyridine.

From [1].

the ability to antagonize and potentiate muscarinic agonists, clearly indicative of the production of different allosteric conformational states. Similarly, the allosterically modified CCR5 receptor demonstrates heterogeneity with respect to sensitivity of antibody binding. In this case, antibodies such as 45531, binding to a specific region of the receptor, reveal different conformations stabilized by aplaviroc and Sch-C, two allosteric modulators of the receptor. This is shown by the different affinity profiles of the antibody in the presence of each modulator (see Figure 7.8). This also has implications for the therapeutic use of such modulators. In the case of Sch-D and aplaviroc in Figure 7.8, the allosterically blocked receptors are similar in that they do not support HIV entry but quite dissimilar with respect to binding of the 45531 antibody. This latter fact indicates that the allosteric conformations produced by each modulator are not the same and this could have physiological consequences. Specifically, it is known that HIV spontaneously mutates [14, 15] and that the mutation in the viral coat protein can lead to resistance to CCR5 entry inhibitors. For example, passage of the virus in the continued presence of the CCR5 antagonist AD101 leads to an escape mutant able to gain cell entry through use of the allosterically modified receptor [16, 17]. It would be postulated that production of a different conformation with another allosteric modulator would overcome viral resistance since the modified virus would not be able to recognize the newly formed conformation of CCR5. Thus, the texture inherent in allosteric modification of receptors (different tertiary conformations of protein) offers a unique opportunity to defeat accommodation of pathological processes to chronic drug treatment.

Texture in antagonism can lead to a unique approach to the therapeutic evaluation of biological targets. For example, if a receptor is required for normal physiological function then eliminating this target pharmacologically is prohibited. This can lead to the elimination of a therapeutic opportunity if that same target is involved in a pathological function. Such a case occurs for the chemokine X-type receptor CXCR4 since loss of normal CXCR4 receptor

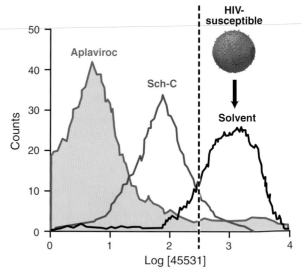

FIGURE 7.8 Binding of the CCR5 antibody 45531 to native receptor (peak labeled solvent) and in the presence of 1 μM Sch-C (blue line) and 1 μM aplaviroc (magenta peak). Different locations of the distributions show different binding sensitivities to the antibody indicative of different receptor conformations. Data courtesy of S. Sparks and J. Demarest, Dept of Clinical Virology, GlaxoSmithKline.

function may be deleterious to normal health. It specifically has been shown that deletion of the genes known to mediate expression of the CXCR4 receptor or the natural agonist for CXCR4 (stromal cell derived factor 1-α, SDF-1α) is lethal and leads to developmental defects in the cerebellum, heart, gastrointestinal tract as well as hematopoiesis [18–20] (i.e., this receptor is involved in normal physiological function and interference with its normal function will lead to serious effects). However, this receptor also mediates entry of the X4 strain of HIV virus leading to AIDS. Therefore, an allosteric modulator that could discern between the binding of HIV and the natural agonist for CXCR4 (SDF-1α) could be a very beneficial drug.

The probe-dependent aspect of allosteric mechanisms could still allow CXCR4 to be considered as a therapeutic target in spite of its crucial role in normal physiology. Suggestions of ligand-mediated divergence of physiological activity and mediation of HIV entry have been reported for CXCR4 in peptide agonists such as RSVM and ASLW. These peptides are not blocked by the CXCR4 antagonist AMD3100, an otherwise potent antagonist of HIV entry suggesting a dissociation of signaling and HIV binding effects [21]. Similar dissociation between HIV and chemokine activity also is observed with other peptide fragments of SDF-1α [22]. These data open the possibility that allosteric molecules can be found that block HIV entry but do not interfere with CXCR4-mediated chemokine function.

Allosteric probe dependence, as well as offering a positive avenue of therapeutic advancement as discussed previously, can have negative effects. For example, allosteric modification of an endogenous signaling system requires the effect to be operative on the physiologically relevant agonist. There are practical circumstances where screening for new drug entities in this mode may not be possible. For example, the screening of molecules for HIV entry theoretically should be done with live AIDS virus but this is not possible for safety and containment reasons. In this case, a surrogate receptor probe, such as radioactive chemokine, must be used and this can lead to dissimilation in activity (i.e., molecules may modify the effects of the chemokine but not HIV). This is discussed specifically in relation to screening in Chapter 8. Another case is the potentiation of cholinergic signaling for the treatment of patients with Alzheimer's disease. It has been proposed that a reduction in cholinergic function results in cognitive and memory impairment in this disease [23]. Therefore, an allosteric potentiation of cholinergic function could be beneficial therapeutically but it would have to be operative for the natural neurotransmitter—in this case, acetylcholine. This agonist is unstable and difficult to use as a screening tool and surrogate cholinergic agonists have been used in drug discovery. However, effects on such surrogates may have no therapeutic relevance if they do not translate to concomitant effects on the natural agonist. For example, the cholinergic test agonist arecoline is potentiated 15-fold by the allosteric modulator eburnamonine but no potentiation, in fact a threefold *antagonism*, is observed with the natural agonist acetylcholine [1]. Such effects underscore the importance of probe dependence in screening for allosteric modulators.

One of the key properties of allosteric modulators is their saturability of effect. With this comes the capability to modulate but not necessarily completely block agonist-induced signals. This stems from the fact that while the allosterically modified receptor may have a diminished affinity and/or efficacy for the agonist the agonist may still produce receptor activation in the presence of the modulator. This submaximal effect on ligand-receptor interaction is shown in Figure 4.10, where it is seen that the displacement of bound ^{125}I-MIP-1α from chemokine C receptor type 1 (CCR1) by allosteric ligand UCB35625

is incomplete (i.e., the ^{125}I-MIP-1α still binds to the receptor but with a lower affinity). An orthosteric antagonist binding to the same binding site as MIP-1α necessarily must completely reverse the binding of MIP-1α. In general, this leads to the possibility that allosteric modulators can modify (i.e., reduce or increase by a small amount) endogenous agonist signals without completely blocking them.

Saturability of the binding to the allosteric site also offers the potential to dissociate duration of effect from magnitude of effect. Since allosteric effects reach an asymptotic value upon saturation of the allosteric site, there is the potential to increase the duration of allosteric effect by loading the receptor compartment with large concentrations of modulator. These large concentrations will have no further effect other than to prolong the saturated allosteric response. For example, consider a system where the therapeutic goal is to produce a tenfold shift to the right of the agonist dose-response curve. A concentration of an orthosteric simple competitive antagonist of $[B]/K_B = 10$ will achieve this and the duration of this effect will be determined by the kinetics of washout of the antagonist from the receptor compartment and the concentration of antagonist. A longer duration of action of such a drug could be achieved by increasing the concentration but this necessarily would increase the maximal effect as well (i.e., $[B]/K_B = 100$ would produce a 100-fold shift of the curve). In contrast, if an allosteric modulator with $\alpha = 0.1$ were to be employed an increased concentration would increase the duration of effect but the antagonism would never be greater than tenfold (as defined by the cooperativity factor α). Thus, the saturability of the allosteric ligand can be used to limit effect but increase the duration.

Another discerning feature of allosterism is the potential for increased selectivity. For example, it could be postulated that it would be difficult for orthosteric antagonists that bind to the acetylcholine recognition site of muscarinic receptors to be selective for muscarinic subtypes (i.e., teleologically these have evolved all to recognize a common agonist). However, the same is not true for the surrounding scaffold protein of the acetylcholine receptor and it is in these regions that the potential for selective stabilization of receptor conformations may be achieved [24].

Finally, the fact that allosteric modulators alter the signaling properties and/or sensitivity of the receptor to physiologically signaling means that their effect is linked to the receptor signal. This being the case, allosteric modulators will augment or modulate function in a reflection of the existing pattern. This may be especially beneficial for complex signal patterns such as those found in the brain. For this reason, the augmentation of the cholinergic system in Alzheimer's disease with cholinesterase inhibitors (these block the degradation of acetylcholine in the synapse and thus potentiate response in accordance with neural firing) has been one approach to treatment of this disease [25]. However, there are practical problems with this idea associated with nonspecific increase in both nicotinic and muscarinic receptor when only selective nicotinic function is

TABLE 7.2
Comparison of properties of orthosteric and allosteric ligands.

Orthosteric Antagonists	Allosteric Modulators
Orthosteric antagonists block all agonists with equal potency.	Allosteric antagonists may block some agonists but no others (at least as well).
There is a mandatory link between the duration of effect and the intensity of effect.	Duration and intensity of effect may be dissociated (i.e., duration can be prolonged though receptor compartment loading with no target overdose).
High concentrations of antagonist block signals to basal levels.	Receptor signaling can be modulated to a reduced (but not to basal) level.
Less propensity for receptor subtype effects.	Greater potential for selectivity.
No texture in effect (i.e., patterns of signaling may not be preserved).	Effect is linked to receptor signal. Thus, complex physiological patterns may be preserved.
All antagonist-bound receptors are equal.	Texture in antagonism where allosterically modified receptors may have different conformations from each other may lead to differences in resistance profiles with chronic treatment.

required. This has opened the field for other strategies such as selective allosteric potentiation of acetylcholine receptor function [26, 27]. In general, as a theoretical approach, allosteric control of function allows preservation of patterns of innervation, blood flow, cellular receptor density, and efficiencies of receptor coupling for complex systems of physiological control in the brain and other organs. The unique properties of allosteric modulators are summarized in Table 7.2.

7.4 Functional Study of Allosteric Modulators

In essence, an allosteric ligand produces a different receptor if the tertiary conformation of the receptor is changed through binding. These different tertiary conformations can have a wide range of effects on agonist function. A different receptor conformation can change its behavior toward G-proteins (and hence the cell and stimulus-response mechanisms) or the agonist, or both. Under these circumstances, there is a wide range of effects that allosteric ligands can have on agonist dose-response curves.

From the point of view of agonist activation, allosteric modulation can be thought of in terms of two separate effects. These effects may not be mutually exclusive and both can be relevant. The first, and most easily depicted, is a change in affinity of the receptor toward the agonist. The most simple system consists of a receptor [R] binding to a probe [A] (a probe being a molecule that can assess receptor behavior; probes can be agonists or radioligands) and an allosteric modulator [B] [28]:

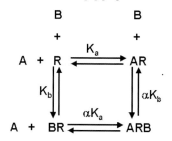

The equation for receptor occupancy for an agonist [A] in the presence of an allosteric ligand [B] is given by (see Section 7.8.1)

$$\frac{[AR]}{[R_{tot}]} = \frac{[A]/K_A(1 + \alpha[B]/K_B)}{[A]/K_A(1 + \alpha[B]/K_B) + [B]/K_B + 1}, \quad (7.1)$$

where K_A and K_B are the equilibrium dissociation constants of the agonist and antagonist receptor complexes respectively and α is the cooperativity factor. Thus, a value for α of 0.1 means that the allosteric antagonist causes a tenfold reduction in the affinity of the receptor for the agonist. This can be seen from the relationship describing the affinity of the probe [A] for the receptor, in the presence of varying concentrations of antagonist:

$$K_{obs} = \frac{K_A([B]/K_B + 1)}{(1 + \alpha[B]/K_B)}. \quad (7.2)$$

It can be seen that a feature of allosteric antagonists is that their effect is saturable (i.e., a theoretically infinite concentration of [B] will cause K_{obs} to reach a maximal asymptote value of K_A/α). This is in contrast to simple competitive antagonists where the degree of antagonism theoretically is infinite for an infinite concentration of antagonist. Therefore, the maximal change in affinity that can be produced by the allosteric modulator is $K_{obs}/K_A = K_A/\alpha K_A = \alpha^{-1}$. Thus, a modulator with $\alpha = 0.1$ will reduce the affinity of the receptor for the agonist by a maximal value of 10.

As well as changing the affinity of the receptor for an agonist, an allosteric effect could just as well change the reactivity of the receptor to the agonist. This could be reflected in a complete range of receptor effects (response production, internalization, desensitization, and so on). This is depicted schematically in the following, below where the agonist-bound receptor goes on to interact with the cell in accordance with the operational model for receptor function [29]. Thus, the receptor bound only to agonist

([AR] complex) interacts with an equilibrium association constant K_e (to yield an efficacy term τ) and the allosterically altered agonist-bound receptor complex ([ABR] complex) interacts with the cell with equilibrium association constant K'_e (to yield an altered efficacy τ'). It is useful to define a ratio of efficacies for the native and allosterically modulated receptor of τ'/τ (denoted ξ, where $\xi = \tau'/\tau$).

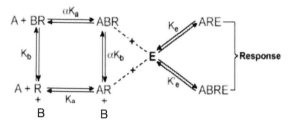

The response to an agonist in the presence of an allosteric modulator that can alter the affinity and efficacy of the receptor is given by (see Section 7.8.2)

Response

$$= \frac{[A]/K_A \tau(1 + \alpha\xi[B]/K_B)E_{max}}{[A]/K_A(1 + \alpha[B]/K_B + \tau(1 + \alpha\xi[B]/K_B)) + [B]/K_B + 1},$$

(7.3)

where K_A and K_B are the respective equilibrium dissociation constants of the agonist [A] and modulator [B] receptor complexes, α is the ratio of affinity of the agonist for the receptor in the presence and absence of the modulator, τ the efficacy of the agonist for the native receptor, and ξ the ratio of τ values of the agonist for the receptor in the presence and absence of modulator. From this general equation, a number of cases can be described.

7.4.1 Surmountable Allosteric Modulation ($\xi = 1$)

The first case to consider is where the modulator affects only the affinity of the receptor for the agonist but does not alter receptor signaling. Under these circumstances, $\xi = 1$ and Equation 7.3 reduces to

$$\text{Response} = \frac{[A]/K_A(1 + \alpha[B]/K_B)\tau\, E_{max}}{[A]/K_A(1 + \tau)(1 + \alpha[B]/K_B) + [B]/K_B + 1}.$$

(7.4)

Equation 7.4 predicts that even when the modulator reduces the affinity of the receptor for the agonist ($\alpha < 1$) the effects will be surmountable with respect to the agonist (i.e., the agonist will produce the control maximal response). This can be seen from Equation 7.4 when $[A] \rightarrow \infty$ and the fractional maximal response $\rightarrow 1$. If the signaling properties of the receptor are not altered by the allosteric modulator, then the concentration-response curve to the agonist will be shifted either to the right (if $\alpha < 1$, see Figure 7.9a) or to the left ($\alpha > 1$, see Figure 7.9b). The distinctive feature of such an allosteric effect is that while the displacements are parallel with no diminution of maxima there is a limiting value (equal to α^{-1}) to the maximal displacement. Figure 7.10a shows an experimentally observed allosteric displacement of acetylcholine effects in cardiac muscle by the allosteric modulator gallamine and the saturable maximal effect (Figure 7.10b).

When an antagonist produces parallel shifts to the right of the dose-response curve with no diminution of the maximal response, the first approach used to quantify potency is Schild analysis (see Section 6.3.1). In cases where the value of α is low (i.e., $\alpha = 0.01$), a tenfold concentration range of the antagonist would cause shifts commensurate with those produced by a simple competitive antagonist.

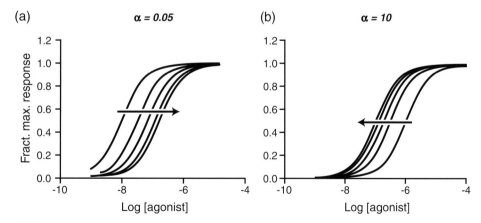

FIGURE 7.9 Functional responses in the presence of allosteric modulators as simulated with Equation 7.4 ($\tau = 30$). (a) Allosteric antagonism. Agonist $K_A = 0.3\,\mu M$, $\alpha = 0.05$, and $K_B = 1\mu M$. Curve furthest to the left is control in absence of modulator. From left to right, concentrations of modulator equal $3\,\mu M$, $10\,\mu M$, $30\,\mu M$, and $100\,\mu M$. Arrow indicates effect of modulator. Note the limited shift to the right. (b) Allosteric potentiation. Agonist $K_A = 30\,\mu M$, $\alpha = 10$, $K_B = 3\,\mu M$. Curve furthest to the right is control in absence of modulator. From right to left, concentrations of modulator equal $3\,\mu M$, $10\,\mu M$, $30\,\mu M$, and $100\,\mu M$. Arrow indicates effect of modulator. Note the limited shift to the left.

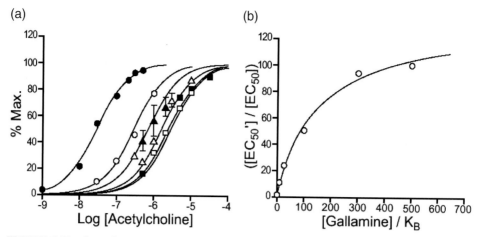

FIGURE 7.10 Operational model fit of the allosteric effects of gallamine on electrically evoked contractions of guinea pig left atrium. (a) Dose-response curves obtained in the absence (filled circles) and presence of gallamine 10 μM (open circles), 30 μM (filled triangles), 100 μM (open triangles), 300 μM (filled squares), and 500 μM (open squares). Data fit to operational model (Equation 7.4) with $K_A = 30$ nM, $E_{max} = 200$, $\tau = 1$. Data fit for gallamine $K_B = 1$ μM and $\alpha = 0.0075$. (b) Ratio of observed EC_{50} values $(EC'_{50}$ for curve in presence of gallamine/EC_{50} control curve) as a function of concentrations of gallamine. Data fit to rectangular hyperbola of max $= 134$ (1/maximum $= \alpha = 0.0075$). Data redrawn from [30].

However, the testing of a wide range of concentrations of an allosteric antagonist would show the saturation of the allosteric binding site as revealed by an approach to a maximal value for the antagonism. The Schild equation for an allosteric antagonist is given by (see Section 7.8.3)

$$Log(DR\text{-}1) = Log\left[\frac{[B](1-\alpha)}{\alpha[B] + K_B}\right]. \qquad (7.5)$$

Expected Schild regressions for allosteric antagonists with a range of α values are shown in Figure 7.11. It can be seen that the magnitude of α is inversely proportional to the

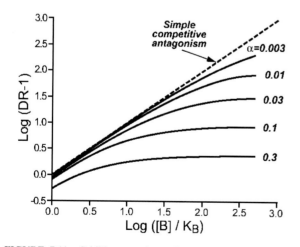

FIGURE 7.11 Schild regressions for allosteric antagonists of differing values of α. Dotted line shows the expected Schild regression for a simple competitive antagonist. With allosteric antagonists of lower values for α, the regression reaches a plateau at higher antagonist concentrations (i.e., curvature occurs at higher antagonist concentrations).

ability of the allosteric antagonist to appear as a simple competitive antagonist (i.e., the lower the value of α the more the antagonist will appear to be competitive). This is discussed further in Section 10.3.1, and an example of this type of analysis is given in Section 12.2.9.

The foregoing discussion has been restricted to allosteric ligands that reduce the affinity of the receptor for the agonist (i.e., allosteric antagonists or modulators). Since allosteric change is the result of a conformational change in the receptor, there is no a priori reason for allosterism to produce only a reduced agonist affinity, and in fact such changes can lead to increases in the affinity of the receptor for the agonist (note the stimulation of the binding of [³H]-atropine by alcuronium in Figure 4.12).

7.4.2 Insurmountable Allosteric Antagonism ($\xi = 0$)

Another possible allosteric effect is to render the receptor insensitive to agonist stimulation (i.e., remove the capacity for agonist response). This may or may not be accompanied by a change in the affinity of the receptor for the agonist. This can be simulated by setting $\xi = 0$ in Equation 7.3 to yield

$$Response = \frac{[A]/K_A \tau E_{max}}{[A]/K_A(1 + \tau + \alpha[B]/K_B) + [B]/K_B + 1}. \qquad (7.6)$$

It can be seen that when there is no effect on the affinity of the receptor for the agonist ($\alpha = 1$) Equation 7.6 is identical to the describing orthosteric noncompetitive antagonism derived by Gaddum and colleagues [31] (see Equation 6.10). However, while the equation is identical and the pattern of concentration-response curves is the same as that for an orthosteric antagonist it should be

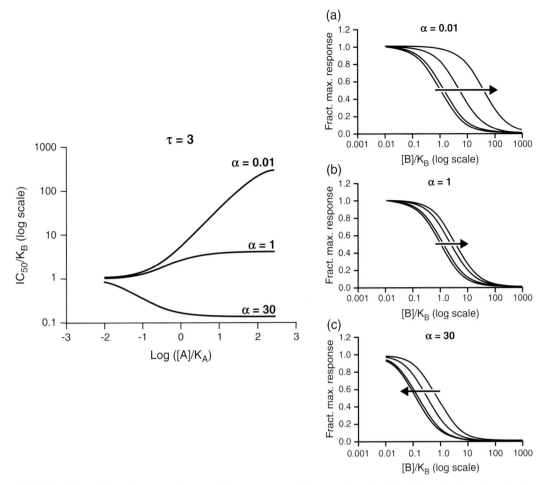

FIGURE 7.17 Effect of varying allosteric effects on agonist affinity on the ratio of the IC_{50} to the true K_B. Graph A shows the effects of a modulator that decreases the affinity of the receptor for the agonist (inset panel a for $\alpha = 0.01$). Shown are inhibition curves in the presence of increasing concentrations of agonist. These shift the inhibition curves to the right and cause an increased ratio for IC_{50}/K_B. Graph b shows the effects of a modulator that has no effect on affinity ($\alpha = 1$). Little shift to the right of the inhibition curve is observed. The slight observable shift is caused by the small receptor reserve due to the value of $\tau = 3$. Graph c shows the effects of a modulator that increases the affinity of the agonist for the receptor ($\alpha = 30$). In this case, the inhibition curves actually shift to the left with increasing concentrations of agonist. The ratio of IC_{50}/K_B decreases with increasing agonist concentrations.

Antagonism can be unclear therefore the concepts of saturability of effect and probe dependence may need to be actively pursued to tease out allosteric mechanisms. If a clear plateau of effect is observed then allosterism is implicated (see Figure 7.15b). If an allosteric antagonism does not interfere with receptor function then surmountable antagonism will be observed (Equation 7.4). A limited Schild analysis may not detect the characteristic curvilinearity of allosteric blockade (Figure 7.11). Therefore, detection of possible allosterism requires extension of normal concentration ranges for testing of blockade (see Figure 7.19).

Differentiation of orthosterism and allosterism also can be made by using different receptor probes. For orthosteric antagonists, the choice of agonist is immaterial (i.e., the same pK_B will result, see Figure 11.21). However, this is not true of allosteric effect where α and ξ values may be unique for every receptor probe. This is a logical consequence of the allosteric model where it can be seen that mathematical terms exist containing the concentration of the antagonist, the α and ξ values for allosterism and the concentration of agonist are together ($[A]/K_A \tau \alpha \xi [B]/K_B$ term in both the numerator and denominator of Equation 7.3). This allows the magnitude of α and ξ to moderate the degree of antagonism. Since these constants are unique for every receptor probe, then the antagonism may also depend on the nature of the receptor probe (agonist). This is in contrast to orthosteric antagonist models where there are no terms containing both $[A]/K_A$ and $[B]/K_B$. In this latter case, there is no possibility of the nature of the agonist determining the magnitude of antagonist effect. Figure 7.20 shows probe dependence on the CCR5 receptor with the

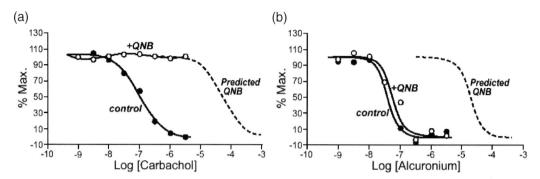

FIGURE 7.18 Ligand-target validation. Lack of sensitivity of putative agonist effect to classical receptor antagonists. (a) Inhibition of cyclic AMP due to activation of muscarinic m2 receptors by the classical muscarinic agonist carbachol in the absence (filled circles) and presence (open circles) of the classical muscarinic antagonist QNB present in a concentration that shifts the agonist curve to the location shown by the dotted line. This concentration of QNB completely blocks the response. (b) Inhibition of cyclic AMP through activation of muscarinic m2 receptors by the allosteric agonist alcuronium in the absence (filled circles) and presence (open circles) of the same concentration of QNB. In this case, the response is insensitive to this concentration of the antagonist. Data redrawn from [32].

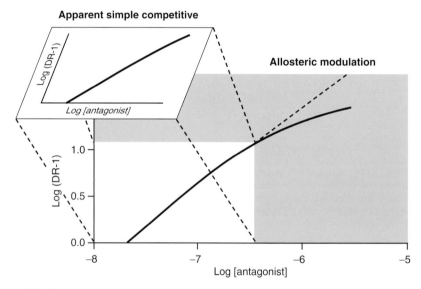

FIGURE 7.19 Schild regression for allosteric modulator of $K_B = 200\,nM$ that has $\alpha = 0.03$ for the agonist. It can be seen that the regression is linear with unit slope at dose ratios < 10. However, extension of concentrations greater than 300 nM reveal saturation of the antagonism and a curvilinear portion of the Schild regression (indicative of allosteric antagonism).

allosteric modulator aplaviroc. It can be seen that the affinity of ^{125}I-MIP-1α is decreased considerably ($\alpha < 0.03$) while the affinity for ^{125}I-RANTES is unchanged (α estimated to be 0.8 [9]).

7.7 Chapter Summary and Conclusions

- Allosteric modulators affect the interaction of the receptor and probe molecules (i.e., agonists or radioligands) by binding to separate sites on the receptor.

These effects are transmitted through changes in the receptor protein.

- Allosteric modulators possess properties different from orthosteric ligands. Specifically, allosteric effects are saturable and probe dependent (i.e., the modulator produces different effects for different probes).
- Allosteric effects can result in changes in affinity and or efficacy of agonists.
- Sole effects on affinity (with no change in receptor function) result in surmountable antagonism.

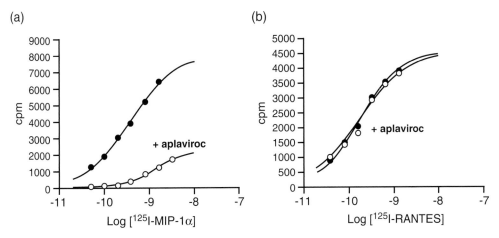

FIGURE 7.20 Effects of aplaviroc, an allosteric modulator of the CCR5 receptor, on the binding of the chemokine ^{125}I-MIP-1α (panel a) and ^{125}I-RANTES (panel b). It can be seen that aplaviroc blocks the binding of MIP-1α but has very little effect on the binding of RANTES. Such probe dependence is indicative of allosteric effect. Data from [32].

The dextral displacement reaches a maximal value leading to a curvilinear Schild regression.

- Allosteric modulators that block receptor function produce insurmountable antagonism. In addition, modulators that block function also can alter (increase or decrease) affinity.

7.8 Derivations

- Derivation of allosteric model of receptor activity (7.7.1)
- Effects of allosteric ligands on response: changing efficacy (7.7.2)
- Schild analysis for allosteric antagonists (7.7.3)
- Relationship of pA_2 and pK_B for insurmountable allosteric antagonism (7.7.4)

7.8.1 Allosteric Model of Receptor Activity

Consider two ligands ([A] and [B]), each with its own binding site on the receptor with equilibrium association constants for receptor complexes of K_a and K_b, respectively. The binding of either ligand to the receptor modifies the affinity of the receptor for the other ligand by a factor α. There can be three ligand-bound receptor species; namely, [AR], [BR], and [ARB]:

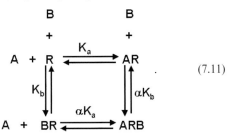

$$. \qquad (7.11)$$

The resulting equilibrium equations are

$$K_a = [AR]/[A][R], \qquad (7.12)$$

$$K_b = [BR]/[B][R], \qquad (7.13)$$

$$\alpha K_a = [ARB]/[BR][A], \quad \text{and} \qquad (7.14)$$

$$\alpha K_b = [ARB]/[AR][B]. \qquad (7.15)$$

Solving for the agonist bound receptor species [AR] and [ARB] as a function of the total receptor species ($[R_{tot}] = [R] + [AR] + [BR] + [ARB]$) yields

$$\frac{[AR] + [ARB]}{[R_{tot}]}$$
$$= \frac{((1/\alpha[B]K_b) + 1)}{((1/\alpha[B]K_b) + (1/\alpha K_a) + (1/\alpha[A]K_a K_b) + 1)}. \qquad (7.16)$$

Simplifying and changing association to dissociation constants (i.e., $K_A = 1/K_a$) yields [28]

$$\rho = \frac{[A]/K_A(1 + \alpha[B]/K_B)}{[A]/K_A(1 + \alpha[B]/K_B) + [B]/K_B + 1}. \qquad (7.17)$$

7.8.2 Effects of Allosteric Ligands on Response: Changing Efficacy

The receptor can bind both the probe (agonist, radioligand, [A]) and allosteric modulator ([B]). The agonist bound receptor signal through the normal operational model ([AR] complex interacting with cellular stimulus-response machinery with association constant K_e) and in a possibly different manner when the allosteric modulator is bound (complex [ABR] interacting with cell with association constant K'_e):

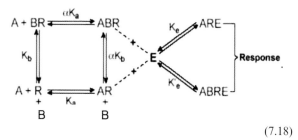

$$(7.18)$$

The equilibrium species are

$$[AR] = [ABR]/\alpha[B]K_b, \qquad (7.19)$$

$$[BR] = [ABR]/\alpha[A]K_\alpha, \quad \text{and} \qquad (7.20)$$

$$[R] = [ABR]/\alpha[B]K_b[B]K_b. \qquad (7.21)$$

According to the operational model, the response producing species activate the response elements of the cell according to

$$\text{Response} = \frac{[AR]/K_E + [ABR]/K'_E}{[AR]/K_E + [ABR]/K'_E + 1}, \qquad (7.22)$$

where $K_E = K_e^{-1}$ and $K'_E = K'_e - 1$. The amount of any receptor species is given by the fractional amount of receptor multiplied by the total receptor number. Thus, Equation 7.22 can be rewritten as

$$\text{Response} = \frac{\rho_{AR}[R_t]/K_E + \rho_{ABR}[R_t]/K'_E}{\rho_{AR}[R_t]/K_E + \rho_{ABR}[R_t]/K'_E + 1}, \qquad (7.23)$$

where ρ_{AR} is the fraction of receptor in the [AR] form given by

$$\rho_{AR} = [A]/K_A/([A]/K_A(1 + \alpha[B]/K_B) + [B]/K_B + 1) \qquad (7.24)$$

and ρ_{ABR} is the fraction of receptor in the [ABR] form given by

$$\rho_{ABR} = \alpha[A]/K_A[B]/K_B/([A]/K_A(1 + \alpha[B]/K_B) \\ + [B]/K_B + 1). \qquad (7.25)$$

Substituting Equations 7.24 and 7.25 into 7.23 and defining τ as $[R_t]/K_E$ and t' as $[R_t]/K'_E$, Equation 7.23 can be rewritten as

Response

$$= \frac{[A]/K_A(\tau + \alpha\tau'[B]/K_B)E_{max}}{[A]/K_A(1 + \alpha[B]/K_B + (\tau + \alpha\tau'[B]/K_B)) + [B]/K_B + 1}. \qquad (7.26)$$

Finally, defining ξ as the ratio of τ values for the agonist bound receptor when it is and is not bound to

modulator ($\xi = \tau'/\tau$), Equation 7.26 becomes

Response

$$= \frac{[A]/K_A\tau(1 + \alpha\xi[B]/K_B)E_{max}}{[A]/K_A(1 + \alpha[B]/K_B + \tau(1 + \alpha\xi[B]/K_B)) + [B]/K_B + 1}. \qquad (7.27)$$

7.8.3 Schild Analysis for Allosteric Antagonists

From Equation 7.3, the observed EC_{50} for the agonist, in the presence of a concentration of allosteric antagonist [B] is given by

$$EC'_{50} = \frac{EC_{50}([B]/K_B + 1)}{(1 + \alpha[B]/K_B)}, \qquad (7.28)$$

where EC_{50} refers to the EC_{50} of the control concentration-respones curve in the absence of modulator. The ratio of the EC_{50} values (concentrations of agonist producing 50% response in the presence and absence of the allosteric antagonist) is given by

$$\frac{EC'_{50}}{EC_{50}} = DR = \frac{([B]/K_B + 1)}{(1 + \alpha[B]/K_B)}. \qquad (7.29)$$

This leads to the logarithmic metameter form of the Schild equation:

$$\text{Log(DR-1)} = \text{Log}\left[\frac{[B](1 - \alpha)}{\alpha[B] + K_B}\right]. \qquad (7.30)$$

7.8.4 Relationship of pA₂ and pK_B for Insurmountable Allosteric Antagonism

As with insurmountable orthosteric antagonists, the shift to the right of concentration-response curves produced by allosteric insurmountable antagonists can be used to calculate a pA_2 value, and in turn this can be related to the pK_B of the antagonist. A concentration of antagonist equal to the pA_2 (i.e., concentration $= 10^{-pA_2}$) causes a dose ratio of 2, leading to the following equality:

$$\frac{2[A]/K_A\tau E_{max}}{2[A]/K_A\left(1 + \tau + \alpha[10^{-pA_2}]/K_B\right) + [10^{-pA_2}]/K_B + 1} \\ = \frac{[A]/K_A\tau E_{max}}{[A]/K_A(1 + \tau) + 1}. \qquad (7.31)$$

The equation for the relationship between the pA_2 and the K_B of an insurmountable allosteric modulator then becomes

$$10^{-pA_2} = K_B/(1 + 2\alpha[A]/K_A) \quad \text{and} \qquad (7.32)$$

$$pK_B = pA_2 - \text{Log}(1 + 2\alpha[A]/K_A). \qquad (7.33)$$

For allosteric modulators that decrease the affinity of the receptor for the antagonist ($\alpha < 1$), the insertion of the α term decreases the error between the observed pA_2 and the true pK_B (thus improving the method). However, if the allosteric modulator *increases* the affinity of the receptor for the agonist ($\alpha > 1$) the error produced by the insurmountable nature of the blockade may become substantial. If the allosteric modulator does not completely block receptor signaling ($\xi \neq 0$), then there is even a closer correspondence between the pA_2 and true pK_B, as shown by

$$10^{-pA_2} = K_B/(1 + 2\alpha[A]/K_A(1 - \xi) - 2\alpha\xi) \quad (7.34)$$

Thus, the non-zero ξ term reduces the effect of α on the pA_2 estimate. It can be seen that when receptor signaling is blocked by the allosteric modulator ($\xi = 0$), Equation 7.34 reduces to Equation 7.32.

References

1. Jakubic, J., Bacakova, I., El-Fakahany, E. E., and Tucek, S. (1997). Positive cooperativity of acetylcholine and other agonists with allosteric ligands on muscarinic acetylcholine receptors. *Mol. Pharmacol.* **52**:172–179.
2. Koshland, D. E. (1960). The active site of enzyme action. *Adv. Enzymol.*, **22**:45–97.
3. Lee, N. H., Hu, J., and El-Fakahany, E. E. (1992). Modulation by certain conserved aspartate residues of the allosteric interaction of gallamine and the m1 muscarinic receptor. *J. Pharmacol. Exp. Ther.* **262**:312–316.
4. Smerdon, S. J., Jager, J., Wang, J., Kohlstaedt, L. A., Chirino, A. J., Friedman, J. M., Rice, P. A., and Steitz, T. A. (1994). Structure of the binding site for nonnucleoside inhibitors of the reverse transcriptase of human immunodeficiency virus type 1. *Proc. Natl. Acad. Sci. USA* **91**:3911–3915.
5. Horn J. R., and Shoichet, B. K. (2004). Allosteric inhibition through core disruption. *J. Mol. Biol.* **336**:1283–1291.
6. Oikonomakos, N. G., Skamnaki, V. T., Tsitsanou, K. E., Gavalas, N. G., and Johnson, L. (2000). A new allosteric site in glycogen phosphorylase b as a target for drug interactions. *Structure* **8**:575–584.
7. Kallen, J. K., Welzenbach, K., Ramage, W. P., Geyl, D., Kriwacki, R., Legge, G., Cottens, S., Weitz-Schmidt, R., and Hommel, U. (1999). Structural basis for LFA-1 inhibition upon lovastatin binding to the CD11a I-domain. *J. Mol. Biol.* **292**:1–9.
8. Teague S. J. (2003). Implications of protein flexibility for drug discovery. *Nature Reviews Drug Discovery* **2**:527–541.
9. Watson, C., Jenkinson, S., Kazmierski,W., and Kenakin, T. P. (2005). The CCR5 receptor-based mechanism of action of 873140, a potent allosteric non-competitive HIV entry-inhibitor. *Mol. Pharmacol.* **67**:1268–1282.
10. Bieniassz, P. D., Fridell, R. A., Aramori, I., Ferguson, S. S., Caron, M. G., and Cullen, B. R. (1997). HIV-1 induced cell fusion is mediated by multiple regions within both the viral envelope and the CCR-5 co-receptor. *EMBO J.* **16**:2599–2609.
11. Kwong, P. D., Wyatt, R., Robinson, J., Sweet, R. W., Sodroski, J., and Hendricks, W. A. (1998). Structure of an HIV gp120 envelope glycoprotein in complex with the CD4 receptor and a neutralizing human antibody. *Nature* **393**:648–659.
12. Doranz, B. J., Lu, Z.-H., Rucker, J., Zhang, T.-Y., Sharron, M., Cen, Y.-H., Wang Z.-X., Guo, H.-H., Du, J.-D., Accavitti,

M. A., Doms, R. W., and Peiper, S. C. (1997). Two distinct CCR5 domains can mediate coreceptor usage by human immunodeficiency virus type 1. *J. Virol.* **71**:6305–6314.
13. Smyth, R. J., Yi, Y., Singh, A., and Collman, R. G. (1998). Determinants of entry cofactor utilization and tropism in a dualtropic human immunodeficiency virus type 1 isolate. *J. Virol.* **72**:4478–4484.
14. Poignard, P., Saphire, E. O., Parren, P. W., and Burton, D. R. (2001). gp120: Biologic aspects of structural features. *Annu. Rev. Immunol.* **19**:253–274
15. Wyatt, R., and Sodroski, J. (1998). The HIV-1 envelope glycoproteins: Fusogens, antigens, and immunogens. *Science* **280**:1884–1888.
16. Trkola, A., Kuhmann, S. E., Strizki, J. M., Maxwell, E., Ketas, T., Morgan, T., Pugach, P., Xu, S., Wojcik, L., Tagat, J., Palani, A., Shapiro, S., Clader, J. W., McCombie, S., Reyes, G. R., Baroudy, B. M., and Moore, J. P. (2002). HIV-1 escape from a small molecule, CCR5-specific entry inhibitor does not involve CXCR4 use. *Proc. Natl. Acad. Sci. USA* **99**:395–400.
17. Kuhmann, S. E., Pugach., P., Kunstman, K. J., Taylor J., Stanfield, R. L., Snyder, A., Strizki, J. M., Riley, J., Baroudy, B. M., Wilson, I. A., et. al. (2004). Genetic and phenotypic analyses of human immunodeficiency virus type 1 escape from small-molecule CCR inhibitor *J. Virol.* **78**:2790–2807.
18. Nagasaw, T., Hirota, S., Tachibana, K., Takakura, N., Nishikawa, S., Kitamura, Y., Yoshida, N., Kikutani, H., and Kishimoto T. (1996). Defects of B-cell lymphopoiesis and bone-marrow myelopoiesis in mice lacking the CXC chemokine PBSF/SDF-1. Nature **382**:635–638.
19. Tachibana, K., Hirota, S., Iizasa, H., Yoshida, H., Kawabata, K., Kataoka, Y., Kitamura, Y., Matsushima, K., Yoshida, N., Nishikawa, S., Kishimoto, T., and Nagasawa, T. (1998). The chemokine receptor CXCR4 is essential for vascularization of the gastrointestinal tract. *Nature* **393**:591–594.
20. Zou, Y. R., Kottmann, A. H., Kuroda, M., Taniuchi, I., and Littman, D. R., (1998). Function of the chemokine receptor CXCR4 in haematopoiesis and in cerebellar development. *Nature* **393**:595–599.
21. Sachpatzidis, A., Benton, B. K., Manfredis, J. P., Wang, H., Hamilton, A., Dohlman, H. G., and Lolis, E. (2003). Indentification of allosteric peptide agonists of CXCR4. *J. Biol. Chem.* **278**:896–907.
22. Heveker, N., Montes, M., Germeroth, L., Amara, A., Trautmann, A., Alizon, M., and Schneider-Mergener, J. (1998). Dissociation of the signalling and antiviral properties of SDF-1-derived small peptides. *Current Biology* **8**:369–376.
23. Bartus, R. T., Dean, R. L., Beer, B., and Lippa, A. S. (1982). The cholinergic hypothesis of geriatric memory dysfunction. *Science* **217**:408–417.
24. Tucek, S., and Proska, J. (1995). Allosteric modulation of muscarinic acetylcholine receptors. *Trends Pharmacol. Sci.* **16**:205–212.
25. Flicker, L. (1999). Acetylcholinesterase inhibitors for Alzheimer's disease. *Br. Med. J.* **318**:615–616.
26. Maelicke A., and Albuquerque, E. X. (1996). New approach to drug therapy of Alzheimer's dementia. *Drug Discovery Today* **1**:53–59.
27. Krause, R. M., Buisson, B., Bertrand, S., Corringer, P.-J., Galzi, J.-L., Changeux, J.-P., and Bertrand D. (1998). Ivermectin: A positive allosteric effector of the a7 neuronal nicotinic acetylcholine receptor. *Mol. Pharmacol.* **53**:283–294.

28. Ehlert, F. J. (1988). Estimation of the affinities of allosteric ligands using radioligand binding and pharmacological null methods. *Mol. Pharmacol.* **33:**187–194.

29. Black, J. W., Leff, P., Shankley, N. P., and Wood, J. (1985) An operational model of pharmacological agonism: The effect of E/[A] curve shape on agonist dissociation constant estimation. *Br. J. Pharmacol.* **84:**561–571.

30. Christopoulos, A. (2000). Overview of receptor alloserism. In *Current Protocols in Pharmacology*, Vol 1, pp. 1.21.21–1.21.45, eds. Enna, S. J., Williams, M., Ferjany, J.

W., Porsolt, R. D., Kenakin, T. P., and Sullivan, J. P., New York: John Wiley and Sons.

31. Gaddum, J. H., Hameed, K. A., Hathway, D. E., and Stephens, F. F. (1955). Quantitative studies of antagonists for 5-hydroxytryptamine. *Q. J. Exp. Physiol.* **40:**49–74.

32. Jakubic, J., Bacakova, L., Lisá, V., El-Fakahany, E. E., and Tucek, S. (1996). Activation of muscarinic acetylcholine receptors via their allosteric binding sites. *Proc. Natl. Acad. Sci. USA* **93:**8705–8709.

8

The Process of Drug Discovery

One doesn't discover new lands without consenting to lose sight of the shore for a very long time.
— ANDRÉ GIDE (1869–1951)

The real voyage of discovery consists not in seeking new landscapes but in having new eyes.
— MARCEL PROUST (1871–1922)

8.1 Pharmacology in Drug Discovery

The drug discovery process can be envisioned as four interconnected phases (see Figure 8.1). Generally, these are the acquisition of chemicals to be tested for biological activity, the determination of the activity of those chemicals on biological systems (pharmacodynamics), the formulation of the most active of these for therapeutic testing in humans (pharmaceutics), and the determination of adequate delivery of the active drug to diseased tissues (pharmacokinetics). Each of these collections of processes is interconnected with the others and failure in any one of them can halt the development process. It is worth considering each process separately, as well as the relationships between them.

8.2 Chemical Sources for Potential Drugs

A starting point to this process is the definition of what the therapeutic end point of the drug discovery process will be; namely, a drug. There are certain properties molecules must have to qualify as therapeutically useful chemicals. While in theory any molecule possessing activity that can be introduced into the body compartment containing the therapeutic target could be a possible drug, in practice therapeutically useful molecules must be absorbed into the body (usually by the oral route), distribute to the biological target in the body, be stable for a period of time in the body, be reversible with time (excreted or degraded in the body after a reasonable amount of time), and be nontoxic. Ideally, drugs must be low-molecular-weight bioavailable

molecules. Collectively, these desired properties of molecules are often referred to as "drug-like" properties. A useful set of five rules for such molecules has been proposed by Lipinski and co-workers [1]. Molecules that fulfill these critieria generally can be considered possible therapeutically useful drugs provided they possess target activity and few toxic side effects. Specifically, these rules state that "drug-like" molecules should have less than five hydrogen-bond donor atoms, have a molecular mass of < 500 Da, have high lipophilicity (cLogP > 5), and that the sum of the nitrogen and oxygen atoms should be < 10. Therefore, when estimating the potential therapeutic drug targets these properties must be taken into consideration.

There are numerous chemical starting points for drugs. Historically, natural products have been a rich source of molecules. The Ebers Papyrus, one of the earliest documents recording ancient medicine, describes 700 drugs—most of them from plants. Similarly, the Chinese *Materia Medica* (100 BCE), the *Shennong Herbal* (100 BCE), *Tang Herbal* (659 AD), the Indian *Ayurvedic* system (1000 BCE), and books of Tibetan medicine *Gyu-zhi* (800 AD) all document herbal remedies for illness. Some medicinal substances have their origins in geographical exploration. For example, tribes indigenous to the Amazon River had long been known to use the bark of the *Cinchona officinalis* to treat fever. In 1820, Caventou and Pelletier extracted the active antimalarial quinine from the bark that provided the starting point for the synthetic antimalarials chloroquine and mefloquine. Traditional Chinese herbal medicine has yielded compounds such as artemisinin and derivatives for the treatment of fever from the *Artemisia annua*. The anticancer vinca alkaloids were isolated from the Madagascar periwinkle *Catharanthus roseus*. Opium is an ancient medicinal substance described by Theophrastus in the third century BCE, used for many years by Arabian physicians for the treatment of dysentery and for the relief of suffering (as described by Sydenham in 1680) in the Middle Ages. Known to be a mixture of alkaloids, opium furnished therapeutically useful pure alkaloids when Serturner isolated morphine in 1806, Robiquet isolated codeine in 1832, and Merck isolated papaverine in 1848. At present, only 5 to 15% of the 25,000 species of higher plants have been studied for possible therapeutic activity. Of prescriptions in the United States written between 1959 and 1980, 25% contained plant extracts or active principals.

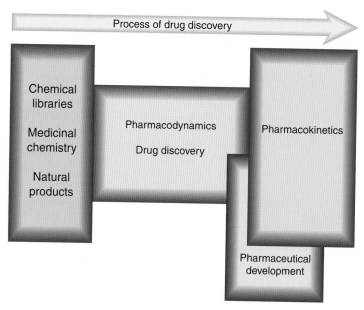

FIGURE 8.1 Schematic diagram of four interactive but separate stages of drug discovery and development.

Marine life can also be a rich source of medicinal material. For example, C-nucleosides spongouridine and spongothymidine isolated from the Caribbean sponge *Cryptotheca crypta* possess antiviral activity. Synthetic analogues led to the development of cytosine arabinoside, a useful anticancer drug. Microbes also provide extremely useful medicines—the most famous case being penicillin from *Penicillium chrysogenum*. Other extremely useful bacterial = derived products include the fungal metabolites, the cephalosporins (from *Cephalosporium cryptosporium*), aminoglycosides and tetracyclines from *Actinomycetales*, immunosuppressives such as the cyclosporins and rapamycin (from *Streptomyces*), cholesterol-lowering agents mevastatin and lovastatin (from *Penicillium*), and antihelmintics and antiparasitics such the ivermectins (from *Streptomyces*). As with plants, less than 1% of bacterial and less than 5% of fungal sources have been explored for medicinal value. In general, the World Health Organization estimates that 80% of the world's population relies on traditional medicine with natural products.

From this perspective, natural products appear to be a great future source of drugs. However, teleologically there may be evolutionary pressure against biological activity of natural products. Thus, while millions of years of selective pressure has evolved molecules that specifically interact with physiological receptors (i.e., neurotransmitters, hormones) with little "cross-talk" to other targets it can be argued that those same years exerted a selective evolutionary pressure to evolve receptors that interact only with those molecules and not the myriad of natural products to which the organism has been exposed. In practical terms, natural products as drugs or starting points for drugs have certain inherent disadvantages as well. Specifically, these tend to be expensive, not chemically tractable (structurally complex and difficult to derivatize), and involve difficult and expensive scale-up procedures (active species tend to be minor components of samples). Natural products also often contain a larger number of ring structures and more chiral centers and have sp^3 hybridization bridgehead atoms present. Natural products are often high in stereo complexity, and in that they contain few nitrogen, halogen, and sulfur atoms and are oxygen rich with many hydrogen donors natural products often are very prone to enzymatic reactions. In addition, a practical problem in utilizing such pharmacophores is the unpredictable novelty and intellectual property that may result. In spite of these shortcomings, between the years 1981 and 2002 of the 67% of 877 synthetic new chemical entities 16.4% utilized pharmacophores derived directly from natural products.

Another approach to the discovery of drugs is "rational design." The basis for this strategy is the belief that detailed structural knowledge of the active site binding the drug will yield corresponding information to guide the design of molecules to interact with that active site. One of the best known examples, yielding rich dividends, is the synthesis of the angiotensin-converting enzyme (ACE) inhibitor captopril from a detailed analysis of the enzyme active site. Similar design of small molecules to fit specific binding loci of enzymes was accomplished for HIV protease (nelfinavir) and Relenza for the prevention of influenza. Other rational design approaches utilize dual pharmacophores from other active drugs to combine useful therapeutic activities. This approach offers the advantage that the dual biological activity will be absorbed, metabolized, and excreted in a uniform manner (i.e., the activity profile of the drug will not change with varying ratios of two

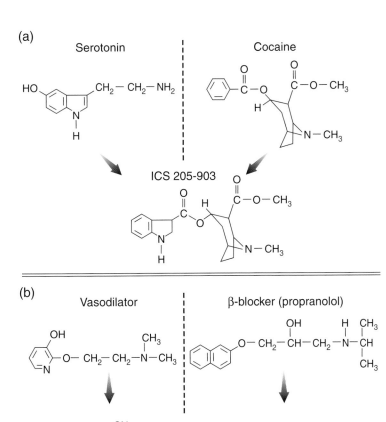

FIGURE 8.2 Examples of drug design through hybridization: combination of two structural types to produce a unique chemical entity. (a) Design of ICS 205-903 [2]. (b) Compound with vasodilating and beta-blocking properties [3].

simultaneously dosed drugs). This also gives medicinal chemists a place to start. For example, ICS 205-903 (a novel and potent antagonist of some neural effects of serotonin in migraine) was made by utilizing the structure of cocaine, a substance known to have seriously debilitating central effects but also known to block some of the neural effects of serotonin with the serotonin structure. The result was a selective serotonin antagonist devoid of the disadvantages of cocaine (Figure 8.2a). Similarly, a beta-adrenoceptor blocker with vasodilating properties has been made by combining the structure of the beta-blocker propranolol with that of a vasodilator (Figure 8.2b). The idea of introducing dual or multi-target activities in molecules is discussed further in Section 9.5.

There are numerous natural substances that have useful therapeutic properties as well as other undesirable properties. From these starting points, medicinal chemists have improved on nature. For example, while extremely useful in the treatment of infection penicillin is not available by the oral route. This shortcoming is overcome in the analogue ampicillin (Figure 8.3a). Similarly, the obvious deleterious effects of cocaine have been eliminated in the local anesthetic procaine (Figure 8.3b). The short activity and weak steroid progesterone is converted to a stronger long acting analogue (+)-norgestrel through synthetic modification (Figure 8.3c). Catecholamines are extremely important to sustaining life and have a myriad of biological activities. For example, norepinephrine produces a useful bronchodilation that has utility in the treatment of asthma. However, it also has a short duration of action, is a chemically unstable catechol, and produces debilitating tachycardia, vasoconstriction, and digital tremor. Synthetic modification to salbutamol eliminated all but the tremorogenic side effects to produce a very useful bronchodilator for the treatment of asthma (Figure 8.3d).

It can be argued that drugs themselves can be extremely valuable starting points for other drugs in that by virtue of the fact that they are tolerated in humans they allow the

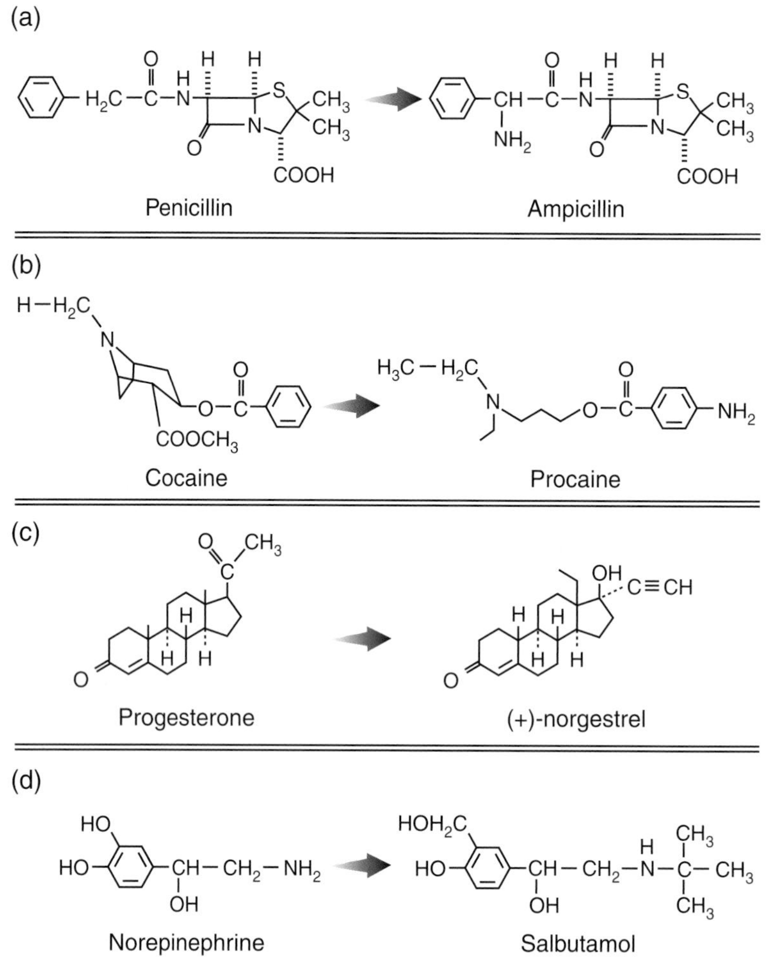

(a)

Penicillin   Ampicillin

(b)

Cocaine   Procaine

(c)

Progesterone   (+)-norgestrel

(d)

Norepinephrine   Salbutamol

FIGURE 8.3 Examples of chemical modification of active drugs that have either unwanted effects (cocaine, norepinephrine) or suboptimal effects (penicillin, progesterone) to molecules with useful therapeutic profiles.

observation of their other effects. Some of those effects ("side effects") may lead to useful therapeutic indications. For example, the observed antiedemal effects of the antibacterial sulfanilamide in patients with congestive heart failure led to the discovery of its carbonic anhydrase inhibitor activity and the subsequent development of the diuretic furosemide (Figure 8.4a). Similarly, the antidiabetic effects of the antibiotic carbutamide led to the development of the antidiabetic tolbutamide (Figure 8.4b). Some of the early antihistamines were found to exert antidepressant and antipsychotic properties. These led to modern psychopharmaceuticals. The immunosuppressant activity of the fungal agent cyclosporine also was exploited for therapeutic utility.

Endogenous substances such as serotonin, amino acids, purines, and pyrimidines all have biological activity and are tolerated in the human body. Therefore, these can be used in some cases as starting points for synthetic drugs. For example, the amino acid tryptophan and neurotransmitter

serotonin were used to produce selective ligands for 5-HT5$_A$ receptors and a selective somatostatin3 antagonist, adenosine A2b receptor antagonists from adenine, and a selective adenosine 2A receptor agonist from adenosine itself (Figure 8.5).

Major pharmaceutical efforts revolve around the testing of large chemical libraries for biological activity. Assuming that most drugs must have a molecular weight of less than 600 (due to desired pharmacokinetic properties as discussed later), there are wide ranges in the estimates of the number of molecules that exist in "chemical space" (i.e., how many different molecules can be made within this size limit). The estimates range from 10^{40} to 10^{100} molecules, although the need for activated carbon centers for the construction of carbon-carbon bonds in synthetic procedures reduces the possible candidates for synthetic congeners. In spite of this fact, the number of possibilities is staggering. For example, in the placement of 150 substituents on mono to 14-substituted hexanes there are 10^{29} possible derivatives.

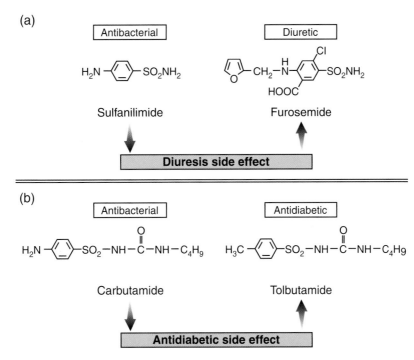

FIGURE 8.4 Examples of case where the side effects of drugs used for another indication led to the discovery and development of a new therapeutic entity for another disease.

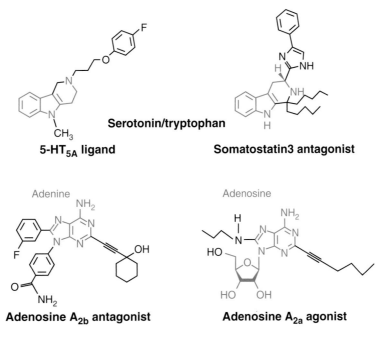

FIGURE 8.5 Examples of natural substances (shown in red) that have been chemically modified to yield therapeutically useful selective drugs.

Considering a median value of 10^{64} possible structures in chemical space clearly indicates that the number of possible structures available is far too large for complete coverage by chemical synthesis and biological screening. It has been estimated that a library of 24 million compounds would be required to furnish a randomly screened molecule with biological activity in the nanomolar potency range. While combinatorial libraries have greatly increased the productivity of medicinal chemists (i.e., a single chemist might have produced 50 novel chemical structures in a year 10 years ago, but with the availability of solid and liquid phase synthesis and other combinatorial techniques a single chemist can produce thousands of compounds in a single month at a fraction of the cost of previous techniques), 24 million compounds per lead is still considerably larger than the practical capability of industry.

One proposed reason for the failure of many high-throughput screening campaigns is the lack of attention to drug-like (namely, the ability to be absorbed into the human body and having a lack of toxicity) properties in the chemical library. The non-drug-like properties of molecules lead to biological activity that cannot be exploited therapeutically. This is leading to improved drug design in chemical libraries incorporating features to improve drug-like properties. One difficulty with this approach is the multifaceted nature of the molecular properties of drug-like molecules (i.e., while drug-like chemical space is more simple than biological target space the screens for drug-like activity are multimechanism based and difficult to predict). Thus, incorporating favorable drug-like properties into chemical libraries can be problematic. Also, different approaches can be counterintuitive to the incorporation of drug-like properties. Thus, rational design of drugs tends to increase molecular weight and leads to molecules with high hydrogen bonding and unchanged lipophilicity. This generally can lead to reduced permeability. A target permeability for drug-like molecules (which should have aqueous solubility minimum of $> 52\,\mu g/ml$) should achieve oral absorption from a dose of $> 1\,mg/kg$. High-throughput screening approaches tend to increase molecular weight, leave hydrogen bonding unchanged from the initial hit, and increase lipophilicity. This can lead to decreases in aqueous solubility with concomitant decrease in drug-like properties.

The assumption made in estimations of the number of molecules that would be required to yield biologically active molecules is that potential drugs are randomly and uniformly distributed throughout chemical space. Analysis of known drugs and biologically active structures indicates that this latter assumption probably is not valid. Instead, drugs tend to cluster in chemical space (i.e., there may be as few as 10,000 drug-like compounds in pharmacological space [4]). The clustering of drug-like molecules in chemical space has led to the concept of "privileged structures" from which medicinal chemists may choose for starting points for new drugs. A privileged structure is defined as a molecular scaffold with a range of binding properties that yields potent and selective ligands for a range of targets through modification of functional groups. Privileged structures can be a part of already known drugs such as the dihydropyridines (known as calcium channel blockers). In this case, inhibitors of platelet aggregation (PAF inhibitors) and neuropeptide Y type 1 receptor ligands have been made form the dihydropyridine backbone (Figure 8.6). Privileged structures also can simply be recurring chemical motifs (such as the indole motif

shown in Figure 8.7) shared by marketed drugs and investigational ligands. Similarly, the 2-tetrazole-biphenyl motif is found in the angiotensin2 receptor antagonist losartan and GHS receptor ligand L-692,429 (Figure 8.8a) and a wide range of biologically active structures are based in spiropiperidines (Figure 8.8b).

8.3 Pharmacodynamics and High-throughput Screening

The history of medicine and pharmacology abound with anecdotes of serendipitous drug discovery. Perhaps the most famous example of this is the discovery of penicillin by Fleming in 1928. This led to the systematic screening of hundreds of microorganisms for antibiotics. However, even in those early discovery efforts the value of screening was appreciated. For example, though Ehrlichs' invention of salvarsan for syphilis has many serendipitous elements it was nevertheless the result of a limited screening of 600 synthetic compounds.

Without prior knowledge of which chemical structure will be active on a particular target, as wide as possible a sampling of chemical space (i.e., diverse choice of chemical structures) must be made to detect biological activity. This is done through so-called high-throughput screening (HTS), whereby a robust biological assay is used to test as large as possible a sample of chemical compounds. Usually robotic automation is employed in this process. Presently, sophisticated liquid-handling devices, extremely sensitive detection devices, and automated assay platforms allow testing of multiple thousands of compounds in very small volumes ($< 100\,\mu L$). The ideal HTS is generic (i.e., can be used for a wide range of targets utilizing formats in which any receptor can be transfected and subsequently expressed), is robust (insensitive to assumptions), is relatively low cost with a low volume (does not require large quantities of substance), is amenable to automation (has a simple assay protocol), ideally is nonradioactive, and has a high tolerance to solvents such as DMSO. Some requirements for functional screening assays are given in Table 8.1.

One of the most negative aspects of drug screening is that it basically is a one-way experiment. The single direction stems from the fact that while activity guides structure activity relationships much less use can be made of lack of activity. This is because of the numerous reasons a compound may not show activity (i.e., there are more defined reasons a molecule is active on a biological target than the reasons it lacks activity [4]). For example, lack of aqueous solubility accounts for a substantial number of potentially false negatives in the screening process.

A major consideration in screening is the detection capability of the screen for both false negatives (lack of detection of an active drug) and propensity to find false positives (detection of a response to the compound not due to therapeutic activity of interest). Ostensibly, false positives might not be considered a serious problem in that secondary testing will detect these and they do not normally interfere with the drug discovery process.

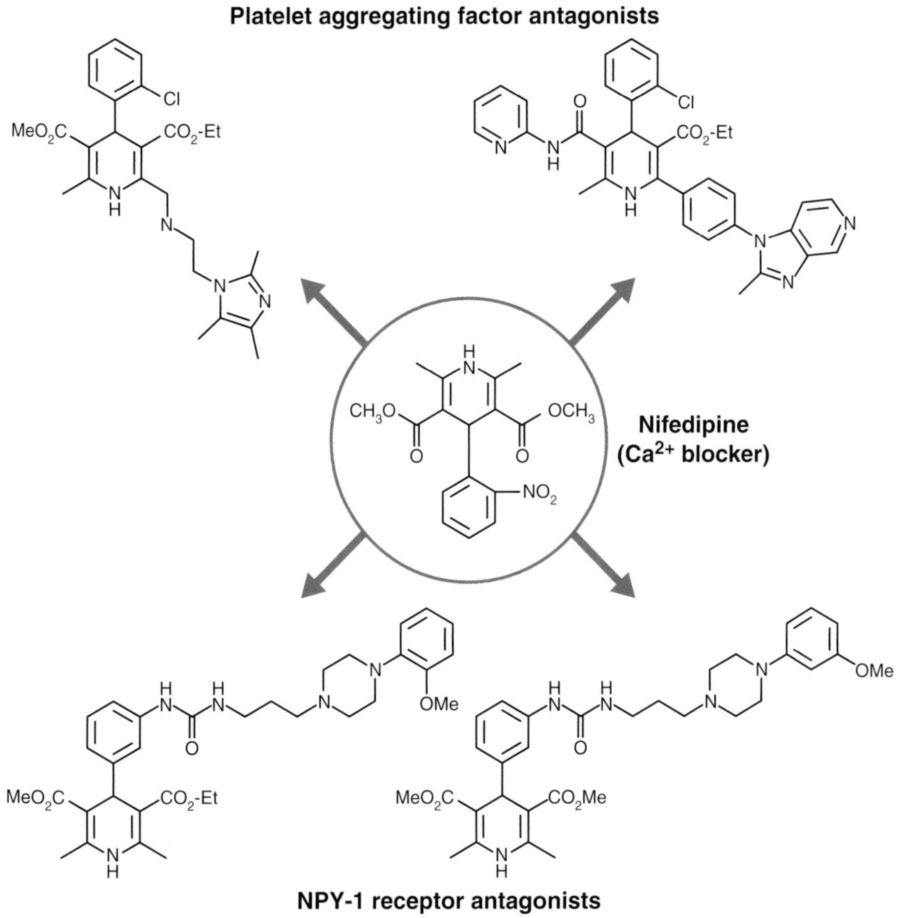

Platelet aggregating factor antagonists

FIGURE 8.6 Example of a preferred structure. In this case, the dihydropyridine scaffold.

However, this can be a serious practical problem if the hit rate of a given HTS is abnormally high due to false positives and the major resource for decoding (following up initial hits) becomes limiting. In this regard, binding assays generally have a lower false positive rate than do functional assays. Also, the false positive rate in functional assays where the exposure time of the assay to the compounds is short (i.e., such as calcium transient studies) is lower than in assays such as reporter assays where the time of exposure is on the order of 24 hours. On the other hand, binding studies require confirmation of primary activity in a functional assay to identify therapeutic activity.

A more serious problem is the one of false negatives, since there is no way of knowing which compounds are active but not detected by the assay. In this regard, binding assays have the shortcoming of detecting only compounds that interfere with the binding of the tracer probe. Within this scenario, allosteric compounds that affect the physiological function of the target but otherwise do not interfere with binding of the tracer are not detected. Since allosterism is probe dependent (i.e., not all molecules are equally

affected by an allosteric ligand, see Chapter 7), the endogenous agonist should be used for screening to detect physiologically relevant activity. For example, the allosteric ligand (alcuronium) for muscarinic receptors produces a tenfold change in the affinity of the receptor for the natural endogenous agonist acetylcholine but only a 1.7-fold change is observed for the synthetic muscarinic agonist arecoline [5]. Therefore, screening with arecoline may not have detected a physiologically relevant (for acetylcholine, the natural agonist) activity of alcuronium.

There are instances where the screen for biologically active molecules cannot be the ideal and appropriate biological test. For example, the screening process for drugs that block against HIV infection theoretically should involve live HIV. However, there are obvious limitations and constraints with using virus that can cause AIDS. Specifically, the containment required with such virulent species is not compatible with HTS. Therefore, a surrogate screen must be done. In this case, a receptor screen of the protein recognition site for HIV (namely, the chemokine receptor CCR5) can be used to screen for drugs that block HIV infection. What is required is a secondary assay to

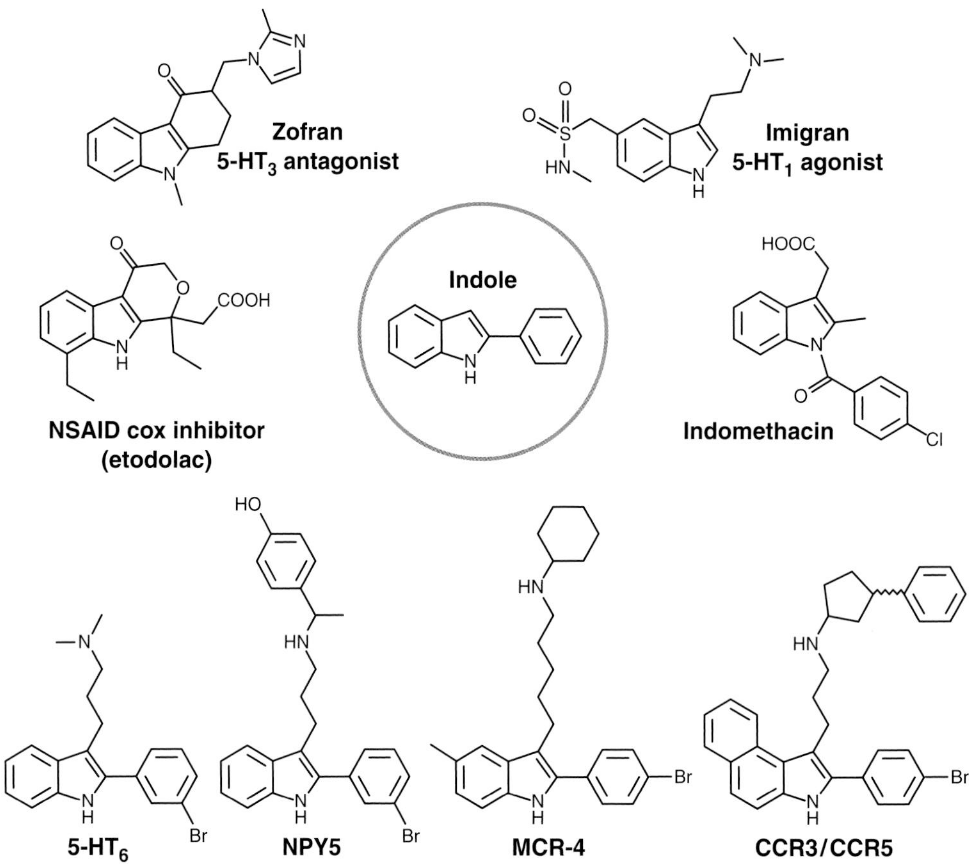

FIGURE 8.7   The preferred indole structure forms the basis of a number of selective ligands for receptors.

ensure that the ligands that block CCR5 also block HIV infection.

The complex protein-protein interactions involved in HIV entry strongly suggest that the blockade of these effects by a small molecule requires an allosteric mechanism (i.e., a specific orthosteric hindrance of a portion of the protein interfaces will not be adequate to block HIV infection). Therefore, the surrogate screen for HIV blockers would be a surrogate allosteric screen. As noted in Chapter 7 and discussed previously, allosteric effects are notoriously probe dependent. Therefore, there is the possibility that the HTS will detect molecules devoid of the therapeutically relevant activity (i.e., block the binding of the probe for screening but not HIV). This also means that the screen may miss therapeutically relevant molecules by using a therapeutically irrelevant allosteric probe. Figure 8.9 shows how usage of a surrogate probe for biological testing can deviate from therapeutic relevance. Initially, a molecule with potent blocking effects on the surrogate probe (radioactive chemokine binding) was shown also to be a potent antagonist of HIV infection (ordinate scale as the IC_{95} for inhibition of HIV infection, see data point for compound A in Figure 8.9). In efforts to optimize this activity through modification of the initial chemical structure, it was found that chemokine blocking

potency could be retained while HIV activity was lost (see data point for compound B in Figure 8.9). In this case, alteration of the chemical structure caused a twofold decrease in chemokine antagonist potency and a disproportionate 3,020-fold decrease in HIV antagonist potency. These compounds clearly show the independence of chemokine binding and HIV binding effects with this molecular series.

The major requirements for a screen are high sensitivity and a large signal-to-noise ratio for detection of effect. This latter factor concerns the inherent error in the basal signal and the size of the window for production of biological effect. A large detection window for response (i.e., difference between basal response and maximal agonist-stimulated response) is useful but not necessary if the random error intrinsic to the measurement of biological effect is low. A smaller maximal detection window, but with a concomitant lower random error in measurement, may be preferable. Since the vast majority of compounds will be exposed to an HTS only once, it is critical that the assay used for screening has a very high degree of sensitivity and accuracy. These factors are quantified in a statistic called the Z' factor [7].

The Z' factor calculates a number that is sensitive to the separation between the mean control values for an HTS

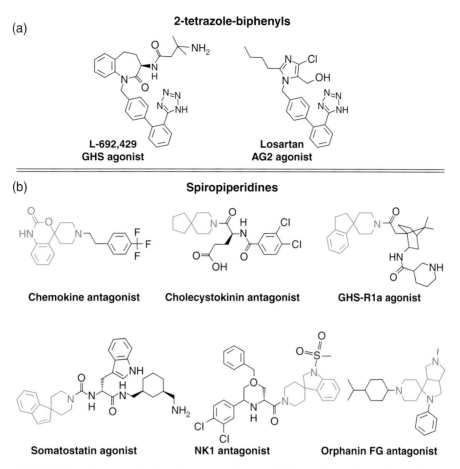

FIGURE 8.8 Examples of preferred structures (2-tetrazole-biphenyls, panel a, and spriopiperidines, panel b) yielding selective ligands for receptors.

(background) and mean of the positive sample as well as the relative standard deviations of both of those means. In validating a screen, a number of negative controls (background signal) and positive controls (wells containing a ligand that gives a positive signal) are run. This process yields a mean value. A positive control mean signal (μ_{c+}) (for example, the maximal response to an agonist for the target receptor) with accompanying standard deviation (denoted σ_{c+}) and negative control signal (background noise, no agonist) denoted μ_{c-} (with σ_{c-}) are generated with a standard positive control drug (i.e., full agonist for the receptor). The bandwidth of values 3 σ units either side of the mean is designated the data variability band, and the width of the spread between the two means ($+3$ σ units) is denoted the separation band (or the dynamic range) of the screen. It is assumed that 3 σ units represents a 99.73% confidence that a value outside of this limit is different from the mean (see Chapter 11 for further discussion). An optimum screen will have a maximum dynamic range and minimum data variability band (see Figure 8.10a). It can be seen that problems can occur with either a large intrinsic standard error of measurement (Figure 8.10b) or small separation band (Figure 8.10c). Interestingly, an efficient

TABLE 8.1

Requirements for a functional screening assay.

Minimal
• Cell line with appropriate receptor is available.
• There is some means of detecting when there is a ligand-receptor interaction taking place.
• Agonist and selective antagonist available.
• Agonist is reversible.

Optimal
• There is a commercial cell line available.
• Response should be sustained, not transient.
• Response should be rapid.

and accurate HTS can be achieved with a low separation band (contrary to intuition) if the data variability band is very small (see Figure 8.10d). The Z′ factor (for a control drug of known high activity for the assay target this is referred to as a Z′ factor) calculates these effects by subtracting the difference between the means from the sum

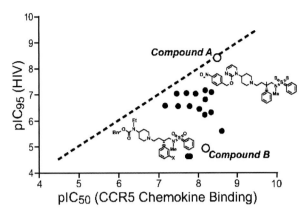

FIGURE 8.9 Correlation between blockade of chemokine binding to CCR5 (abscissae as pK_i values) and 95% inhibition of HIV infection as pIC_{95} (ordinates) for a series of CCR5 antagonists. It can be seen that compound A is nearly equiactive as a blocker of chemokine binding ($pK_i = 8.5$) and HIV infection ($pIC_{95} = 8.4$; ratio of affinities $= 1.3$), whereas structural analogues (filled circles) clearly differentiate these activities. For the structure B shown, the chemokine-blocking activity has been somewhat retained ($pK_i = 8.2$), whereas the HIV-blocking activity has largely been lost ($pIC_{95} = 4.9$; ratio of affinities $= 3,020$). Data drawn from [6].

of the difference of the standard deviations of the means divided by the difference between the means:

$$Z' = \frac{|\mu_{c+} - \mu_{c-}| - (3\sigma_{c+} + 3\sigma_{c-})}{|\mu_{c+} - \mu_{c-}|} = 1 - \frac{(3\sigma_{c+} + 3\sigma_{c-})}{|\mu_{c+} - \mu_{c-}|}.$$
(8.1)

Table 8.2 shows the range of possible Z' values, with comments on their meaning in terms of high-throughput screening assays.

The calculation of Z values for experimental compounds can yield valuable data. Values of Z for test compounds are calculated in the same way as Z' values except that the μ_{c+} and σ_{c+} values are the signals from the test compounds (denoted μ_s and σ_s for test sample) and μ_{c-} and σ_{c-} from the assay with no test compounds run (i.e., controls for noise, denoted μ_c and σ_c for controls). While the Z' indicates the robustness and detection capability of the screen (calculated with known active compounds), a value of Z for a set of unknown compounds can also test other factors related to the screen such as the concentration at which the compounds are tested and/or the chemical makeup of the compound set. For example, Figure 8.11a shows a screen with an excellent Z' value ($Z' = 0.7$) and Z values for a set test compounds run at two concentrations. It can be seen that the higher concentration yields a higher signal and variation (possibly due to toxic effects of the high concentration). This, in turn, will lead to a lower Z factor. Similarly, Figure 8.11b shows distributions for two chemical libraries. It can be seen that there is a clear difference in the quality of the assay with these two sets of compounds, indicating a possible inherent property of one of the chemical scaffolds leading to variability in the screen.

In effect, the quality of the compound set can be quantified for this assay with a value of Z [7].

Of major importance for an HTS is sensitivity to weak ligands. As discussed in Chapter 2, functional systems generally amplify responses as the signal is measured distal to the agonist-receptor interaction. For this reason, agonist screens utilizing end organ response are preferred (i.e., melanophore function, reporter assays). In contrast, the sensitivity of antagonist screening can be controlled by adjustment of the magnitude of the agonism used to detect the blockade. Clearly, the lower the amount of stimulation to the receptor of the system the more sensitive it will be to antagonism. This effect is inversely proportional to the window of detection for the system. On one hand, as large a window of agonist response as possible is preferred to maximize signal-to-noise ratios. On the other hand, too large a window may require a strong agonist stimulation that in turn would create insensitivity to antagonism. This can be offset by screening at a higher concentration of antagonist, but this can introduce obfuscating factors such as toxic effects of high concentrations of weakly active compounds. Thus, for antagonist screening it becomes a trade-off of strength of agonist stimulation against concentration of antagonist. An optimal screening assay must adjust for maximal sensitivity and minimal variability. Figure 8.12 shows some potential scenarios for single concentration inhibition of different levels of agonist stimulation by different concentrations of an antagonist. It can be seen that the maximal sensitivity to antagonism is observed with low levels of receptor stimulation (Figure 8.12a, see $[A]/K_A = 0.3$). However, the standard deviation of the signal is large enough to interfere with the determination of antagonism. As the magnitude of the receptor stimulation increases ($[A]/K_A = 1.5$, and 10), the standard deviation of the signal ceases to be a problem but there is less inhibition of the signal. This can be overcome by increasing the concentration of antagonist (Figure 8.12a). Figure 8.12b shows the relationship between the initial level of receptor stimulation and the percent inhibition of that signal by an antagonist. If it assumed that a 40% or greater inhibition of the signal is unequivocal for detection of antagonism, then it can be seen from this figure that the initial level of receptor stimulation cannot exceed 33% maximum for screening antagonist concentrations at the equilibrium dissociation constant (K_B) and <90% maximum stimulation for antagonist concentration $= 10 \times K_B$.

From the standpoint of sensitivity to antagonist, a receptor stimulation level of 50% is optimal for functional studies. However, in view of signal-to-noise factors and the need for a clear window of inhibition an 80% level of stimulation often is employed. In this regard, binding may hold some advantages since the window of detection for a binding assay with a low level of nsb may be greater than that for a functional assay. Figure 8.13 shows the antagonism by a concentration of antagonist of $[B] = K_B$, of a dose-response curve for receptor stimulation of 80% (function, see Figure 8.13a),

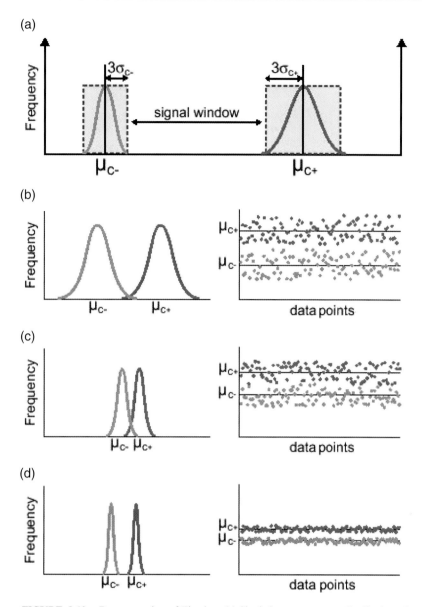

FIGURE 8.10 Representation of Z' values (a) Shaded areas represent distribution of values for control readings (no drug) and the distribution for readings from the system obtained in the presence of a maximal concentration of standard active drug. The signal window for this assay is the separation between the distributions at values 3× the standard deviation of the mean away from the mean. (b) A representation of an assay with a low Z' value. Although there is a separation, the scatter about the mean values is large and there is no clear window between the lower and upper values. (c) An assay with a low signal window. This assay has a low Z' value. (d) An assay with a low signal window but correspondingly low error leading to a better Z' value.

and receptor binding level of 10%. It is assumed that both of these initial levels of receptor stimulation yield adequate windows of detection for the respective assay formats. It can be seen that the concentration of antagonist produces 50% inhibition of the binding and only 23% inhibition of the functional signal (i.e., the binding assay format is more sensitive to the antagonism). A reexpression of this effect in terms of the minimal

potency of antagonist that each screen could detect (assuming that a 40% inhibition is required for detection) indicates that the binding assay would be capable of detecting antagonists with a $K_B \geq 8\,\mu M$ while the functional assay would only detect antagonists of $K_B \geq 3\mu M$ (a 2.7-fold loss of sensitivity). It should be stressed that binding and function have been somewhat arbitrarily assigned these two levels of receptor stimulation.

TABLE 8.2
Z′ values and high-throughput screening assays.

Z′ Value	Description of Assay	Comments
Z′ = 1	No variation ($\sigma = 0$) or infinite band of separation	Ideal assay
1 > Z′ ≥ 0.5	Large dynamic range	Excellent assay
0.5 > Z′ > 0	Small dynamic range	Adequate assay
0	No band of separation, σ_{c+} and σ_{c-} touch	Dubious quality
< 0	No band of separation, σ_{c+} and σ_{c-} overlap	Impossbible for screening

From [7].

The association of an assay format need not be associated with the sensitivity. In practice, if the functional signal-to-noise level is high there would be no need to turn to radioligand binding to increase sensitivity of the screen. Similarly, if the nsb levels of the binding screen were high the level of initial B_o values for screening would need to be increased to levels comparable to functional assays (i.e., 50% stimulation) and the advantage of binding over function would be lost. In general, sensitivity is not the major factor in the choice of screening format.

The process of tracking screening hits and determining which chemical series is likely to produce a fruitful lead involves the verification of activity within a series of related

structures. While the absolute potency of the hit is clearly important, it is recognized that factors such as selectivity, favorable physicochemical properties, absence of toxophores (pharmacophores leading to toxicity), and the capability for the rapid production of chemical analogues are also very important features of lead molecules. For this reason, the concept of "ligand efficiency" has been used to evaluate the worth of screening hits. This idea converts ligand affinity to the experimental binding energy per atom (so-called Andrews binding energy [8]) to normalize the activity of ligand to its molecular weight [9]. It has been estimated that a maximum affinity per atom for organic compounds is $-1.5 \, kcal \, mol^{-1}$ per non-hydrogen atom (Δg (free energy of binding) $= -RT \, lnK_d$/number of non hydrogen atoms) [10].

Before discussion of the drug discovery process following lead identification, it is relevant to discuss variations on the theme of hit identification. Screening traditionally has been based on finding a defined primary biological activity (i.e., receptor-based agonism or antagonism of physiological effect). Such an approach presupposes that all potentially useful receptor activity will be made manisfest through these effects. However, some receptor activities may not be mediated through G-protein activation. For example, the CCK antagonist D-Tyr-Gly-[(Nle^{28,31},D-Trp^{30})cholecystokinin-26-32]-phenethyl ester actively induces receptor internalization without producing receptor activation [11]. This suggests that screening assays

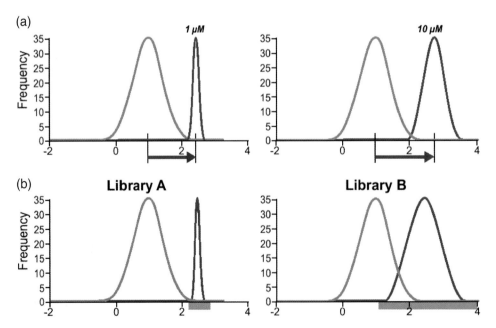

FIGURE 8.11 Distributions for various screens. (a) The larger distribution represents inactive compounds while the smaller one shows a small sample with values greater than the mean of the total compound library. Distributions are shown for two concentrations tested from this library. It can be seen that while the mean of the higher concentration is slightly further away from the control distribution the error is also much greater, leading to a lower Z′ value. (b) The results of single concentration of two compound libraries are shown. It can be seen that library A has a smaller standard error about the mean and therefore is a higher-quality library for potentially active molecules.

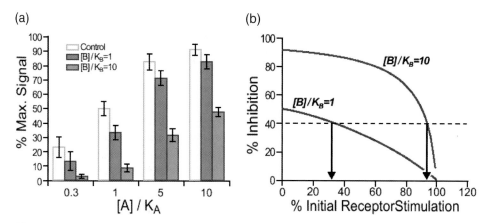

FIGURE 8.12 Antagonism of single concentration stimulation (either functional or radioligand binding) by two concentrations $[B]/K_B = 1$ and 10) of a simple competitive antagonist in screening experiments. (a) Various levels of receptor stimulation in the absence of antagonist (open bars), in the presence of a concentration equal to the K_B (cross-hatched bars), and in the presence of $10 \times K_B$ antagonist (shaded bars). (b) Percent inhibition (ordinates) of initial receptor stimulation (abscissae) produced by two concentrations of antagonist. If it is assumed that a minimum of 40% inhibition of initial signal is required for adequate detection of antagonism then the receptor stimulation levels must not be greater than those that produce 33% and 90% receptor activation (or initial radioligand binding B_o value) in the HTS for antagonist concentrations of $[B]/K_B = 1$ and 10, respectively.

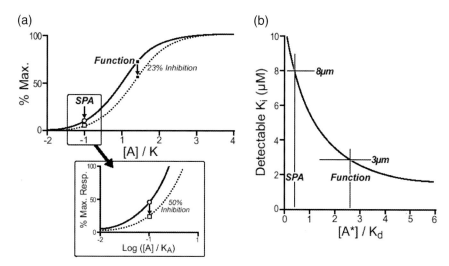

FIGURE 8.13 Windows of detection for antagonism. A twofold shift in a dose response curve (either to an agonist in a functional study or a radioligand in a saturation binding study) will be perceived differently in different regions of the dose-response curve. Thus, a concentration that produces 80% response will be blocked 23%, while a concentration that produces only 10% will be blocked by a factor of 50%. Therefore, the lower the initial signal input to an antagonist assay the more sensitive it will be to antagonists. In general, functional assays require stronger input signals to achieve acceptable windows (usually an EC_{80} agonist concentration) than do binding studies (such as scintillation proximity assays SPA). Inset shows where a 10% maximal initial radioligand binding signal can still yield a useful window for observation of antagonism. (b) Ordinate axis shows the lowest potency of hypothetical antagonists that are detectable in an assay (assume 50% blockade of initial signal) as a function of the signal strength used for the assay. If it is assumed that a minimal signal strength for functional assays is $[A]/K_s = 2.5$ while that for an SPA can be lower ($[A]/K_d = 0.5$), it can be seen that the binding assay will detect weaker antagonists ($IC_{50} < 8 \mu M$) than will the functional assay (must be $IC_{50} < > 3 \mu M$).

other than simple agonism and/or antagonism may be useful for the detection of ligand activity.

A similar idea involves the modification of screening assays for the detection of special ligands. For example,

certain inhibitors of enzyme function trap the enzyme in dead-end complexes that cannot function. This is referred to as *interfacial inhibition* [12]. Thus, inhibitors such as brefeldin A and camptothecin target a transient kinetic intermediate that is not normally present in a nonactivated protein. Screening assays designed to detect these types of inhibitors have a small concentration of substrate in the medium to produce the enzyme transition state (the target of the interfacial inhibitor). Similarly, topoisomerase assays have been designed to identify transient trapping of catalytic-cleavage complexes. Interestingly, such inhibitors may offer an added measure of selectivity since they are active only when both partners of a physiological interaction are present and target only this interaction.

This has particular relevance to allosteric modification of receptors. As described in Chapter 7, the fraction of receptor bound to an agonist [A], expressed in terms of the presence of an allosteric modulator [B], is given as

$$\frac{[AR]}{[R_{tot}]} = \frac{[A]/K_A(1 + \alpha[B]/K_B)}{[B]/K_B(\alpha[A]/K_A + 1) + [A]/K_A + 1}. \qquad (8.2)$$

This leads to the expression for the observed affinity (expressed as equilibrium dissociation constant of the ligand-receptor complex) of the modulator as

$$K_{obs} = \frac{K_B([A]/K_A + 1)}{\alpha[A]/K_A + 1}. \qquad (8.3)$$

It can be seen from Equation 8.3 that the concentration of the probe molecule ($[A]/K_A$) affects the observed affinity of the modulator. This can have practical consequences, especially when allosteric potentiators are the desired chemical target. Just as an allosteric potentiator will increase the affinity of the probe molecule (agonist, radioligand), the reciprocal also is true; namely, that the agonist will increase the affinity of the receptor for the modulator. This can be used in the screening process to make an assay more sensitive to potentiators. For example, for a potentiator that increases the affinity of the agonist 30-fold ($\alpha = 30$), the observed affinity of the modulator will increase by a factor of 15.5 when a small concentration of agonist ($[A]/K_A = 1$) is present in the medium. Such modification of screening assays can be used to tailor detection for specific types of molecules.

Finally, as a corollary to the screening process, there are thermodynamic reasons for supposing that any ligand that has affinity for a biological target may also change that target in some way (i.e., have efficacy). This is because the energetics of binding involve the same forces responsible for protein conformation (i.e., as discussed in Section 1.10, a ligand will bias the natural conformational ensemble of the receptor). This can be simulated with a probabilistic model of receptor function [13, 14].

To describe this model quantitatively, it is simplest to arbitrarily begin with one receptor state (referred to as $[R_o]$) and define the affinity of a ligand [A] and a G-protein [G] for that state as

$$^Ak_o = [AR_o]/[R_o][A] \qquad (8.4)$$

and

$$^Gk_o = [GR_o]/[R_o][G], \qquad (8.5)$$

respectively. The probability of the receptor being in that state is denoted p_o. The probability of the receptor forming another conformation $[R_1]$ is defined as p_1, and the ratio of the probabilities for forming state R_1 vs R_o is given as j_1, where $j_1 = p_1/p_o$. The value j controls the energy of transition between the states. The relative probability of forming state $[R_1]$ with ligand binding is denoted $^Aj_1 = {}^Ap_1/{}^Ap_o$, and with G-protein binding as $^Gj_1 = {}^Gp_1/{}^Gp_o$. An important vector operating on this system is defined as b, where b refers to the fractional stabilization of a state with binding of either ligand (defined as $^Ab_1 = {}^Aj_1/j_i$) or G-protein ($^Gb_1 = {}^Gj_1/j_i$). Every ligand and G-protein has characteristic values of b for each receptor state and it is these b vectors that constitute ligand affinity and efficacy. With these probabilities and vectors, the following operators are defined:

$$\Omega = 1 + \sum j_i, \qquad (8.6)$$

$$\Omega_A = 1 + \Omega \sum {}^Ab_ip_i, \qquad (8.7)$$

$$\Omega_G = 1 + \Omega \sum {}^Gb_ip_i, \quad \text{and} \qquad (8.8)$$

$$\Omega_{AG} = 1 + \Omega \sum {}^Ab_i{}^Gb_ip_i, \qquad (8.9)$$

where i refers to the specific conformational state and the superscripts G and A refer to the G-protein and ligand bound forms, respectively. With these functions defined, it can be shown that macroaffinity is given by

$$\text{Macroaffinity } (K) = {}^Ak_o\Omega_A(\Omega)^{-1}, \qquad (8.10)$$

where k_o is related to the interaction free energy between ligand and a reference microstate of the receptor. A measure of efficacy is given by

$$\text{Efficacy } (\alpha) = (\Omega\Omega_{AG})(\Omega_A\Omega_G)^{-1}. \qquad (8.11)$$

Figure 8.14 shows calculated values for affinity (ordinates) and efficacy (abscissae) for 5,000 simulated ligands. The probabilities are random but it can seen that there is a correlation between affinity and efficacy. The calculations show that the energy vectors that cause a ligand to associate with the protein will also cause a shift in the bias of protein conformations (i.e., the act of binding will cause a change in the nature of the protein ensemble). This suggests that if a ligand binds to a receptor protein it will in some way change its characteristics toward the system. This has implication in screening since it suggests that all compounds with measured affinity should be tested for all aspects of possible biological activity, not just interference with the binding of an endogenous agonist [15].

Once hits have been identified, they must be confirmed. The test data obtained from a screen form a normal distribution. One criterion for determining possible active molecules is to retest all initial values > 3 σ units away from the mean. This will capture values for which there is $> 99.3\%$ probability of being significantly greater than the

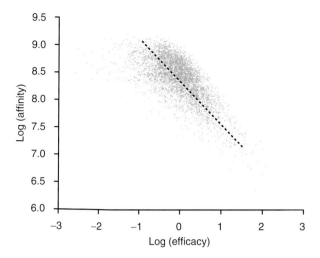

FIGURE 8.14 Simulation for 5,000 theoretical ligands with calculated efficacy (Equation 8.11) and affinity (Equation 8.10). It can be seen that efficacy and affinity are correlated suggesting that all ligands that have been shown to bind to a receptor should be extensively tested for possible efficacy effects on the receptor directly, through agonist-effects on the receptor or changes in constitutive behavior of the receptor itself. Redrawn from [15].

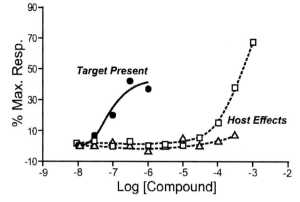

FIGURE 8.16 Ligand-target validation. Dose-response curves to a putative agonist for a therapeutic target on cell line transfected with the the target receptor (filled circles) and on a cell line not transfected with the target receptor (dotted lines, open circles, and open triangles). The open symbol curves reflect nonspecific and non-target-related effects of the compound on the host cell line. The clear differentiation between the target curves and the host curves indicate a specific effect on the therapeutically relevant target.

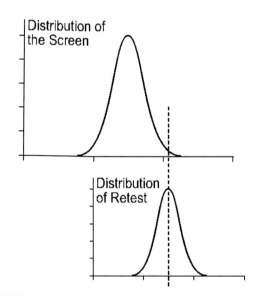

FIGURE 8.15 Confirmation of initial hits in the HTS. Top panel shows the distribution of values from a single test concentration of a high-throughput screen. The criteria for activity and subsequent retest is all values >3 standard error units away from the mean (dotted line). The process of retesting will generate another distribution of values, half of which will be below the original criteria for activity.

mean of the population (see Figure 8.15). The distribution of the apparently active compounds, when retested, will have a mean centered on the 3 σ value for the distribution of the total compound set. It can be seen that 50% of these will retest as active (be greater than 3 σ units away from the

initial total compound set mean). Therefore, the compounds that retest will have a 99.85% probability of having values greater than the mean of the original data set. The criteria for retest may be governed by practical terms. If the hit rate is inordinately high, then it may be impractical to test all hits that give values >3 σ units from the mean. A lower (having a greater probability of retest) number of "hits" (>4 σ or 5 σ units away from the mean) may need to be tested to reduce the retest load.

Another important concept in the process of early confirmation of lead activity is ligand-target validation. The first and most obvious criterion for selective target interaction is that the ligand effect is observed in the host cell only when the target is present. Thus, in a cell-based assay using cells transfected with receptor the response to a putative agonist should be observed only in the transfected cell line and not in the host cell line (or at least a clearly different effect should be seen in the host cell line, see Figure 8.16).

There are two general types of observable biological responses: agonism and antagonism. The lead optimization process is the topic of Chapter 10, where specifics of the methods and theory of determining molecular activity are outlined. For the remainder of this chapter, it is assumed that the hit from screening has been through the lead optimization process to the point where it can be considered a drug candidate. As shown in Figure 8.1, the next stages involve the developability of the molecule(s) in terms of pharmacokinetics, pharmaceutics, and propensity for adverse drug reactions.

The preceding discussion involves the elucidation of the primary hit and lead activity, obviously a crucial step in the drug discovery process. However, there are numerous other reasons a molecule with good primary activity may still fail

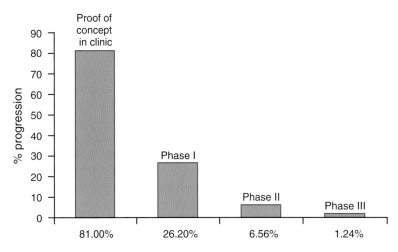

FIGURE 8.17 Attrition of molecules as they are taken through the clinical testing procedure. It can be seen that very few become drugs (1.34%). Redrawn from [16].

as a drug and it is becoming increasingly clear that the factors that lead to this failure need to be addressed as early as possible in the lead optimization process. Figure 8.17 shows the outcome of a risk analysis for the probability of a new compound emerging as a drug. It can be seen that attrition is extremely high. An active molecule must be absorbed into the body, reach the biological target, be present for a time period sufficient for therapeutic activity, and not produce untoward side effects. It will be seen that an important part of the lead optimization process is to incorporate these properties into the primary lead molecule early on in the process [16]. One reason this is important is that the concepts involved are, in some cases, diametrically opposed. For example, while low molecular weight is a known positive property of drugs the lead optimization process generally results in increased molecular weight as pharmacophores are added to increase potency. For this reason, the concept of "lead likeness" [17] can be used to determine the suitability of lead molecules for beginning the lead optimization process (*vide infra*). The problems involved in introducing lead likeness into screening hits is exacerbated by the fact that as analogues become more potent there is less tolerance for chemical analogueing to improve physicochemical properties. In fact, it is a general observation that there often are relatively minor differences between leads and launched drug candidates (see Figure 8.18) [24]. On the other hand, there is abundant evidence to show that apparently very minor changes in chemical structure can impose large effects on biological activity (see Figure 8.19).

In general, there are three milestones for the drug discovery process. The first is the identification of a verified hit series (primary activity in a related series of molecules), the second the determination of a lead series (series with primary activity and drug-like properties), and the third a clinical candidate (activity, positive pharmaceutical, and pharmacokinetic properties devoid of toxicity). An example

of developability being a key factor in the emergence of a drug from an active molecule can be found in the histamine H2 receptor antagonist molecules. The first active histamine active H2 antagonist burimamide, while active by the parenteral route, did not have the oral absorption properties required for an oral drug (Figure 8.20). The second in the series, metiamide, was active by the oral route but had fatal bone marrow toxicity (thereby precluding clinical utility). The third in the series fulfilled the requirements of target activity, acceptable absorption, and toxicity profile and thus became a prototype blockbuster drug in the new series (Figure 8.20).

8.4 Pharmacokinetics

The essence of pharmacology is the relationship between the dose of a drug given to a patient and the resulting change in physiological state (the response to the drug). Qualitatively, the type of response is important, but since (as put by the German pharmacologist Walter Straub in 1937) "there is only a quantitative difference between a drug and poison" the quantitative relationship between the dose and the response is paramount. Thus, the concentration (or dose) of drug is the *independent variable* (that set by the experimenter) and the pharmacological effect returned by the therapeutic system is the *dependent variable*. The value of the dependent variable only has meaning if the value of the independent variable is correct (i.e., if the experimenter truly knows the magnitude of this variable). Pharmacokinetics furnishes the tools for the clinician to determine the true value of the independent variable.

Drugs can only be effective if enough is present at the target site and they can be harmful if too much is present so as to produce toxic side effects. Any attempt to draw conclusions about the clinical efficacy of a drug in a clinical trial without knowledge of the concentration at the target site is premature. The science of pharmacokinetics basically

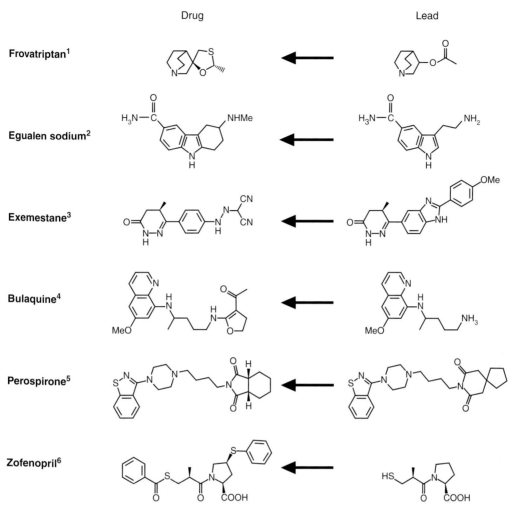

Drug   Lead

Frovatriptan[1]

Egualen sodium[2]

Exemestane[3]

Bulaquine[4]

Perospirone[5]

Zofenopril[6]

FIGURE 8.18 Structural relationships between the initial lead for a molecule and the eventual drug. It can be seen that changes in structure are in some cases not extensive. Data shown for frovatriptan[1] [18], egualen sodium[2] [19], exemestane[3] [20], bulaquine[4] [21], perospirone[5] [22], and zofenopril[6] [23]. Drawn from [24].

seeks to answer the questions: how much of the drug that is given to the patient actually reaches the target organ?, where in the body does the drug go?, and how long does it stay in the body? Therefore, as a prerequisite to pharmacodynamics (study of drug-receptor interactions), pharmacokinetics examines the journey of drugs into the body and toward their intended therapeutic target organ. For example, a drug taken by the oral route is absorbed from the stomach into the systemic circulation and carried by the bloodstream throughout the body. Thus, an antiarrhythmic drug intended to prevent fatal arrhythmia of the heart must travel through the systemic circulation and the coronary arteries and be absorbed through the wall of capillaries and into the heart muscle. As it diffuses through layers of cells it finally encounters the sinus node and interacts with specific sites on the cell membrane to mediate electrical activity of the cell. Each barrier to this distribution can affect the concentration of the drug reaching the target site. A useful acronym to describe pharmacokinetics is ADME.

This generally describes the process of drug *absorption* into the body, *distribution* throughout the body, *metabolism* by degradative and metabolizing enzymes in the body, and finally *elimination* from the body. It is useful to consider each of these steps because together they summarize pharmacokinetics.

There are numerous routes of administration of drugs into the body. The choice of which route to use in a given therapeutic situation can be determined by convenience, maximization of compliance (for example, a drug taken once a day by the oral route is much easier to sustain on a chronic basis than one that needs to be injected twice a day), and attainment of concentration bias to gain advantage therapeutically. For example, β-adrenoceptor agonists such as salbutamol are very useful for rapid relaxation of constricted bronchioles in asthma. However, these drugs also can produce some tachycardia and notably a debilitating digital tremor. However, if taken by aeorsol salbutamol reaches the target organ first (bronchioles) for

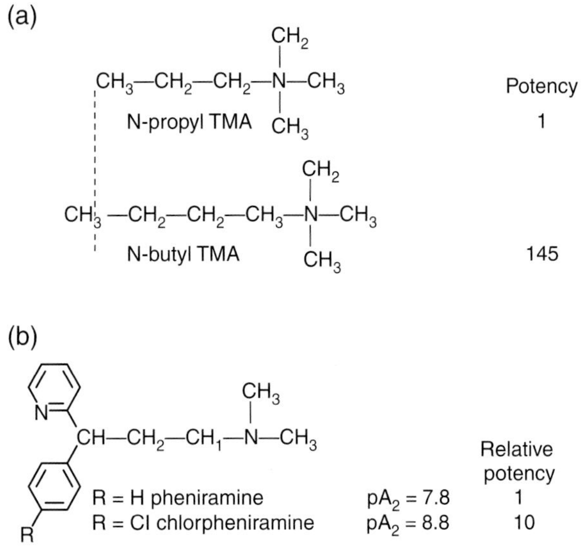

(a)

N-propyl TMA   Potency
                               1

N-butyl TMA   145

(b)

R = H pheniramine $pA_2 = 7.8$ Relative
R = Cl chlorpheniramine $pA_2 = 8.8$ potency
                                            1
                                           10

FIGURE 8.19 Small changes in the chemical structure of N-propyl tetramethylammonium and pheniramine produce 145-fold and tenfold increases in potency, respectively.

maximal effect and then diffuses throughout the bloodstream in a reduced concentration for minimal effect on the heart and skeletal muscle. Thus, side effects are minimized. Similarly, ocular drugs for glaucoma can be introduced as eyedrops directly into the eye for maximal concentration effect and minimal cardiovascular side effects.

While there is interstitial space between cells, drugs generally must go through, not around, cells to penetrate membranes and gain access to internal organs. Under these circumstances, the ability of molecules to pass through cell membranes is a very important determinant of absorption. A passive concentration gradient may be the only driving force behind the passage of a drug from the central compartment into secondary specialized compartments (i.e., organs, the brain). In these cases, the lipophilicity

of the molecule is important (i.e., a non-lipophilic molecule will not pass through a lipid bilayer easily). The state of ionization also is relevant (ionized charged molecules do not pass easily), as is the size (a general target maximal size for most orally available drugs is m.w. <600). For some molecules, active processes of transport into the cell are operative, and in these cases general lipophilicity and size issues are less prominent. As discussed in Section 8.2, some guidelines for medicinal chemists are contained in a set of rules known as the "rule of 5" derived by Lipinski [1]. In general, any molecule that violates any two of these rules would be predicted to yield poor absorption in vivo. In this analysis, drug-like properties (leading to good ADME properties) were found in molecules of molecular weight less than 500, with less than 10 total nitrogen and oxygen atoms, containing less than 5 hydrogen bond donors and less than 10 hydrogen bond acceptors. In addition, more lipophilic molecules (Log P > 5, where P is the partition coefficient in the aqueous versus organic phase) were absorbed to a greater extent.

As discussed in Section 8.3, the process of lead optimization often leads to increased molecular weight and lipophilicity. Therefore, the idea that initial leads should possess lead-likeness properties has been proposed [24]. For example, assuming that the lead optimization process will increase both the lipophilicity and molecular weight of a lead, good lead-likeness values for a screening hit would be molecular weight < 350 and cLogP < 3 with a primary affinity for the biological target of approximately 0.1 μM.

If the entry of a molecule into the body were simply a temporally restricted absorption process, then a steady-state concentration would be achieved given enough time for complete absorption. However, what in fact is observed in drug pharmacokinetics is a complex curve reflecting absorption of the drug into the body and the diminution of the concentration that is absorbed back down to negligible levels. The reason for this complex pattern of rise and fall in

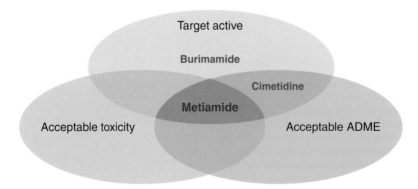

FIGURE 8.20 Drugs as subsets of clinical profiles. While burimamide, cimetidine, and metiamide are all active histamine H2 antagonists with ulcer healing activity burimamide lacks a suitable toxicity and pharmacokinetic profile and cimetidine is adequately absorbed but still toxic. Only metiamide fulfills the requirements of a clinically useful drug.

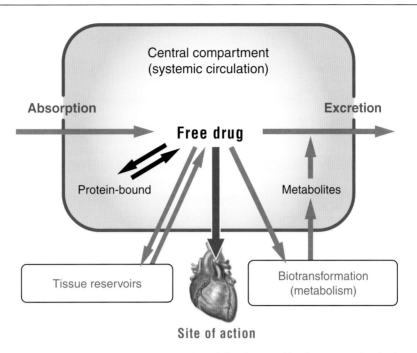

FIGURE 8.21 Schematic representation of the pharmacokinetic processes involved in drug absorption, distribution, and elimination.

drug concentration in vivo is due to the number of processes that impinge on the drug concentration as it passes into and out of the body. These are summarized in Figure 8.21. First, the drug must pass into the systemic circulation via the chosen route of administration. Once it is in the circulation, it is subject to a number of processes that reduce the concentration of freely accessible drug. One of these is binding to proteins in the blood, usually albumin for acidic drugs and alpha1-acid glycoprotein for basic drugs. The complex between proteins and drug can cause a sequestration of free drug into a pool not readily accessible for therapeutic purpose (i.e., only free drug can cross plasma membranes). Drugs are also excreted unchanged, mainly by the kidneys. In addition, the liver metabolizes drugs, thereby effectively eliminating them from the circulation. As well as converting drugs to inactive metabolites, the liver converts many drugs to polar metabolites that are then rapidly excreted. Various tissues (such as fat) can act as reservoirs for drugs also sequestering them into compartments. For example, the antimalarial drug quinacrine can concentrate several-thousand-fold in the liver. All of these elimination processes compete for the target organ for concentration of drug.

The liver (and other organs) removes active drugs through two general processes. One is through the conversion of biologically active to inactive molecules. The other is to the conversion into polar metabolites that are readily excreted (to a greater extent than the parent drug). The actual metabolic processes can be biochemically classified into two types of reactions—so-called phase I and phase II metabolism. Phase I metabolic reactions place a functional group on the parent molecule to render them biologically inactive

(in some rare instances, retention or even enhancement of activity can result). Phase II reactions (conjugation reactions) covalently link between a functional group on the molecule and glucuronic acid, sulfate, amino acids, glutathione, or acetate to create highly polar metabolites that are rapidly excreted in urine. In general, the liver and other organs in the body participate in the removal of foreign chemicals from the body through conjugation, hydrolysis, oxidation, reduction, and finally excretion.

While pharmacokinetics is the science of drug disposition in the body, the field of clinical pharmacokinetics is concerned with the practical presentation of therapeutic drugs to the target organ(s) for the therapy of disease. The main consideration here is the attainment of a consistent concentration of drug freely accessible to the biological target for a sustained period in order to nullify, reverse, or ameliorate a pathological process. There are four general parameters that are of paramount importance to clinicians in the study of clinical pharmacokinetics. Thus, the *clearance* yields a measure of the body's efficiency to eliminate the drug. The *volume of distribution* of the drug is the apparent volume of fluid containing the drug in the body. A measure of the length of time the drug stays in the body can be gained from the *half time*. Specifically, this is the length of time it takes for the concentration of the drug to be reduced by half its initial value. Finally, the *bioavailability* of the drug is a measure of the efficiency of absorption and presentation to the systemic circulation. For example, a drug taken by the oral route may have a bioavailability of only 20%; that is, only 20% of the orally ingested amount reaches the general systemic circulation after ingestion. Clearance is measured as the volume of

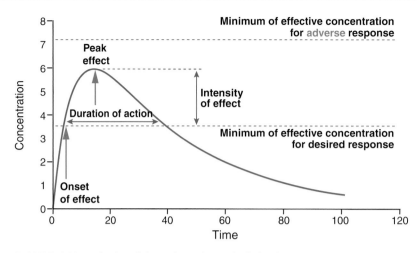

FIGURE 8.22 Kinetics of drug absorption and elimination as viewed by the plasma concentration of an orally administered drug with time.

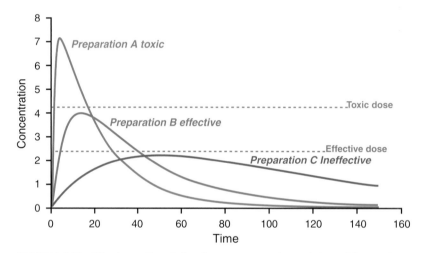

FIGURE 8.23 Kinetic profiles of the plasma concentrations of three different drugs taken by the oral route. If absorption is rapid, toxic effects may ensue (red line). If too slow, a therapeutically effective level may not be attained (blue line).

fluid per unit time from which the drug would have to be completely removed to account for the elimination from the body. The efficiency of clearance is dependent on the ability of the organ to remove the drug and also the rate of blood flow through the organ.

Referring to the observed temporal relationship between concentration of an ingested drug in a central compartment such as the systemic circulation, there are various parameters that can be used to describe a drug's pharmacokinetic performance. These are summarized in Figure 8.22. There is a required level of drug needed for therapeutic effect (minimal effective concentration for desired response), and usually a toxic level of drug as well (minimum effective concentration for adverse effects). Thus, the therapeutic aim is to exceed the first limit but stay below the second. The time at which the level of drug achieves the minimal

therapeutic level describes the time to onset of effect. The difference between the minimal effective concentration for response and highest concentration (peak effect) is referred to as the intensity of effect. The length of time that the concentration exceeds the minimal effective therapeutic concentration is called the duration of effect.

A measure of the actual amount of drug in the body can be obtained from the area under the curve of the temporal concentration curve (calculated by integration). Interestingly, the temporal behavior of a drug can be extremely important in therapeutics. For example, consider three preparations of a drug that present identical values for area under the curve (i.e., amount of drug absorbed) but have different kinetics of absorption (Figure 8.23). As shown, preparation B produces a useful profile whereby the concentration exceeds the minimal effective concentration

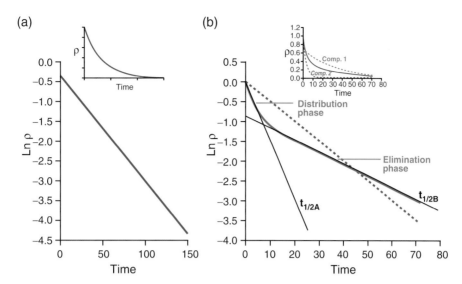

FIGURE 8.24 Kinetics of drug elimination/redistribution. (a) First-order elimination from a single compartment. Inset shows fractional concentration with time. Larger graph shows the same with natural logarithmic ordinates. (b) Plasma concentration in a two-compartment system. The initial rapid elimination from the plasma in all probability represents redistribution of the drug out of the plasma to portions of the body. The slower phase represents elimination from the single body compartment. Inset shows elimination from each compartment as dotted lines with the observed combined effects shown in the solid blue line. This results in a curvilinear semilogarithmic plot.

but stays below the toxic level. In contrast, preparation A exceeds this level (to produce toxic effects) and preparation C never achieves the minimal effective concentration though very similar amounts of the drug are absorbed. In general, unfavorable pharmacokinetics can completely preclude the effective therapeutic of an active molecule. The discipline of clinical pharmacokinetics quantifies the ADME properties of new drugs.

It is worth considering the actual mechanics of clinical pharmacokinetics to get an idea of what data actually drives the conclusions around determining ADME properties of drugs. A basic and important process is the measurement of the concentration of drug in the bloodstream at various times after administration. The elimination of a drug from the body can be approximated by the exit of a substance from a single compartment via a first-order elimination process. Thus, the fraction of drug at any time (denoted by ρ_t) is given by $e^{-k_e t}$, where k_e is an elimination rate constant and t is time. One very useful feature of exponential relationships is the fact that they are linear when plotted on a semilogarithmic scale. Thus, ln ρ_t as a function of time yields a straight-line, the slope of which can readily be estimated for an estimate of the elimination rate constant (Figure 8.24a). This is the fraction of drug eliminated from the body per unit time. The slope of the linear plot of ln ρ_t on time can be used to measure the elimination half time ($t_{1/2}$) for a drug. This is time it takes for the concentration to be reduced to half its initial value. It is calculated by dividing -0.693 (ln 0.5) by the k_e. It is inversely proportional to the rate of elimination (i.e., a drug

with a $t_{1/2}$ of 4 hours is present in the body approximately twice as long as one with a half time of $t_{1/2}$ of 2 hours). Relating the temporal concentration of a drug in the body during the elimination phase with k_e values is not intuitive. Therefore, it is frequently expressed in terms of $t_{1/2}$. Thus, a period of one $t_{1/2}$ is the time required for the drug concentration to fall to 50% of its original value and 97% of the drug is eliminated after five periods of $t_{1/2}$. For the kinetic process of elimination, the velocity of elimination is first order unless the eliminating mechanism is saturated. In the first-order phase, the velocity of the process is linearly related to the concentration. However, at high doses of drug where the elimination is saturated the process is zero order. This means that a constant amount, not a constant fraction, of drug is eliminated until the process is no longer saturated.

Drug elimination may not be first order at high doses due to saturation of the capacity of the elimination processes. When this occurs, a reduction in the slope of the elimination curve is observed since elimination is governed by the relationship $V_{max}/(K_m + [conc])$, where V_{max} is the maximal rate of elimination, K_m is the concentration at which the process runs at half maximal speed, and [conc] is the concentration of the drug. However, once the concentration falls below saturating levels first-order kinetics prevail. Once the saturating levels of drugs fall to ones eliminated via first-order kinetics, the half time can be measured from the linear portion of the ln ρ_t versus time relationship. Most elimination processes can be estimated by a one compartment model. This "compartment" can

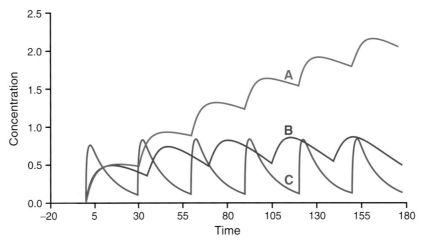

FIGURE 8.25 Repeated oral administration of drugs leads to steady-state plasma concentrations. If elimination is rapid and administration not often enough, then an elevated and therapeutically effective steady-state concentration may not be achieved (green lines). In contrast, if elimination is very slow (or administration too often), then an accumulation of the drug may be observed with no constant steady state (red line). Blue line shows a correct balance between frequency of administration and elimination.

actually be physically separated (as in blood versus a tissue depot) and still behave as one compartment kinetically if the rate constants for exchange between them are relatively fast. Using this approach, clinical data on temporal concentration in the blood can be used to estimate parameters useful for tailoring of dosage. For example, blood samples can be taken at regular intervals from a patient who received a 3-mg dose of a drug and used to determine the half time and volume of distribution of the drug by assuming a first-order elimination process. A semi-logarithmic (natural logarithms) of the data yields a straight line, which is then extrapolated to time zero. The ordinate intercept yields the concentration of the drug in the blood at time zero. Knowing the dose given allows a calculation to be made of the volume of distribution for this drug. This is the volume containing the drug assuming complete absorption. For example, a 3-mg dose giving a calculated concentration at time zero of 0.25 mg/L gives a volume of distribution of 12 liters, approximately the complete volume of fluid in a 70-kg male. Therefore, it appears from these data that the drug is reasonably distributed throughout the body. An estimate of the clearance for a drug (in volume that the body can completely clear per unit time) can be made from the rate constant for elimination (k_e) and the extrapolated volume of distribution. It should be noted that there can be exceptions to this simple rule and it is based on a one-compartment assumption.

There are instances where the observable kinetics of elimination clearly are not due to first-order exit from a single compartment. For example, a steep curve relating drug concentration and time may indicate a two-compartment system in which the drug exits in two phases, one fast and one slow (Figure 8.24b). In practical cases, usually the second slower half time is clinically relevant. However, the first elimination is often indicative of distribution into an important body compartment for the drug. Thus, the drug may be sequestered in various organs upon absorption, thereby giving rise to two-compartment kinetics. The first phase is one of distribution to the sites of sequestration, while the second represents elimination from the entire central compartment. This can be seen in a semilogarithmic plot that is clearly nonlinear. Under these circumstances, two half times for elimination are calculated (Figure 8.24b). The rapid phase is usually due to redistribution of the drug from the central compartment to other specialized compartments such as organs, protein, or fat. The second reflects elimination from the central compartment.

This type of analysis can yield insights into the distribution of some drugs. For example, the administration of 500 mg of the cardiac glycoside digoxin yields a zero time concentration of 0.75 ng/ml. The calculation of volume of distribution in this case provides a volume of 665 liters, nearly ten times the possible volume of fluid in a human being—suggesting a gross miscalculation in the volume of distribution. These types of data reveal how sequestration of drugs in tissue depots can severely skew volume of distribution estimates. The calculation for volume of distribution assumes a uniform distribution in the plasma. In this case, it is known that digoxin selectively concentrates in muscle and fat. Therefore, the free concentration in the plasma is inordinately low. This, in turn, provides the serious overestimation of the volume of distribution.

As noted previously, the aim of clinical pharmacokinetics is to achieve a steady concentration of blood at the site of action over a prolonged period of time. Unless a drug has extraordinary pharmacokinetics, repeated administration of the drug is necessary. This reveals another complication in the practical dosing for therapeutic effect. Just as a uniform area under the curve does not ensure uniform presentation of drug to tissues, the kinetics of absorption and elimination can confound multiple dose regimens. As shown in Figure 8.25, assume that the pattern shown for drug B is the desired pharmacokinetic profile.

The same dosing regimen with another drug (drug A) that is not eliminated as quickly as drug B shows that accumulation to dangerous levels occurs with repeated administration. In contrast, the same regimen with a drug that is cleared more quickly (drug C) shows that the desired sustained drug level is never attained even though the peak concentration for drug C is higher than for drug B.

An important concept in clinical pharmacokinetics is the bioavailablility of a drug. This is the actual fraction of drug that enters the central systemic circulation upon administration via the chosen therapeutic route. For example, drugs taken by the oral route must be absorbed either through the stomach or most likely the small intestine into the bloodstream. The blood preferentially flows through the liver. Thus, the drug is subjected to metabolism before it enters the general circulation. This first barrier of metabolism is referred to as the first-pass effect. Bioavailability is calculated as the ratio of area under the curves when the drug is given intravenously (assume 100% bioavailability) versus the chosen route of administration.

It is useful to finish with a restatement of the driving questions in clinical pharmacokinetics and how the various tools previously discussed can be used to answer them. Thus, to the question of how much of the drug administered reaches the therapeutic site an indirect answer can be obtained by measuring the bioavailability of the molecule. As to determining where in the body the drug goes, compartmental analysis can sometimes show patterns of distribution. Also, the volume of distribution can be used to detect sequestration. Finally, how long the drug stays in the body can readily be determined by measuring $t_{1/2}$, although it should be noted that this would be from the central compartment and not necessarily from the therapeutic site of action. In general, it can be seen that pharmacokinetics can completely control the outcome of a clinical trial, and as such deserves as much attention as the pharmacodynamic process.

A drug candidate must be adequately absorbed, reside in the body for a time sufficient to reach its target organ(s), and be excreted or degraded completely. There are general guidelines that can be used to determine early in the process whether or not a given molecule will fulfill these criteria. For example, a molecule with a clearance of $>25\%$ of liver blood flow by the intravenous route or $<10\%$ oral availability (assuming it is designed to be a drug taken by the oral route) would not augur well for further development. In contrast, a molecule with $<25\%$ liver blood flow clearance and $>30\%$ oral bioavailability would be a good candidate. In addition to pharmacokinetics, the chemical

form of the candidate also is important. This issue can be addressed by pharmaceutical studies.

8.5 Pharmaceutical Development

Pharmaceutics is the process of determining the best form for use in the study of the molecule in toxicological and clinical studies and the most stable preparation for dispensability as a drug product. It is a complete discipline within itself, the full discussion of which is beyond the scope of this present book. The pharmaceutical development of drug candidates is an important step that must go on in partnership with the study of pharmacokinetics. Ideally, the oral absorption of the molecular substance in capsule form should be equal to or greater than its absorption when administered as an soluble aqueous solution. The substance should be stable in a crystalline form as well. If stable crystals are not evident, nanomilled solid suspensions or spray-dried preparations can be made. Alternatively, polyethylene-glycol-surfactant-enhanced solutions can be used to model soft gel caps. In general, while these techniques can assist in the presentation of molecules for in vivo study pharmaceutical preparation is limited in terms of making a molecule suitable as a drug substance. Absorption via the oral route (preferably in a capsule) should be adequate to allow $30\times$ to $100\times$ dosing for toxicological studies. There are guidelines for these developmental steps. For example, the DCS (Developability Classification System) classifies compounds in terms of possessing high/low solubility and high/low permeability from studies of aqueous and gastrointestinal tract solubility and calculated or cell-based permeability measures. Figure 8.26 shows a graphical depiction of various regions of developability based on the aqueous solubility (abscissae) and permeability of the substance (ordinates). Ideal developability resides in the top left-hand quadrant (DCI), passable profiles in the green DCIIa area, and problematic profiles elsewhere on this grid (DCIIb, DCIII, and DCIV).

8.6 Adverse Drug Effects

New drugs must be efficacious, reach the site of action, and do no harm. This latter condition is the subject of drug liability studies. For the decade 1991 to 2000, new drug registration was a mere 11% of compounds submitted for first in human studies with toxicity and safety issues accounting for approximately 30% of the failures. There are clear "zero-tolerance" toxicities, and there are those that are tolerable—with tolerance depending on the indication, patient population (i.e., age and gender), length of treatment, and seriousness of illness. Table 8.3 shows some of most common clinically observed side effects of drugs.

Side effects commonly arise from exaggerated effects at the primary target (mechanism-based toxicity), problems with dosing, prolonged use, or cytotoxicity (i.e.,

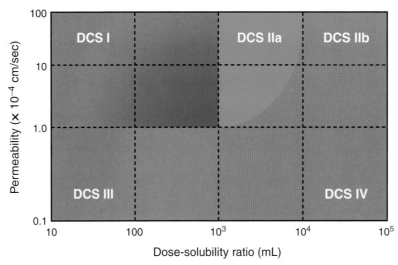

FIGURE 8.26 Developability criteria based on a grid showing the aqueous solubility of compounds (abscissae) and permeability through a lipid bilayer (ordinates). Adequate properties for development of a useful clinical entity with drug-like properties reside in the blue and portions of the green sectors. Compounds in the red region pose problems that can preclude progression to clinical candidate status.

TABLE 8.3
Major adverse side effects associated with clinical use of drugs.

Cardiovascular
Arrhythmias
Hypotension
Hypertension
Congestive heart failure
Angina/chest pain
Pericarditis
Cardiomyopathy

Dermatological
Erythema
Hyperpigmentation
Photodermatitis
Eczema
Urticaria
Acne
Alopecia

Endocrine
Thyroid dysfunction
Sexual dysfunction
Gynecomastia
Addison syndrome
Galactorrhea

Gastrointestinal
Hepatitis/hepatocellular damage
Constipation
Diarrhea
Nausea/vomiting
Ulceration
Pancreatitis
Dry mouth

Hematological
Agranulocytosis
Hemolytic anemia
Pancytopenia
Thrombocytopenia
Megaloblastic anemia
Clotting/bleeding
Eosinophilia

Musculoskeletal
Myalgia/myopathy
Rhabdomyolysis
Osteoporosis

Metabolic
Hyperglycemia
Hypoglycemia
Hyperkalemia
Hypokalemia
Metabolic acidosis
Hyperuricemia
Hyponatremia

Neurological
Seizures
Tremor
Sleep disorders
Peripheral neuropathy
Headache
Extrapyramidal effects

Renal
Nephritis
Nephrosis
Tubular necrosis
Renal dysfunction
Bladder dysfunction
Nephroliythiasis

Respiratory
Airway obstruction
Pulmonary infiltrates
Pulmonary edema
Respiratory depression
Nasal congestion

Ophthalmic
Disturbed color vision
Cataract
Optic neuritis
Retinopathy
Glaucoma
Corneal opacity

Otological
Deafness
Vestibular disorders

Psychiatric
Delirium/confusion
Depression
Hallucination
Drowsiness
Schizophrenia/paranoia
Sleep disturbances

TABLE 8.4
Cardiovascular targets assoicated with adverse drug effects.

Target	Possible Adverse Drug Effects
Adenosine A1	Bradycardia/AV-block/renal vasoconstrict
Adenosine A2a	Hypotension/coronary vasodilation/platelet aggreg
Adenosine A3	Mediator release
α_{1a}-adrenoceptor	Hypertension/orthostatic hypotension/inotropy
α_{1b}-adrenoceptor	Othostatic hypotension
α_{2a}-adrenoceptor	Hypertension/possible hyperglycemia
α_{2b}-adrenoceptor	Hypertension /cardiac ischemia/vasoconstriction/central $\downarrow$ blood pressure
α_{2c}-adrenoceptor	Hypertension/cardiac ischemia/skel. muscle blood flow
β_1-adrenoceptor	Cardiac inotropy bronchospasm/heart rate/ventricular fibrillation
β_2-adrenoceptor	Fascil. cardiac arrest/impairs cardiac perform
Angiotensin AT$_1$	Hypertension/cell proliferation and migration/tubular Na$^+$ resorption
Bradykinin B$_1$	Nociception/inflammation/cough
Bradykinin B$_2$	Nociception/inflammation/cough
CGRP	Hypocalcemia/hypophosphatemia
Ca^{2+} channel	Hypotension
Dopamine D$_1$	Induces dyskinesia/vasodilatation, schizophrenia/$\downarrow$ coordination
Endothelin ET$_a$	Vasoconstriction/cell proliferation/aldosterone secretion
Endothelin ET$_b$	Vasoconstriction/cell proliferation/bronchoconstriction
Histamine H$_3$	$\downarrow$memory, sedation/vasodilatation/$\downarrow$ GI motility
Muscarinic m1	Δ blood pressure/$\downarrow$ GI secretion
Muscarinic m2	Vagal effects/Δ blood pressure/tachycardia
Muscarinic m3	Vagal effects, salivation/Δ blood pressure, dry mouth/$\downarrow$ ocular accommodation
Muscarinic m4	Vagal effects, salivation adrenergic/Δ blood pressure/facilitates D1 stim
NE transporter	Hyperreactivity/facilitates α-activation
Nicotinic Ach	Autonomic functions/palpitations, nausea, sweating $\downarrow$ gut motility, gastric/tremor, ganglionic function
NPY$_1$	Venous vasoconstriction/emptying/anxiogenic
K$^+$ channel (hERG)	Cardiac QTc prolongation
K$^+$ channel [ATP]	Hypotension/hypoglycemia
5-HT$_{2b}$	Cardiac valvulopathy
5-HT$_4$	Facilitates GI transit/mechanical intestinal allodynia
Na$^+$ channel (site 2)	Cardiac arrhythmia
Thromboxane $_{a2}$	Vascular constriction/bronchial constriction/allergic inflamm. platelet ag
Vasopressin V$_{1a}$	Vasopressor
Vasopressin V$_{1b}$	Vasopressor, anxiogenic

Taken from [19].

hepatoxicity and bone marrow toxicity). Effects on other biological targets (i.e., GPCRs), ion channels, and liver metabolic enzymes account for major drug liabilities. In most cases, such as effects on GPCRs, the untoward effects are a direct result of the receptor activation (or blockade). Table 8.4 shows some liability effects commonly associated with some GPCRs [25]. In some cases, the receptor activity belies effects that are not obvious. For example, muscarinic m3 receptor activity has been associated with type 2 diabetes [26]. Promiscuous GPCR activity is a potential problem with GPCR active drugs. Similarly, hydrophobic drugs have been shown to have affinity for calcium channels, and notably potassium channels. This latter activity is a clear liability since blockade of the hERG potassium channel can lead to cardiac QTc prolongation and a condition called *torsades de pointes*, a potentially fatal cardiac arrhythmia. Other promiscuous targets are the pregnane X-receptor, a nuclear receptor associated with regulation of cytochrome P450 enzymes. Induction of PXR

can have large effects on metabolism, drug-drug interactions, multi-drug resistance, and transport mechanisms. Cytochrome P450 enzymes are particularly susceptible to drug activity due to their broad substrate specificity. Four of these enzymes (CYP3A4, CYP2C9, CYP2C19, and CYP2D6) account for 80% of known oxidative drug metabolism [27]. Blockade of these enzymes can lead to detrimental interactions with other drugs. For example, the antihistamine terfenadine was high-affinity for the hERG channel (leading to serious liability). This drug is rapidly metabolized and the metabolite fexofenadine is weakly active at the hERG channel. However, in the presence of other drugs that interfere with terfenadine metabolism (ctyochrome enzymes) this antihistamine poses a serious risk of life-threatening arrhythmia.

In general, the detection of adverse drug reactions early in the drug discovery process is becoming commonplace. So-called "liability panels" of receptors, hERG channel activity, and cytochrome enzymes are utilized to identify liability activity in chemical series. Pharmacophore modeling of "anti-targets" [28] can also be used to virtually screen for potential problematic drug activity.

8.7 Chapter Summary and Conclusions

- The drug discovery process can be divided into four subsets: acquisition of chemical drug candidates, pharmacodynamic testing of large numbers of compounds (screening), and the optimization of pharmacokinetic and pharmaceutical properties.
- Potential chemical structures for drug testing can originate from natural products, design from modeling the active site of the biological target, modification of natural substances, hybridization of known drugs, or random screening of chemical diversity.
- There is evidence to suggest that drug-like structures exist in clusters in chemical space (privileged structures). Identification of these can greatly enhance success in screening.
- Large-scale sampling of chemical space can be achieved with high-throughput screening. This process involves the design of robust but sensitive biological test systems and the statistical sifting of biological signals from noise. The Z statistic can be useful in this latter process.
- Surrogate screening (utilizing similar but not exact therapeutically relevant targets) can lead to dissimulation in screening data, especially for allosteric molecules. For this reason, frequent reality testing with a therapeutically relevant assay is essential.
- Lack of favorable ADME properties (absorption, distribution, metabolism, elimination) can preclude therapeutic use of an otherwise active molecule. The clinical pharmacokinetic parameters of clearance, half-life, volume of distribution, and bioavailability can be used to characterize ADME properties.
- Active molecules also must not have toxic side effects and must have favorable pharmaceutical properties for qualification as useful drugs.

References

1. Lipinski, C., Lombardo, F., Dominy, B., and Feeney, P. Experimental and computational approaches to estimate solubility and permeability in drug discovery and development settings. *Adv. Drug. Deliv. Rev.* **23**:3–25.
2. Richardson, B. P., Engel, G., Donatsch, P., and Stadler, P. A. (1985). Identification of serotonin M-receptor subtypes and their specific blockade by a new class of drugs. *Nature* **316**:126–131.
3. Baldwin, J. J., Lumma, W. C. Jr., Lundell, G. F., Ponticello, G. S., Raab, A. W., Engelhardt, E. L., Hirschmann, E. L., Sweet, C. S., and Scriabine, A. (1979). Symbiotic approach to drug design: Antihypertensive β-adrenergic blocking agents. *J. Med. Chem.* **22**:1284–1290.
4. Lipinski, C. A. (2000). Drug-like properties and the causes of poor solubility and poor permeability. *J. Pharmacol. Tox. Meth.* **44**:235–249.
5. Jakubic, J., Bacakova, I., El-Fakahany, E. E., and Tucek, S. (1997). Positive cooperativity of acetylcholine and other agonists with allosteric ligands on muscarinic acetylcholine receptors. *Mol. Pharmacol.* **52**:172–179.
6. Finke, P. E., Oates, B., Mills, S. G., MacCoss, M., Malkowitz, L., Springer, M. S., Gould, S. L., DeMartino, J. A., Carella, A. Carver, G., et al. (2001). Antagonists of the human CCR5 receptor as anti-HIV-1 agents. Part 4: synthesis and structure—Activity relationships for 1-[N–(Methyl)-N-(phenylsulfonyl)amino]-2-(phenyl)-4-(4-(N-(alkyl)-N-(benzyloxycarbonyl)amino)piperidin-1-yl)butanes. *Bioorg. Med. Chem. Lett.* **11**:2475–2479.
7. Zhang, J.-H., Chung, T. D. Y., and Oldenburg, K. R. (1999). A simple statistical parameter for use in evaluation and validation of high throughput screening assays. *J. Biomeolecular Screening* **4**:67–72.
8. Andrews, P. R., Craik, D. J., and Martin, J. L. (1984). Functional group contributions to drug-receptor interactions. *J. Med. Chem.* **27**:1648–1657.
9. Hopkins, A. L., Groom, C. R., and Alex, A. (2004). Ligand efficiency: A useful metric for lead selection. *Drug Disc. Today* **9**:430–431.
10. Kuntz, I. D., Chen, K., Sharp, K. A., and Kollman, P. A. (1999). The maximal affinity of ligands. *Proc. Natl. Acad. Sci. USA* **96**:9997–10002.
11. Roettger, B. F., Ghanekar, D., Rao, R., Toledo, C., Yingling, J., Pinon, D., and Miller, L. J. (1997). Antagonist-stimulated internalization of the G protein-coupled cholecystokinin receptor. *Mol. Pharmacol.* **51**:357–362.
12. Pommier, Y., and Cherfils, J. (2005). Interfacial inhibition of macromolecular interactions: Nature's paradigm for drug discovery. *Trend. Pharmacol. Sci.* **26**:138–145.
13. Onaran, H. O., and Costa, T. (1997). Agonist efficacy and allosteric models of receptor action. *Ann. N Y Acad. Sci.* **812**:98–115.
14. Onaran, H. O., et al. (2000). A look at receptor efficacy: From the signaling network of the cell to the intramolecular motion of the receptor. In: *The pharmacology of functional, biochemical, and recombinant systems handbook of experimental pharmacology*, Vol. 148, edited by Kenakin, T. P. and Angus, J. A., pp. 217–280. Springer.
15. Kenakin, T. P., and Onaran, O. (2002). The ligand paradox between affinity and efficacy: Can you be there and not make a difference? *Trends Pharmacol. Sci.* **23**:275–280.
16. Tang, Z., Taylor, M. J., Lisboa, P., and Dyas, M. (2005). Quantitative risk modeling for new pharmaceutical compounds. *Drug Disc Today* **22**:1520–1526.
17. Teague, S. J., Davis, A. M., Leeson, P. D., and Oprea, T. J. (1999). The design of leadlike combinatorial libraries. *Angewandte Chemie International Edition* **38**:3743–3748.
18. King, F. D., Brown, A. M., Gaster, L. M., Kaumann, A. J., Medhurst, A. D., Parker, S. G., Parsons, A. A., Patch, T. L., and Rava, P. (1993). (.+−.)-3-Amino-6-carboxamido-1,2,3,4-tetrahydrocarbazole: A conformationally restricted analog of 5-carboxamidotryptamine with selectively for the serotonin 5-HT1D receptor. *J. Med. Chem.* **36**:1918.
19. Yanagisawa, T., Wakabayashi, S., Tomiyama, T., Yasunami, M., and Takase, K. (1998). Synthesis and anti-ulcer activities of sodium alkylazulene sulfonates. *Chem. Pharm. Bull.* **36**:641.
20. Giudici, D., Ornati, G., Briatico, G., Buzzetti, F., Lombardi, P., and di Salle, E. (1988). 6-Methylenandrosta-1,4-diene-3,17-dione (FCE 24304): A new irreversible aromatase inhibitor. *J. Steroid Biochem.* **30**:391.

21. Bhat, B., Seth, M., and Bhaduri, A. P. (1981). *Indian J. Chem.* **20B:**703.

22. Krapcho, J., Turk, C., Cushman, D. W., Powell, J. R., DeForrest, J. M., Spitzmiller, E. R., Karanewsky, D. S., Duggan, M., Rovnyak, G., Schwartz, J., Natarajan, S., Godfrey, J. D., Ryono, D. E., Neubeck, R., Atwal, K. S., and Petrillo, E. W. (1988). Angiotensin-converting enzyme inhibitors. Mercaptan, carboxyalkyl dipeptide, and phosphinic and inhibitors incorporating 4-substituted prolines. *J. Med. Chem.* **31:**1148.

23. Sham, H. L., Kempf, D. J., Molla, A., Marsh, K. C., Kumar, G. N., Chen, C.-M., Kati, W., Stewart, K., Lal, R., Hsu, A., Betebenner, D., Korneyeva, M., Vasavanonda, S., McDonald, E., Saldivar, A., Wideburg, N., Chen, X., Niu, P., Park, C., Jayanti, V., Grabowski, B., Granneman, G. R., Sun, E., Japour, A. J., Leonard, J. M., Plattner, J. J., and Norbeck, D. W. (1998). ABT-378, a highly potent inhibitor of the human immunodeficiency virus protease Antimicrob. *Agents Chemother.* **42:**3218.

24. Proudfoot, J. R. (2002). Drugs, leads, and drug-likeness: An analysis of some recently launched drugs. *Bioorgan. Med. Chem. Lett.* **12:**1647–1650.

25. Whitebread, S., Hamon, J., Bojanic, D., and Urban, L. (2005). In vitro safety pharmacology profiling: An essential tool for successful drug development. *Drug Disc. Today* **10:**1421–1433.

26. Silvetre, J. S., and Prous, J. (2005). Research on adverse drug events: Muscarinic m3 receptor binding affinity could predict the risk of antipsychotics to induce type 2 diabetes. *Meth. Find. Exp. Clin. Pharmacol.* **27:**289–304.

27. Wienkers, L. C., and Heath, T. G. (2005). Predicting in vivo drug interactions from in vitro drug discovery data. *Nature Rev. Drug Discovery* **4:**825–833.

28. Klabunde, T., and Evers, A. (2005). GPCR antitarget modeling: Pharmacophore models for biogenic amine binding GPCRs to avoid GPCR-mediated side effect. *Chem. Bio. Chem.* **6:**876–889.

9

Target- and System-based Strategies for Drug Discovery

I am interested in physical medicine because my father was. I am interested in medical research because I believe in it. I am interested in arthritis because I have it.

— BERNARD BARUCH (1959)

New techniques may be generating bigger haystacks as opposed to more needles.

— D. F. HORROBIN (2000)

9.1 Some Challenges for Modern Drug Discovery

The identification of primary biological activity on the target of interest is just one of a series of requirements for a drug. The capability to screen massive numbers of compounds has been ever-increasing over the past 10 to 15 years yet no corresponding increase in successfully launched drugs has ensued. As discussed in Chapter 8, there are required pharmacokinetic properties and absence of toxic effects that must be features of a therapeutic entity. In the 1990s, 40% of the attrition in drug discovery was due to lack of bioavailability and pharmacokinetics. As more attention was paid to ADME, properties of chemical screening libraries, toxicity, lack of therapeutic efficacy, and differentiation from currently marketed drugs have become the major problems. As shown in Figure 9.1, the number of new drug entities over the years has decreased. This particular representation is normalized for the increasing costs of drug discovery and development, but it does reflect some debilitating trends in the drug discovery process. Undue reliance on robotic screening with simplistic single-gene-target approaches (inappropriate reliance on the genome as an instruction booklet for new drugs) coupled with a deemphasis of pharmacological training may have combined to cause the current deficit in new drugs [2]. The lack of success in drug discovery is reflected in the number of drugs that have failed in the transition from phase II clinical trials (trial in a small number of patients designed to determine efficacy and acute side effects) to phase III clinical trials (larger trials meant to predict effects in overall populations and to determine overall risk-to-benefit ratio of drug). (See Figure 9.2.) While the 62 to 66% of the new drugs entering phase I

passed from phase II to phase III in the years 1995 to 1997, this percentage fell to 45% in 2001 through 2002 [3]. In view of the constantly increasing number of new drugs offered for clinical trial, this suggests that the quality of molecules presented to the clinic is diminishing from that seen 10 years ago.

At the heart of the strategies for drug discoveries are two fundamentally different approaches, one focusing on the target whereby a molecule is found to interact with a single biological target thought to be pivotal to the disease process and one focusing on the complete system. It is worth considering these separately.

9.2 Target-based Drug Discovery

A target-based strategy for drug discovery has also been referred to as a "reductionist approach." The term originates in physics, where it describes complex matter at the level of fundamental particles. In drug discovery, target-based refers to the fact that the responsible entity for a pathological process or disease is thought to be a single gene product (or small group of defined gene products) and is based on the premise that isolation of that gene product in a system is the most efficient and least ambiguous method of determining an active molecule for the target. Reductionist approaches are best suited for "me-too" molecules with well-validated targets when first-in-class already exists. They are also well suited to Mendelian diseases such as cystic fibrosis and sickle cell anemia, where the inheritance of a single gene mutation can be linked to the disease.

Reductionist systems are most often recombinant ones with the target of interest (for example, human GPCR) expressed in a surrogate cell. The nature of the cell is thought to be immaterial since the cell is simply a unit reporting activation of the target of interest. For example, belief that peptic ulcer healing is facilitated by blockade of histamine-H2-receptor-induced acid secretion suggests a reductionist system involving antagonism of histamine response in surrogate cells transfected with human histamine H2 receptors. In this case, refining primary activity when the target-based activity disease relationship has been verified is a useful strategy. It can be argued that

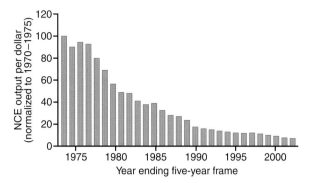

FIGURE 9.1 Histograms show the number of new drugs (normalized for the cost of drug discovery and development in the years they were developed) as a function of the years they were discovered and developed. Adapted from [1].

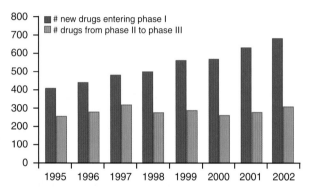

FIGURE 9.2 Histograms showing the number of new drug entities entering phase I clinical development (blue bars), and concomitantly the number entering phase III development as a function of year. Adapted from [2].

considerable value may be mined in this approach since first-in-class is often not best-in-class.

Focusing in on a single target may be a way of treating a disease but not necessarily curing it. The interplay of multiple genes and the environment leads to complex diseases such a diabetes milletus, coronary artery disease, rheumatoid arthritis, and asthma. To consider this latter disease, it is known that bronchial asthma is the result of airway hyperreactivity that itself is the result of multiple system breakdowns involving allergic sensitization, failure of neuronal and hormonal balance to airway smooth muscle, and hyperreactivity of smooth muscle. Bronchial spasm can be overcome by a system override such as powerful β-adrenergic muscle relaxation, providing a life-saving treatment, but this does not address the origins of the disease nor does it cure it. The divergence in phase II from phase III studies (shown in Figure 9.2) is cited as evidence that the target approach is yielding molecules, but that they may be wrong molecules for curing (or even treating) the disease.

Whereas in physics the path from the fundamental particle to the complex matter is relatively linear

(reductionism requires linearity and additivity), in biology it often is extremely nonlinear. This can be because of system-specific modifications of genes and highly complex interactions at the level of the cell integration of the genes. This can lead to some impressive disconnections (i.e., the principal defect is known in type I diabetes but targeted approaches have still been unable to cure the disease). In theory, pathways can be identified in disease processes, critical molecules in those pathways identified, prediction of the effects of interference with the function of those molecules determined, and the effect of this process on the disease process observed. However, this simple progression can be negated if many such pathways interact in a nonlinear manner during the course of the disease. In fact, in some cases the design of a surrogate system based on the target may be counterproductive. For example, for anticancer drugs the test system tumors are sometimes chosen or genetically manipulated for sensitivity to drugs. This can make the models overpredictive of drug activity in wild-type tumors where multiple pathways may be affected by numerous accumulated mutations and/or chromosomal abnormalities used to maintain their phenotype. A classic example of where a single target fails to emulate the properties of diseases is in the therapy of psychiatric disorders. These diseases have a shortage of validated targets (it is unlikely that there are single-gene lesions accounting for psychiatric disorders) and the high-through-put screening systems bear little resemblance to the in vivo pathology. Genetic approaches in psychiatry are problematic since the effects of "nurture" and epigenetic changes (identical genotypes yielding different phenotypes) are prevalent. In addition, animal models cannot be transposed to phase I and phase II clinical testing. In the clinic, placebo effects can approach 60% (in anxiety and depression studies), and inappropriate inclusion of patients clouds interpretation of data. In general, it is extremely difficult to use a single gene product as a target for psychiatric diseases, making a reductionist approach in this realm impractical [4].

In a target-based strategy, the preclinical process of drug discovery can be roughly divided into three stages. The first is the discovery phase. This involves the identification of a valid therapeutic target (i.e., receptor), the development of a pharmacological assay for that target, and the screening of large numbers of molecules in the search for initial activity. The next phase is the optimization phase, where chemical analogs of the initial lead molecule are made and tested in either the screening assay or a related assay thought to reflect the therapeutically desired activity. From this stage of the process comes the optimized lead molecule that has sufficient activity and also no obvious non-drug-like properties that would preclude development to a candidate for clinical study. The third phase is the development phase, where the pharmacokinetic properties of the lead molecules are optimized for maximal drug-like properties in the clinic. In terms of strategies for drug development, the latter two steps are common to all modes (i.e., screening and lead optimization are required).

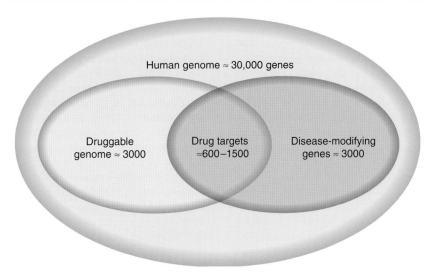

FIGURE 9.3 Venn diagram indicating the human genome and the subsets of genes thought to mediate disease and those that are druggable (thought to be capable of influence by small molecules such as proteins). The intersection of the subsets comprises the set that should be targeted by drug discovery. Adapted from [5].

However, the target validation step is unique to target-based drug discovery.

Once a target-based approach is embarked upon, the choice of target is the first step. In biological systems, there are generally four types of macromolecules that can interact with drug-like molecules: proteins, polysaccharides, lipids, and nucleic acids. As discussed in Chapter 1, by far the richest source of targets for drugs are proteins. The sequencing of the human genome was completed in April 2003 and the outcome predicts that of the estimated 30,000 genes in the human genome approximately 3,000 code for proteins that bind drug-like molecules [5]. Of the estimated 3,000 to 10,000 disease-related genes [6, 7], knockout studies (animals bred devoid of a specific naturally occurring gene) indicate that 10% of these genes have the potential to be disease modifying. From these estimates, it can be proposed that there are potentially 600 to 1,500 small-molecule drug targets as yet undiscovered (see Figure 9.3) [5].

9.2.1 Target Validation and the Use of Chemical Tools

A detailed discussion of the science of target validation is beyond the scope of this book, but some of the general concepts will be illustrated by example. Evidence in estimating a given target's relevance in a disease can be pharmacologic and/or genetic. For example, the chemokine receptor CCR5 has been described as the critical target for M-tropic HIV entry into healthy cells (*vide infra*). It is useful to examine the data supporting this idea as an illustration of how these lines of evidence converge to validate a target. One line of evidence to support this is co-location of the target with sensitivity to the disease. Thus, it is known that CCR5 receptors must be present on the cell membrane for HIV infection to occur [8, 9]. Similarly,

removal of CCR5 from the cell membrane in vitro leads to resistance to M-tropic HIV [10]. Another line of evidence is in vitro data to show that ligands for CCR5, such as natural chemokines and chemokine small-molecule antagonists, interfere with HIV infection [11–15]. This effect extends in vivo, where it has been shown that individuals with high levels of circulating chemokines (ligands for CCR5) have a decreased progression to AIDS [16, 17]. Similarly, patients with herpes virus 6 (HHV-6) have increased levels of chemokine and this leads to suppression of HIV replication [18].

Genetic evidence can be powerful for target validation. For example, an extremely useful finding from genetic evidence are data to indicate the effects of a long-term absence of the target. For CCR5, this is the most compelling evidence to show this protein is the target for HIV. Specifically, individuals with a mutation leading to lack of expression of operative CCR5 receptors (Δ32 CCR5 allele) are highly resistant to HIV infection. These individuals are otherwise completely healthy, indicating that this drug therapy to render this target inoperative should not be detrimental to the host [19–23]. Often these types of data are obtained in genetically modified animals; for example, a knockout mouse where genetic therapy leaves the mouse devoid of the target from birth. In the case of CCR5, the knockout mouse is healthy, indicating the benign consequences of removal of this receptor [24]. Complementary genetic evidence also is available to show that AIDS patients possessing a CCR5 promoter (−2450 A/G leading to high cellular expression levels of CCR5) have a highly accelerated progression toward death [25]. In general, the data for CCR5 serve as an excellent example of where pharmacological and genetic evidence combine to highly validate a therapeutic target. Genetic knockout animals can also be used to identify pathways relevant to pathological phenotypes. For example, a number of inbred strains of mice fed a diet that promotes hyperlipidemia develop lesions and lipid plaques. However, knockout mice lacking the major carrier of plasma cholesterol, apolipoprotein E, spontaneously form plaques on a *normal* diet—thereby implicating a role for cholesterol in cardiovascular disease. Gene knockout animals can be used to explore phenotypes resulting from the removal of a given target. Thus, CNS-target expression of RGS-I $G_q\alpha$ protein leads to tremulousness, decreased body mass, heightened response to the 5-HT_{2C} receptor agonist RO60-0175 (which induces anorexia), and convulsions to the the 5-HT_{2A} receptor agonists 2,5-dimethoxy-4-iodoamphetamine and muscarinic agonist pilocarpine (at concentrations that are ineffective in normal mice) [26].

Another approach to target validation is through chemical tool compounds. A reductionist view of drug discovery is premised on the fact that a single gene product (or small collection of identifiable gene products) is responsible for a given disease. There are numerous untestable assumptions made in this process, and if these are unchecked the final test becomes a very expensive one (namely, the clinical testing of a drug molecule). A large part of the expense of this process results from the fact that the test molecule must be a drug (i.e., there are numerous criteria that a molecule must pass to be become a "drug" candidate, and this constitutes much effort and expense en route to the final testing of the reductionist hypothesis). The use of chemical tools that may not qualify as drug candidates may substantially reduce the effort and expense of this process (i.e., use of a molecule with target activity that does not qualify as a drug per se to test the disease target-link hypothesis). Such hypothesis-testing molecules may be parenterally administered (obviating the need for oral absorption) and the results assessed on a timescale that may avoid longer-term toxicity problems. For example, the natural product staurosporine (not a drug in its own right) provided useful information regarding tyrosine kinase inhibition in cancer, leading to the anticancer drug imatinib (inhibitor of BCR-ABL tyrosine kinase). A classic example of tool compound validation (although unintended) is the progression of histamine H2 receptor antagonists for the treatment of ulcer (see Figure 8.20). In this case, the data obtained with the ultimately unsuitable compounds burimamide and cimetide led to the clinically useful drug metiamide. Chemical tools have intrinsic advantages over genetic approaches since the latter can adequately answer questions of removal of gene function, but not gain of function. Chemical tools can approach both loss and gain of function. To determine whether the addition of gene activity is involved in disease, an agonist of the gene product is required—a role that can be fulfilled by a chemical tool. This has lead to the terms *chemical genetics* and *chemical genomics* for the use of molecules to determine the relevance of gene products in disease. A shortcoming of this approach is that molecules are usually not exquisitely selective (as genetic knockouts are), leading to some ambiguity in the analysis of results.

The requirement for target validation can be a serious limitation of target-based strategies. In addition to being a high-resource requirement (estimates suggest three years and US$390 million per target), target validation has intrinsic hazards in terms of equating the data with a conclusion that the given target is the causative factor of (or even intimately related to) a disease. One of the mainstays of target validation is the observation of animal health and behavior after the gene controlling the target of interest is knocked out. However, a problem with this strategy is the different genomic background the organism is exposed to when the gene is eliminated from birth as opposed to when it is eliminated by a drug in adult life. Removing the gene from birth may bring into effect compensating mechanisms that allow the organism to survive. These may not be operative (or there may not be enough time for them to compensate) in adult life upon sudden elimination of the target. For example, while it is known that humans containing the $\Delta32$ CCR5 mutation (which prevents cell surface expression of CCR5) are otherwise healthy it still is not certain that elimination of CCR5 with CCR5-based HIV entry inhibitors to adult AIDS patients will not cause abnormalities in chemotaxis. The induction of compensatory mechanisms can be substantially overcome by the construction of conditional knockouts whereby inducible promoters are used to produce tissue dependent and/or time-dependent knockout after animal development.

In general, systems achieve robustness with redundancy (i.e., several isoenzymes catalyze the same reaction), making the interaction with a single target of questionable value. Also, the use of mouse knockouts brings in obvious questions as to species-dependent differences between humans and mice ("mice are not men" [27]). Animal studies in general have been shown not to be infallible predictors of clinical activity in humans. For example, preclinical studies in animals indicated that antagonists of the neurokinin NK_1 receptor attenuate nociceptive responses. Studies with nonsteroidal anti-inflammatory drugs (NSAIDs) indicate that this should be a predictor of analgesic activity in humans. However, unlike NSAIDs the NK_1 activity in animals does not transfer into an analgesic activity in humans [28].

It is prudent to not treat target validation as a single-answer type of experiment (i.e., if the appropriate data indicates that the target is "validated" then no further examination is required). As with all hypothesis testing, theories cannot be proven correct only incorrect. The fact that data is obtained to support the notion that a given target is involved in a disease does not prove that interference with that target will influence the disease. Target validation is an ongoing process that really does not end until the drug is tested in the actual disease state in patients with a properly controlled clinical trial.

Finally, another consideration in target selection and subsequent prosecution of a biological target is random variation in gene expression leading to slightly modified proteins. These could be devastating to drug activity. As discussed in Section 8.3, an antagonist of the chemokine

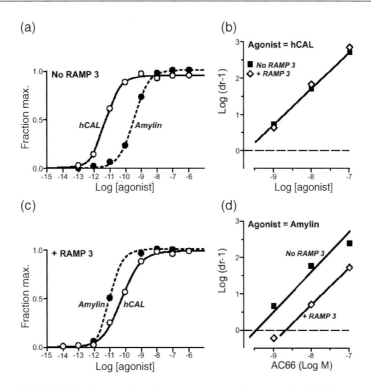

FIGURE 9.8 Assumption of a new receptor phenotype for the human calcitonin receptor upon coexpression with the protein RAMP3. (a) Melanophores transfected with cDNA for human calcitonin receptor type 2 show a distinct senstivity pattern to human calcitonin and rat amylin. hCAL is 20-fold more potent than rat amylin. (b) A distinct pattern of sensitivity to the antagonist AC66 is also observed. Both agonists yield a pK_B for AC66 of 9.7. (c) Coexpression of the protein RAMP3 (receptor activity modifying protein type 3) completely changes the sensitivity of the receptor to the agonists. The rank order is now changed such that amylin has a threefold greater potency than human calcitonin. (d) This change in phenotype is carried over into the sensitivity to the antagonist. With coexpression of RAMP3, the pK_B for AC66 changes to 8.85 when rat amylin is used as the agonist. Data redrawn from [41].

(Table 9.2b). Increasing also is the list of phenotypes associated with these dimerization processes. With the emergence of receptor dimers as possible therapeutic targets have come parallel ideas with dimerized ligands (see Section 9.5).

Drug targets can be complexes made up of more than one gene product (i.e., integrins, nicotinic acetylcholine ion channels). Thus, each combination of targets could be considered a target in itself [40]. Some of these phenotypes may be the result of protein-protein receptor interactions [41–43]. For example, the human calcitonin receptor has a distinct profile of sensitivity to and selectivity for various agonists. Figure 9.8a shows the relative potency of the human calcitonin receptor to the agonists human calcitonin and rat amylin. It can be seen that human calcitonin is a 20-fold more potent agonist for this receptor than is rat amylin [41]. When the antagonist AC66 is used to block responses, both agonists are uniformly sensitive to blockade

$(pK_B = 9.7$, Figure 9.8b)., However, when the protein RAMP3 (receptor activity modifying protein type 3) is coexpressed with the receptor in this cell the sensitivity to agonists and antagonists completely changes. As seen in Figure 9.8c, the rank order of potency of human calcitonin and rat amylin reverses such that rat amylin is now threefold more potent than human calcitonin. Similarly, the sensitivity of responses to AC66 is reduced by a factor of 7 when amylin is used as the agonist $(pK_B = 8.85$, Figure 9.8d). It can be seen from these data that the phenotype of the receptor changes when the cellular milieu into which the receptor is expressed changes. RAMP3 is one a family of proteins that affect the transport, export, and drug sensitivity of receptors in different cells. The important question for the drug development process is, if a given receptor target is thought to be therapeutically relevant, what is the correct phenotype for screening? As can be seen from the example with the human calcitonin receptor, if a RAMP3 phenotype for the receptor is the therapeutically relevant phenotype then screening in a system without RAMP3 coexpression would not be useful.

9.3 Systems-based Drug Discovery

With a target-based approach, the activity of molecules interacting with the previously identified target of interest can readily be assessed. As discussed previously, such an approach requires a linear relationship between targets and cellular activity. If pathways interact in a complex and nonlinear fashion, then redundancy and feedback effects may make predictions from single targets difficult and erroneous. A major criticism of target approaches is that they stray from a relatively tried and true successful historical strategy in drug research, whereby discovery relied upon proven physiology and/or pathophysiology and appropriate models. Another more pragmatic criticism of target-based strategies is that while they yield drugs for targets, this activity does not necessarily translate to overall clinical utility (see Figures 9.1 and 9.2).

An alternative to target-based strategies is referred to as "systems-based" drug discovery. The study of the assembled cellular system has evolved into "systems biology," whereby natural cells are used for screening with complex outputs obtained ranging from secreted cellular products to genomic data utilized to measure system responses to drugs. The term originated in engineering, where it describes a theoretical framework for controlling a complicated system (i.e., flying an airplane). The assembly of genes into living cells creates an infinitely richer palette for potential intervention.

> Move over human genome, your day in the spotlight is coming to a close. The genome...contains only the recipes for making proteins...it's the proteins that constitute the bricks and mortar of cells and that do most of the work.
> — Carol Ezzell, *Scientific American* (April 2002)

Systems approaches may yield more abundant opportunities for drug discovery. In organs under the control of pathological mechanisms, genes can interact to provide multifactorial phenotypes. This can greatly expand the possible targets for drugs. Therefore, the study of the same target in its therapeutic environment can enrich the recognition possibilities for new drugs, in essence increasing the biological space of that target [40].

There is a fundamental difference between the target-based approach (where a very large number of compounds are screened against one target) versus a systems approach, where a smaller number of compounds (but perhaps higher quality more drug-like molecules) are screened in a system that has many targets. Systems can have a great many (possibly hundreds) small-molecule intervention sites and can be engineered to incorporate many disease relevant pathways. The output of such systems can be extremely complex and requires high-throughput genomic tools and technologies to process. The development of sophisticated computing tools as well as the advancement of genetic

technology has facilitated the construction of biological systems for screening and the study of structure-activity relationships. Specifically, short interfering RNA duplex molecules (siRNA) can be used to silence specific genes in the cell, which allows the observation of their relevance to total cellular function (Figure 9.9a). This approach is vulnerable to biological redundancy in the system but overexpression of targets in the cell also can be used in conjunction with siRNA approaches to identify and characterize pathways. Analysis of multiple readouts of cellular function then act as fingerprints for the particular silenced portion of a pathway. As multiple histograms viewed from the top and color coded for response, these outputs form a heat map for cell function that can be used to compare control conditions and the effects of drugs (Figure 9.9b).

In general, systems allow the identification of unknown (and previously hidden) drug activity and/or can add texture to known drug activity. This can lead to the identification of new uses for existing targets, identification of new targets (so-called "therapeutic target space" involving discovery of a molecular phenotype in a system and subsequent determination of the molecular target), and determination of an entry point into signaling cascades that may be amenable to drug intervention (optimize efficacy and minimize side effects). Comparison of normal and diseased samples can be used to determine disease-specific signaling as a target for drug intervention. The complexity of the systems response output allows discrimination of subtle drug activities. For example, Figure 9.10 shows three levels of output from a system and the results observed for three hypothetical compounds. Compound A is inactive in the system, whereas compounds B and C block different points on the integrated pathway cascades. The first level of output (i.e., second-messenger production) does not indicate activity in any of the three ligands. It can be seen that the second level of output does not discriminate between the activity seen for compounds B and C, whereas the third (and more complex) level of output shows them to be different. In general, systems are designed to provide maximally complex outputs in different contexts (different milieu of cellular activating agents) to yield complex heat-map fingerprints of drug activity. Statistical methods such as multidimensional scaling are then used to associate similar profiles (define functional similarity maps) and determine differences. This gives added levels of power to screening systems and subsequent lead-optimization assays. In general, integrated systems can be used to correlate functional responses with mechanistic classes of compounds, identify secondary activities for molecules, provide insight into the mechanism of action of compounds that give clinical activities, and characterize pathways and correlate them with functional phenotypes [44, 45].

Cellular screening systems can be developed with primary human cells cultured in biologically relevant contexts. The outputs of these systems are focused sets of biologically relevant parameters (gene transcription, protein

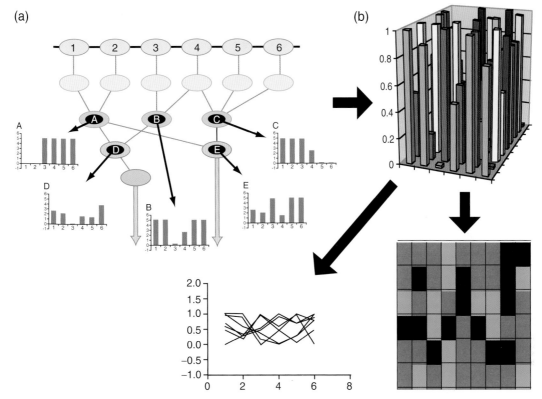

FIGURE 9.9 Study of integrated cellular pathways. (a) The activation of six extracellular targets is represented by histograms. Selective inhibition of various points along the pathways (by application of siRNA) yields characteristic patterns for the activity histograms. The letters on the sets of histograms refer to the effect of blocking the corresponding letter intersecting point of the pathway. (b) A collection of such histograms are combined into a 3D array. Multiple readouts of such arrays can be made either by showing lines for related sets of readings (for example, readings obtained for the same cellular context) or by coding the height of the various histograms with colors to form a heat map. These 2D representations become characteristic fingerprints for a given biological activity in the system.

production). For example, vascular endothelium cells in different contexts, defined by stimulation with different proinflammatory cytokines, are used to screen for drugs of possible use in inflammatory diseases such as rheumatoid arthritis (Figure 9.11). Cellular outputs can be enhanced by overexpression to constitutively active levels. For asthma (TH2-mediated inflammation), arthritis and autoimmune diseases (TH1-mediated disease), transplantation (T-cell driven), and cardiovascular disease-related (monocyte and endothelial cell driven) inflammatory responses, four complex cell systems can be utilized [44]. With this approach, the NF-kB signaling pathway, phosphatidylinositol 3-kinase (PI3K/Akt pathway), and RAS/mitogen-activated protein kinase (MAPK) pathways can be used to model proinflammatory activity. Measurement of surface proteins such as VCAM-1, ICAM-1, and E-selectin (vascular adhesion molecules for leukocytes); MIG/CXCL9 and IL-8/CXCL8 (chemokines that mediate selective leukocyte recruitment); platelet-endothelial cell adhesion molecule 1/CD31 (controls leukocyte transmigration); and HLA-DR (MHC class II; the protein responsible for antigen presentation) are then used to monitor drug effect. Figure 9.11 shows the components of the system.

Integrated systems are useful to differentiate intracellular targets such as kinase. The kinome is large, and the targeted ATP binding sites are very similar. In this regard, systems can show texture where there is none in isolated target assays. For example, general tyrosine kinase inhibitors with poor target specificity (such as AG126 and genisten); nonspecific JAK inhibitors ZM39923, WHI-P131, and AG490; and the nonselective 5-lipoxygenase inhibitors AA861 and NGDA are quite dissimilar when tested in an integrated system [44]. Systems are also useful in detecting off-target or secondary activities. For example, differences can be seen among Raf1 inhibitors BAY 43–9006, GW5074, and ZM336372 and among casein kinase inhibitors apigenin, DRB (5,6-dichloro-1-b-D-ribofuranosylbenzimidazole, and TBB 4,5,6,7-tetrabromo-2-aza-benzimidazole. The selective p38 MAPK inhibitors PD169316 and SB2033580 have similar potency for the primary target p38a. However, testing in an integrated system reveals significant differences between the two drugs consistent with newly detected inhibiton of P-selectin expression and strong inhibition of VCAM-1, E-selectin, and IL-8 for SB203580 (consistent with an off-target activity for this compound).

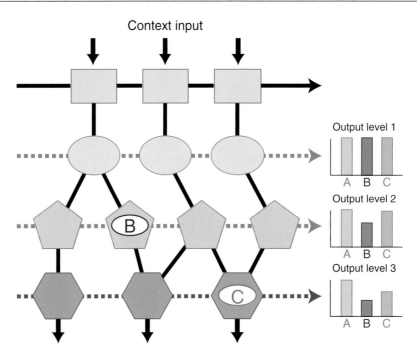

FIGURE 9.10 Levels of complexity for response readouts of cellular systems. Extracellular targets (light blue boxes) activate intracellular networks to produce biological response. Histograms show the activity of three hypothetical compounds (coded green, blue, and red). The green compound is inactive, the blue componud blocks an intracellular target (green pentangle labeled with oval marked B), and the red compound blocks another intracellular target (blue hexagram), labeled with oval marked C. If the response is read at a primary level of response (for example, levels of intracellular second messenger), the three compounds all appear to be inactive. Readings further down the cellular cascade detect one active compound (output level 2), and even further down detect the other active compound and differentiate the activity of the two active compounds (output level 3).

Systems can also reveal similarity in functional responses by mechanistically distinct drugs. For example, the activity of the mTOR antagonist rapamycin correlates with that of general PI-3 kinase inhibitors LY294002 and wortmannin. Similarly, nonsteroidal fungal estrogen receptor agonists zearalenone and β-zearalenol cluster activity with many p38 MAPK inhibitors. In fact, some striking mechanistic dissimilarities show like behavior in integrated systems. For example, phosphodiesterase IV inhibitors Ro-20–1724 and rolipram cluster with glucocorticoids dexamethasone, budesonide, and prednisolone. Both classes of drug have shown involvement in suppression of leukocyte function.

Studying established drugs in systems can yield new biological insights into mechanisms. For example, statins targeting HMG-CoA reductase for lipid lowering show anti-inflammatory effects (reduction in the leukocyte activation antigen CD69), activity shared by other HMG-CoA inhibitors. Subsequent studies have shown that the integrated activity is the consequence of HMG-Co-A inhibition and not an off-target activity. Interestingly, experiments in systems-based assays have shown different ranking of potency from isolated target potency. Specifically, the anti-inflammatory potency of statins in an integrated cellular system is cerivastatin ≫ atorvastatin ≫ simvastatin ≫ lovastatin ≫ rosuvastatin ≫ pravastatin. However, the most potent target-based HMg-acetyl coA (cholesterol-lowering) compounds are atorvastatin and rosuvastatin.

Clearly, as testing of candidate molecules progresses toward the clinical therapeutic end point the complexity increases. Thus, complications ensue along the journey from biochemical studies (isolated receptors, enzymes) through recombinant cellular systems to natural whole systems. The next level of complexity beyond these involve assays in context and in vivo systems (Figure 9.12). It should be noted that while the veracity of data to the true clinical profile for a molecule increases as the testing enters into these realms so too does the resource requirement and risk. For this reason, a paramount need in drug discovery is the collection of quality data capable of predicting failure in these expensive systems, as early as possible in the drug discovery process. It is worth discussing some unique applications of complex conditions in testing systems for drug screening and the determination of surrogate markers for prediction of successful therapeutic activity.

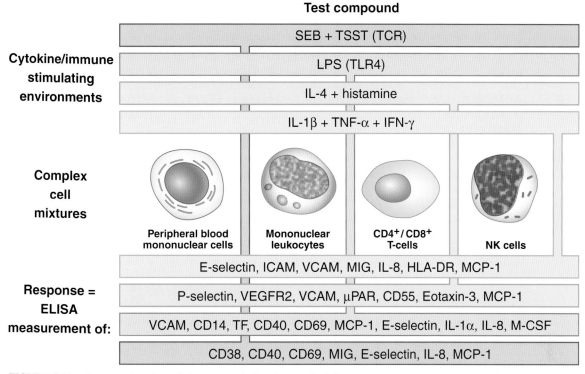

FIGURE 9.11 An example of a cellular system designed to study inflammatory processes related to asthma and arthritis. Multiple readouts (ELISA measurements) from each of four cell types are obtained under conditions of four contexts (mixture of stimulating agents). This results in a complex heat map of basal cellular activities that can be affected by compounds. The changes in the heat map (measured as ratios of basal to compound-altered activity) are analyzed statistically to yield associations and differences.

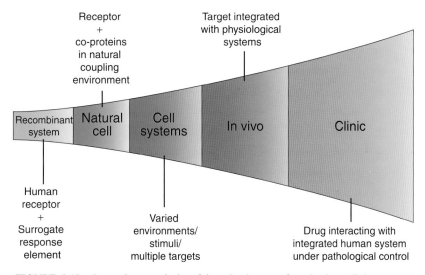

FIGURE 9.12 Increasing complexity of drug development from in vitro cellular systems to the clinic.

Cellular context refers to the physiological conditions present for the particular tissue of interest in a therapeutic environment. It can be important in determining the effects of drugs, and therefore in how drugs are screened and tested. For example, the signaling molecule TGF-β helps prevent malignant transformation of cells in breast epithelium. However, if the cells are already transformed TGF-β enhances blood vessel formation and tumor-cell invasiveness—thereby *promoting* tumor growth and dispersion [46]. Context can be especially important in vivo, and this may be critical to the therapeutic use of new drugs. Some context can be discerned with knockout animals. For example, the role of β-adrenoceptors, bradykinin B_2, prostanoid EP_2, and dopamine D_3 receptors in the control of blood pressure becomes evident only if physiological stress is applied (i.e., salt-loading or exercise). For these reasons, it is important that cell models mimic conditions in vivo and incorporate environmental effects and cell-cell interactions.

Through "context-dependent" biological effect, increased breadth of function can be detected. Additional discrimination (context-dependent activity) can be obtained by changing conditions. For example, as discussed previously, PDE-IV inhibitors and glucocorticoids cluster in leukocyte dependent systems. However, they can be differentiated in lipopolysaccharide systems under different cell stimulus. For drugs that produce effect by modifying signaling, context can be critical. For example, the phosphodiesterase inhibitor fenoximone produces positive cardiac inotropy and can be useful for congestive heart failure. The positive inotropic effects can be observed in vivo [47] in a working myocardium under hormonal and transmitter control (Figure 9.13a). However, in an isolated heart in vitro with no such neural tone fenoximone has no visible effect (Figure 9.13b). Fenoximone blocks the degradation of intracellular cyclic AMP. Therefore, increased inotropy is observed only under conditions where cyclic AMP is being produced by transmitter tone. These conditions can be simulated by adding a very low concentration of weak β-adrenoceptor agonist (in this case, prenalterol). Figure 9.13b shows the positive inotropic

effect to fenoximone observed in the presence of subthreshold levels of prenaloterol [48]. This defines a possible context for assays designed to potentiate cyclic AMP levels; namely, the presence of a subthreshold of β-adrenoceptor agonism. Similarly, adenosine receptors mediate renal vascular tone but mainly through the modification of the existing renal tone. Figure 9.14 shows the relative lack of effect of the adenosine agonist 2-chloroadenosine on vascular tone in a perfused kidney in vitro. In a different context—namely, subthreshold α-adrenoceptor vasoconstriction with methoxamine and vasodilation with forskolin (elevated cyclic AMP)—2-chloroadenosine vascular effects become evident (Figure 9.14). In this case, a context of physiological vascular tone increases the effect of the modifying adenosine agonism [49].

The interplay of levels of low-intrinsic-efficacy compounds with levels of physiological tone is very important. For example, the effects of β-adrenoceptor partial agonists/antagonists pirbuterol, prenalterol, and pindolol are quite different in conditions of high-basal and low-basal physiological tone (as altered by types of anesthesia, Figure 9.15) [50]. It can be seen that the partial agonist with the highest intrinsic efficacy (pirbuterol) produces elevated heart rate under conditions of low basal tone, and little effect on heart rate with anesthesia producing high basal tone (Figure 9.15a). Prenalterol has a lower intrinsic efficacy and produces less tachycardia under conditions of low tone and a slight bradycardia with high tone (Figure 9.15b). Finally, the very-low-intrinsic-efficacy β-adrenoceptor partial agonist pindolol produces very little tachycardia with high tone and profound bradycardia in conditions of high tone (Figure 9.15c). Such changes in the effects of drugs with low levels of intrinsic efficacy make prediction of

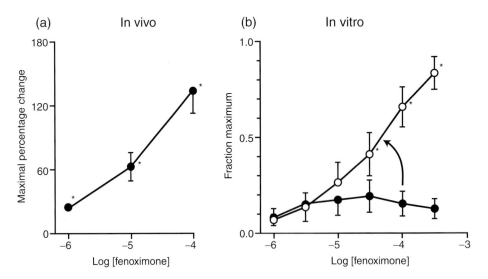

FIGURE 9.13 Cardiovascular responses to the PDE inhibitor fenoximone in different contexts. (a) In vivo effects of fenoximone in anesthetized dogs. Ordinates reflect positive inotropy. Redrawn from [47]. (b) In vitro effects of fenoximone in guinea pig untreated isolated left atria (filled circles) and atria in the presence of subthreshold β-adrenoceptor stimulation with prenalterol (open circles). Redrawn from [48].

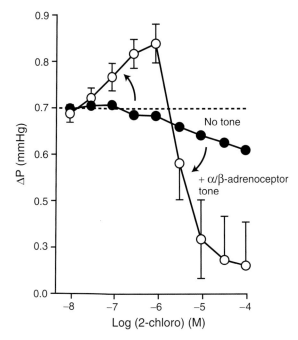

FIGURE 9.14 Effects of adenosine receptor agonist 2-chloro-adenosine on vascular perfusion pressure of isolated perfused rat kidneys. Minor effects seen in untreated kidneys (filled circles) and pronounced vasoconstriction while vasodilatation in kidneys co-perfused with subthreshold concentrations of α-adrenoceptor vasoconstrictor methoxamine and vasodilatatory activation of adenylyl cyclase with forskolin (open circles). Redrawn from [49].

therapeutic response in vivo difficult without data obtained in cellular context.

9.4 In Vivo Systems, Biomarkers, and Clinical Feedback

Pharmacological hypotheses are the most rigorously tested in all of biological science. A potential drug molecule must emerge through the entire drug discovery and development process and be tested in humans to give a desired therapeutic effect before the initial hypothesis beginning the process can be negated. In keeping with the notion that systems are more predictive of eventual therapeutic worth than isolated target assays, the next step in complexity are in vivo models of normal physiological function and disease (Figure 9.12). Historically, drug discovery was based on animal models and natural cell systems. On one hand, the differences in species (humans and animals) was a hurdle and a potential stopping point for the development of drugs for humans in such systems. On the other hand, it could be argued that testing was done in systems of proven physiology and pathology. The system was more like what the drug would encounter when it was used in the therapeutic environment. In vivo systems also allow observation of what a small drug molecule usually is designed to do; namely, perturb the diseased state to cause it to return to a normal state or at least to alleviate symptoms.

The relevant phenotype for complex multifaceted diseases such as obesity, atherosclerosis, heart failure, stroke, behavioral disorders, neurodegenerative diseases, and hypertension can only be observed in vivo. Historically, in vivo animal testing has led to the initiation of some classical treatments for disease. For example, the mode of action of the antihypertensive clonidine and subsequent elucidation of presynaptic α_2-adrenoceptors resulted from in vivo experimentation. Similarly, the demonstration of an orally active ACE inhibitor showing reduced blood pressure in spontaneously hypertensive rats led the emergence of captopril and other clinically active ACE inhibitors for hypertension. While investigation of drug effect is more complicated in vivo there are tools and techniques that can be used to better derive this information. Thus, protein-specific antibodies, gene knockouts and knock-ins, RNA interference, and imaging techniques can

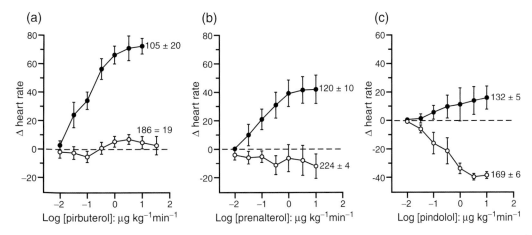

FIGURE 9.15 In vivo effects of β-adrenoceptor partial agonists of differing intrinsic efficacy. Changes in heart rate (increases in beats/min) shown in anesthetized cats. Chloralose/pentobarbital anesthesia (filled circles) yield low basal heart rates, while urethane/pentobarbital anaesthesia (open circles) yields high basal heart rates. Responses to (in order of descending relative intrinsic efficacy) (a) pirbuterol, (b) prenalterol, and (c) pindolol. Redrawn from [50].

provide rich information on in vivo processes and validation of pathways. In vivo experimentation can show integrated response from multiple sources, reveal unexpected results, determine therapeutic index (ratio between efficacious and toxic concentrations), help assess the importance of targets and processes identified in vitro, and assess pharmacokinetics and help predict clinical dosing. These obvious advantages come with the price tag of high resource requirement (Figure 9.12).

While the obvious value of in vivo animal models is clear, there also are instances—especially in cases of inflammatory arthritis, behavior, and tumor growth—where they have failed to be predictive of useful clinical activity in humans [51]. For example, leukotriene B_4 (LTB_4) antagonists showed activity in animal models of inflammatory arthritis yet failed to be useful in rheumatoid arthritis [52]. Similarly, dopamine D4 antagonists showed activity in animal behavior models previously predictive of dopamine D2 antagonists in schizophrenia. However, testing of dopamine D4 antagonists showed no efficacy in humans [53].

The ultimate in vivo model is humans in a controlled clinical environment, and there are considerable data to show that even complex models fail to predict clinical utility [40, 54]. Increasingly it is becoming evident that the complexities of disease states modify, cancel, and change target-based drug effects sometimes in unpredictable ways. Clinical data is extremely valuable in the assessment of both the drug in question and understanding of the relationship between the target and the disease state. Therefore, clinical feedback of these data is an essential part of the drug discovery process. The emerging field that relates to the use of clinical data in the drug discovery process is translational medicine. The metaphor used to describe the translational medicine process of information utilization from the clinic is that of a highway. The insights and information gained have led to the idea that whereas in the past the drug discovery process has been a one-way highway (from the bench to the clinic) it now needs to be a two-way highway where the learnings in the clinic should directly be applied to the criteria used early on in discovery. Furthermore, the lanes of this highway need to be expanded and much more information from the clinic needs to be regarded earlier.

The next question is therefore what tools are available to obtain such clinical data. Imaging techniques can be used to gain insight into drug activity in a noninvasive manner. Similarly, surrogate (from the Latin surrogare, "to substitute") end points are increasingly used, especially in cancer research where monitoring of effects such as cell cycle, mitotic spindle separation, apoptosis, angiogenesis, and tumor invasion are relevant to the assessment of clinical value. Thus, readings of tumor shrinkage and time-to-disease progression can be better predictors of long-term survival. Another increasingly valuable avenue of efficacy assessment is through biomarkers. These are especially useful in the treatment of diseases requiring long-term administration of drugs. The impact of drugs on cellular processes requires metabolite data predictive of subtle changes in molecular networks not accessible in target studies. In cancer, serum biochemical tumor markers can be useful predictors of outcome. Biomarkers are especially useful in cases where the precise mechanism of the drug is known. This can open the possibility of restricting clinical testing to those patients expressing the marker. In cancer patients, this includes HER2b overexpression I breast cancer (Herceptin), bcr-abl translocation in chronic myeloid leukemia (Gleevec), and expression of CD20 in non-Hodgkin's lymphoma (Rituximab) [55]. In general, a biomarker can be a physiological by-product (i.e., hypotension, platelet aggregation) or a biochemical substance (tumor markers). In this latter case, serum cholesterol or glycated hemoglobin can be a useful biomarker for statin therapy, control of diabetes, or antihypertensive treatment. A biomarker also can be a change in image (i.e., positron emission tomography). Thus, functional imaging can be used to visualize mitosis, apotosis, inflammation, structural changes in tumor regression, and blood flow. Immunohistology also can be used to furnish predictive markers of success of a given treatment.

9.5 Types of Therapeutically Active Ligands

In addition to diversity in biological targets, there is emerging diversity in the types of chemicals that can be used therapeutically to interact with these targets. Before the advent of widespread functional high-throughput screening (HTS), the majority of new therapeutic entities could be classed as full agonists, partial agonists, or antagonists. Since the screening mode used to discover these often was orthosterically based (i.e., displacement of a radioligand in binding), the resulting leads were usually correspondingly orthosteric. With HTS in functional mode, there is the potential to cast a wider screening net to include allosteric modulators. The changing paradigm of biologically active molecules found in HTS is shown in Figure 9.16. With the use of the cellular functional machinery in detecting biologically active molecules comes the potential to detect allosteric antagonists (modulators), where $\alpha < 1$ or potentiators ($\alpha > 1$). As discussed in Chapter 7, there are fundamental differences between orthosteric and allosteric ligands that result in different profiles of activity and different therapeutic capability (see Section 7.3). As more allosteric ligands are detected in functional HTS, the ligand-target validation issues may become more prominent. In general, the requirement of target presence in the system to demonstrate an effect is the first, and most important, criterion to be met. In cases where sensitivity of the effect to known target antagonists is not straightforward, demonstration of the target effect when the target is transfected into a range of host cells is a useful confirmation (see Figure 8.16).

Another variation on a theme for biological targets involves a concept known as polypharmacology (pharmacology involving, ligands with activity at more than one target within the same concentration range). The unique therapeutic profiles of such molecules rely on the interplay

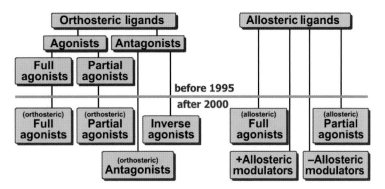

FIGURE 9.16 The use of new screening techniques employing functional assays promises a richer array of biologically active molecules that will not only mimic natural endogenous ligands for the targets but modify existing physiological activity.

of activities on multiple biological targets. There are increasing examples of clinically active drugs in psychiatry that have multiple target activities. For example, olanzapine (a useful neuroleptic) has highly unspecific antagonist activity at 10 different neurotransmitter receptors. Similarly, there is evidence that this may be an important aspect of kinase inhibitors in oncology. The unique value of the antiarrhythmic drug amiodarone is its activity on multiple cardiac ion channels [56].

Introducing multiple activities into molecules can be a means of maximizing possible therapeutic utility. Figure 9.17 shows the theoretical application for activity at two types of receptors; namely, α- and β-adrenoceptors. Depending on the dominant activities, molecules from a program designed to yield dual α- and β-adrenoceptor ligands could be directed toward a range of therapeutic applications. Chemical strategies for introducing multiple activities into a single molecule range from dimerization of structures known to possess the single activities to utilization of structures known to possess multiple activities. In the latter case, it is known that drug activity is seldom specific (i.e., the molecule possesses only one single activity at all concentration ranges) but rather is selective possessing a range of activities over a broad concentration range. Figure 9.18 shows the numerous activities found in the α_2-adrenoceptor antagonist yohimbine (Figure 9.18a) and antidepressant amitriptyline (Figure 9.18b). The linkage of known active chemical structures for multiple activity has been described as a strategy in Chapter 8 (see Section 8.2), but an even more obvious amalgam of structures (joined with a linker) can be used to target receptor homo- and heterodimers [57]. Dimeric ligands can show increased potency. For example, a dimer of the 5-HT_{1B} receptor sumatriptan used for the treatment of migraine shows a 100-fold increase in potency over monomeric sumatriptan [58]. Dimerization of ligands is a way to introduce mixtures of activity. One example of this is a dimeric linking of a δ-opioid antagonist (naltrindole) and κ_1-opioid agonist (ICI-199,441) to yield a molecule of

greater potency and mixed activity [59]. (See Figure 9.19.) Dimeric ligands need not be obvious amalgams of active structures. For example, in view of clinical data suggesting that a mixture of histamine and leukotriene antagonism was superior to either agent singly in asthma—as well as in view of the finding that the antihistamine cyproheptadine was a weak antagonist of LTD4—a molecule based on cyproheptadine and modified with features from the endogenous leukotriene agonist LTD4 yielded a molecule with better activity in asthma [60]. (See Figure 9.20a.) Dual activity has also been designed from knowledge of similar substrates. The treatment of hypertension with the angiotensin-converting enzyme (ACE) inhibitor captopril is established. The enzyme-neutral endopeptidase (NEP) is a metalloprotease that degrades atrial natriuretic factor, a peptide known to cause vasodilatation and oppose the action of angiotensin. These activities led to the postulate that a combined ACE/NEP inhibitor would be efficacious in hypertension, and one approach to this utilizes the notion that these two enzymes cleave similar dipeptide fragments. From this, a constrained anti-phenylalanine dipeptide mimetic (designed to mimic a low-energy conformation of the His-Leu portion of angiotensin bound to ACE) and the Phe-Leu portion of leu-enkephalin bound to NEP were used to produce a dual inhibitor of both ACE and NEP (Figure 9.20b). This formed the basis for the synthesis of a potent ACE/NEP inhibitor of nanomolar potency (Figure 9.20b).

One of the practical problems involved with ligands yielding polypharmacology is that their therapeutic profiles of action can often only effectively be tested in vivo. For example, debilitating concomitant tachycardia seen with beneficial increases in cardiac performance is a common finding for standard β-adrenoceptor agonist catecholamines such as isoproterenol (see Figure 9.21a). However, β-adrenoceptor agonist dobutamine produces much less tachycardia for the same increased cardiac performance. This interesting differentiation has been shown to be due to a low-level pressor effect of dobutamine

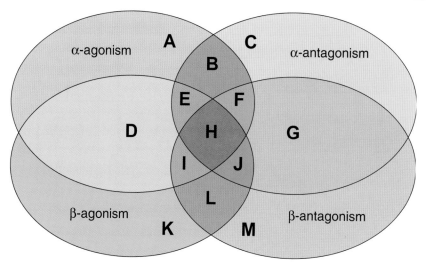

	α-Adrenoceptors	β-Adrenoceptors	Possible Indication
A	Full agonist		
B	Partial agonist		Shock, trauma
C	Antagonist		Hypertension
D	Full agonist	Full agonist	
E	Partial agonist	Full agonist	Acute cardiac decompensation
F	Partial agonist	Antagonist	Nasal decongestion, glaucoma, shock, cardiopulmonary resuscitation
G	Antagonist	Antagonist	Hypertension
H	Partial agonist	Partial agonist	Lipolysis
I	Full agonist	Partial agonist	
J	Antagonist	Partial agonist	Asthma, hypertension, congestive heart failure
K		Full agonist	Asthma
L		Partial agonist	Asthma
M		Antagonist	Hypertension, angina, glaucoma

FIGURE 9.17   Venn diagram consisting of the various possible activities (agonism and antagonism) on two receptor subtypes (α- and β-adrenoceptors). Letters label the areas of intersection denoting joint activity. The table shows possible therapeutic application of such joint activity.

(which opposes tachycardia through a reflex vagal stimulation) caused by a weak α-adrenoceptor agonism [61]. Blockade of α-adrenoceptors in vivo greatly reduces the difference between isoproterenol and dobutamine (see Figure 9.21b). This inotropic (over chronic selectivity) cannot be seen in isolated organs; only in the in vivo system. In this case, the whole animal is needed to detect the beneficial properties of dobutamine polypharmacology (α- + β-agonism).

Another type of therapeutically active molecule is one designed primarily with pharmacokinetics in mind (designed to be well absorbed and to enter the central compartment readily), which can then be converted to the therapeutically active molecule in the body. These are referred to as pro-drugs. This process, called latentiation, consists of the conversion of hydrophilic drugs into lipid-soluble drugs (usually by masking hydroxyl, carboxyl, and

primary amino groups). A concentrating effect can be achieved once the pro-drug enters a compartment and the active moiety is released and trapped by enzymatic hydrolysis. This can be a useful strategy for drug therapy in the central nervous system, which is protected by the blood-brain barrier (an obstacle relatively impervious to polar molecules). For example, a lipid-soluble diacetyl derivative of morphine crosses the blood-brain barrier at a rate 100 times faster than morphine. Once in the brain, pre-capillary pseudocholinesterase deacylates the molecule to morphine (Figure 9.22a). Similarly, the delivery of gamma-aminobutyric acid (GABA) into the CNS for treatment of depression, anxiety, Alzheimer's disease, parkinsonism, and schizophrenia is difficult due to the presence of the blood-brain barrier. However, the Schiff-base progamide crosses into the CNS to release gabamide and then GABA (Figure 9.22b). A particularly effective

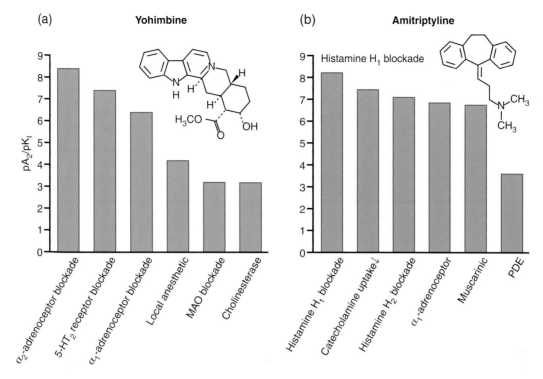

FIGURE 9.18 Multiple receptor effects (ordinates denote pK values for antagonism or receptor occupancy) of (a) yohimbine and (b) amitriptyline.

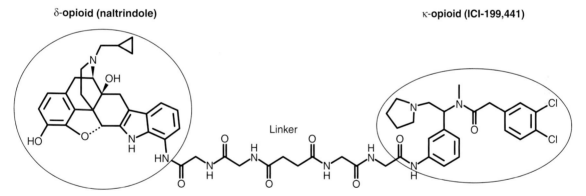

FIGURE 9.19 Dimeric antagonist formed by oligoglycyl-based linkage of two opioid receptor subtype antagonists naltrindole and ICI-199,441. From [59].

pro-drug strategy is the use of the dual ester dipivalyl epinephrine for the treatment of glaucoma. Epinephrine reduces intraocular pressure and is an effective treatment for the disease. However, it does not readily penetrate the cornea (it is unstable and short acting). Dipivalyl epinephrine easily penetrates the cornea, and active epinephrine is released in the eye through enzymatic hydrolysis, making the pro-drug 17 times more potent than the parent by the ocular route. Since epinephrine itself is metabolically unstable, it degrades before reaching the general circulation (thereby eliminating side effects

(Figure 9.23)). The use of the pro-drug optimally produces a maximally effective concentration of the active drug in the eye, the target organ.

9.6 Summary and Conclusions

- There is evidence to suggest that while more drugs are being discovered there is no commensurate increase in the number of novel treatments for disease.
- A major approach to discovery is the target-based approach, whereby a single biological target is

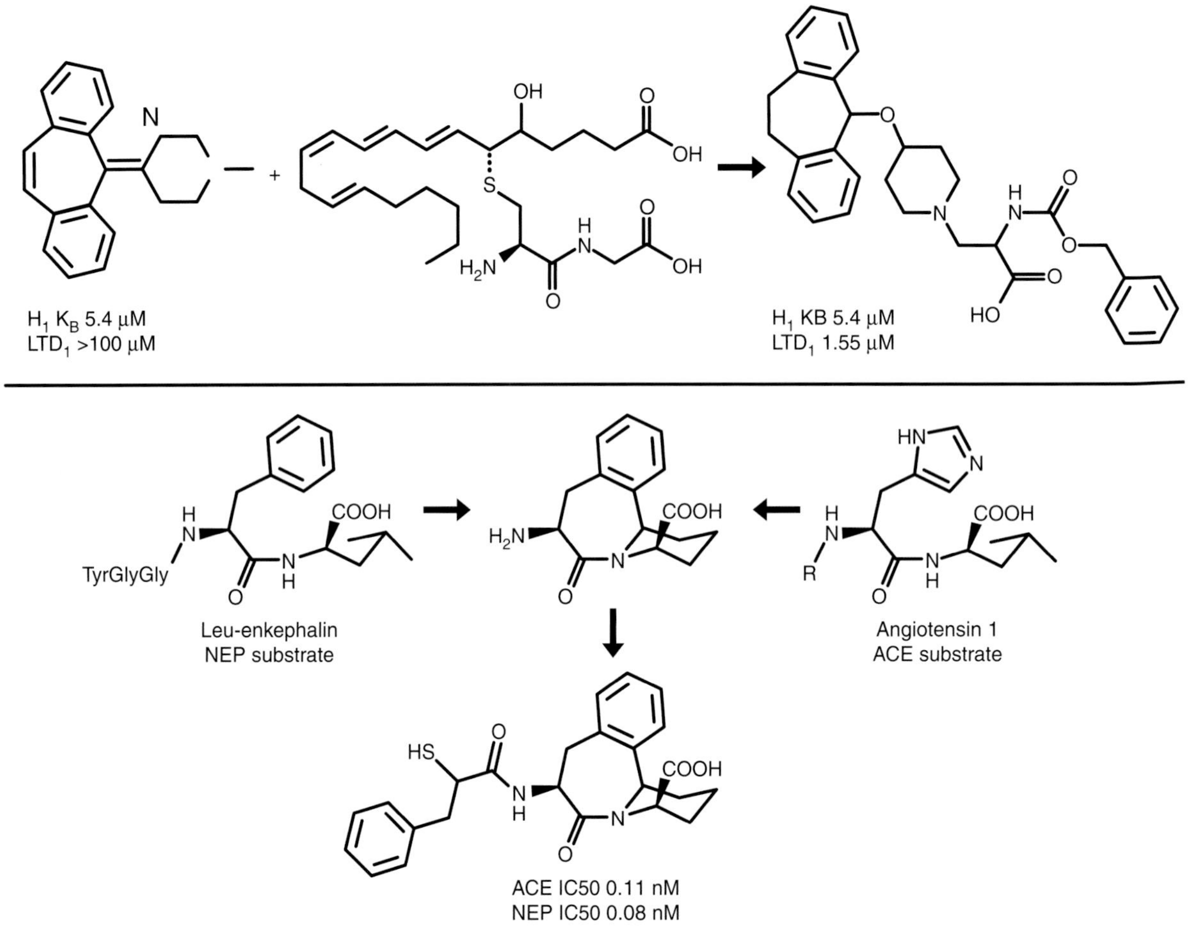

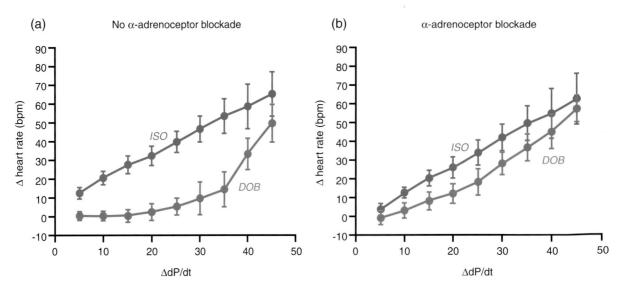

FIGURE 9.20 Design of multiple ligand activity. (a) Dual histamine H1 receptor and leukotriene receptor antagonist incorporating known antihistaminic properties of cyproheptadine and LTD$_4$. (b) Joint ACE/NEP inhibitor formed from incorporating similarities in substrate structures for both enzymes. From [57].

FIGURE 9.21 Changes in heart rate (ordinates) for agonist-induced changes in cardiac inotropy (changes in rate of ventricular pressure) in anesthetized cats. Responses shown to isoproterenol (filled circles) and dobutamine (open circles). (a) Response in normal cats shows inotropic selectivity (less tachycardia for given changes in inotropy) for dobutamine over isoproterenol. (b) The inotropic selectivity of dobutamine is reduced by previous α-adrenoceptor blockade by phentolamine. From [61].

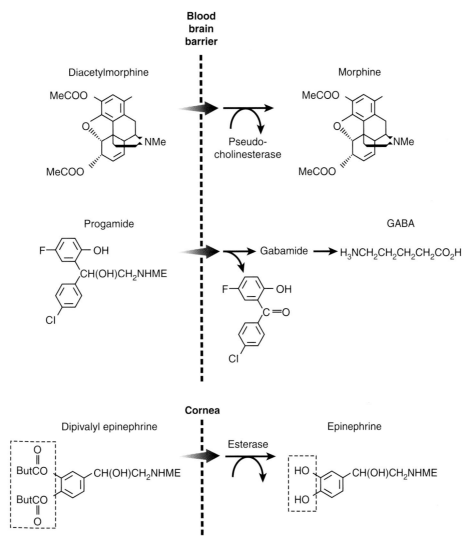

FIGURE 9.22 Latentiation of morphine and gamma amino-butyric acitd (GABA) allows entry through the blood-brain barrier and subsequent trapping by enzymatic hydrolysis. Diacetylmorphine is converted to morphine by pseudocholinesterase, while progamide is converted to gabamide and subsequently to the active drug GABA.

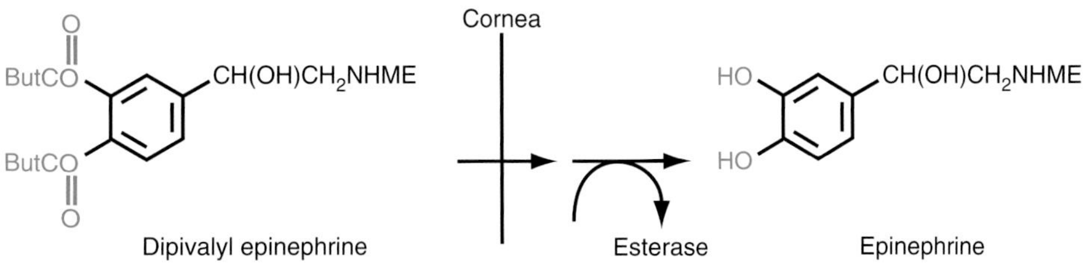

FIGURE 9.23 The pro-drug dipivalyl epinephrine enters the cornea of the eye to allow esterase to produce epinephrine in the eye to alleviate high pressure in glaucoma.

identified (and validated) as a primary cause of disease. Ligands that produce a defined action at the target (i.e., agonism, antagonism) are therefore expected to alleviate the disease in the therapeutic situation.

- Recombinant systems are the main tools of target-based approaches. These can be manipulated, but information is lacking for complete modeling of therapeutic systems.

- Biological targets may consist of single-entity proteins, complexes of receptors (dimers), or receptors plus accessory proteins. Mixtures of gene products can produce unique phenotypic biological targets.
- An alternative approach involves testing of new drug entities on whole-cell systems and measuring effects on integrated cellular pathways. Favorable phenotypic responses are identified with this approach that may better produce alteration of multicomponent disease processes.
- An added complexity, but one that may better predict therapeutic activity, is the testing of drugs in assays with different contexts (i.e., basal stimulation).
- Testing in vivo can further produce therapeutic model systems. Certain multicomponent disease conditions can only be adequately modeled in vivo.
- The ultimate model is the human in the clinical situation. Translational medicine with noninvasive imaging techniques and biomarkers is now able to furnish valuable information that can be used in the initial discovery process to produce better defined drugs.
- As well as complex biological targets, complex chemical targets (drugs with multiple activity, prodrugs) can be used to produce therapeutically useful phenotypic responses.

References

1. Booth, B., and Zemmel, R. (2004). Prospects for productivity. *Nature Rev. Drug Disc.* **3:**451–456.
2. Williams, M. (2004). A return to the fundamentals of drug discovery. *Curr. Opin. Investigational Drugs* **5:**29–33.
3. Walker, M. J. A., Barrett, T., and Guppy, L. J. (2004). Functional pharmacology: The drug discovery bottleneck? *Drug Disc. Today* **3:**208–215.
4. Spedding, M., Jay, T., Cost de Silva, J., and Perret, L. (2005). A pathophysiological paradigm for the therapy of psychiatric disease. *Nature Rev. Drug Disc.* **4:**467–476.
5. Hopkins, A. L., and Groom, C. R. (2002). The druggable genome. *Nature Rev. Drug Disc.* **1:**727–730.
6. Claverie, J.-M. (2001). What if there were only 30,000 human genes? *Science* **291:**1255–1257.
7. Drews, J. (2000). Drug discovery: A historical perspective. *Science* **287:**1960–1964.
8. Luster, A. D. (1998). Mechanisms of disease: Chemokines, chemotactic cytokines that mediate inflammation. *N. Engl. J. Med.* **338:**436–445.
9. Zaitseva, M., Blauvelt, A., Lee, S., Lapham, C. K., Klaus-Kovtun, V., Mostowski, H., Manischewitz, J., and Golding, H. (1997). Expression and function of CCR5 and CXCR4 on human langerhans cells and macrophages: Implications for HIV primary infection. *Nature Medicine* **3:**1369–1375.
10. Cagnon, L., and Rossi, J. J. (2000). Downregulation of the CCR5 beta-chemokine receptor and inhibition of HIV-1 infection by stable VA1-ribozyme chimeric transcripts. *Antisen. Nuc. A. Drug Dev.* **10:**251–261.
11. Baba, M., Nishimura, O., Kanzaki, N., Okamoto, M., Sawada, H., Iizawa, Y., Shiraishi, M., Aramaki, Y., Okonogi, K., Ogawa, Y., Meguro, K., and Fujino, M.

(1999). A small-molecule, nonpeptide CCR5 antagonist with highly potent and selective anti-HIV-1 activity. *Proc. Natl. Acad Sci. USA* **96:**5698–5703.
12. Cocchi, F., De Vico, A. L., Garzino-Demo, A., Arya, S. K., Gallo, R. C., and Lusso, P. (1995). Identification of RANTES, MIP-1, and MIP-1 as the major HIV-suppressive factors produced by CD8+ T cells. *Science* **270:**1811–1815.
13. Finke, P. E., Oates, B., Mills, S. G., MacCoss, M., Malkowitz, L., Springer, M. S., Gould, S. L., DeMartino, J. A., Carella, A., Carver, G., et al. (2001). Antagonists of the human CCR5 receptor as anti-HIV-1 agents. Part 4: synthesis and structure-activity relationships for 1-[N-(Methyl)-N-(phenylsulfonyl) amino]-2-(phenyl)-4-(4-(N-(alkyl)-N-(benzyloxycarbonyl)amino) piperidin-1-yl) butanes. *Bioorg. Med. Chem. Lett.* **11:**2475–2479.
14. Mack, M., Luckow, B., Nelson, P. J., Cihak, J., Simmons, G., Clapham, P. R., Signoret, N., Marsh, M., Stangassinger, M., Borlat, F., Wells, T. N. C., Schlondorff, D., and Proudfoot, A. E. I. (1998). Aminooxypentane-RANTES induces CCR5 internalization but inhibits recycling: A novel inhibitory mechanism of HIV infectivity. *J. Exp. Med.* **187:**1215–1224.
15. Simmons, G., Clapham, P. R., Picard, L., Offord, R. E., Rosenkilde, M. M., Schwartz, T. W., Buser, R., Wells, T. N. C., and Proudfoot, A. E. I. (1997). Potent inhibition of HIV-1 infectivity in macrophages and lymphocytes by a novel CCR5 antagonist. *Science* **276:**276–279.
16. Garzino-Demo, A., Moss, R. B., Margolick, J. B., Cleghorn, F., Sill, A., Blattner, W. A., Cocchi, F., Carlo, D. J., DeVico, A. L., and Gallo, R. C. (1999). Spontaneous and antigen-induced production of HIV-inhibitory chemokines are associated with AIDS-free status. *Proc. Natl. Acad. Sci. USA* **96:**11986–11991.
17. Ullum, H., Lepri, A. C., Victor, J., Aladdin, H., Phillips, A. N., Gerstoft, J., Skinhoj, P., and Pedersen, B. K. (1998). Production of beta-chemokines in human immunodeficiency virus (HIV) infection: Evidence that high levels of macrophage in inflammatory protein-1-beta are associated with a decreased risk of HIV progression. *J. Infect. Dis.* **177:**331–336.
18. Grivel, J.-C., Ito, Y., Faga, G., Santoro, F., Shaheen, F., Malnati, M. S., Fitzgerald, W., Lusso, P., and Margolis, L. (2001). Suppression of CCR5- but not CXCR4-tropic HIV-1 in lymphoid tissue by human herpesvirus 6. *Nature Medicine* **7:**1232.
19. Dean, M., Carrington, M., Winkler, C., Huttley, G. A., Smith, M. W., Allikmets, R., Goedert, J. J., Buchbinder, S. P., Vittinghoff, E., Gomperts, E., Donfield, S., Vlahov, D., Kaslow, R., Saah, A., Rinaldo, C., and Detels, R. (1996). Genetic restriction of HIV-1 infection and progression to AIDS by a deletion allele of the *CKR5* structural gene. *Science* **273:**1856–1862.
20. Huang, Y., Paxton, W. A., Wolinsky, S. M., Neumann, A. U., Zhang, L., He, T., Kang, S., Ceradini, D., Jin, Z., Yazdanbakhsh, K., Kunstman, K., Erickson, D., Dragon, E., Landau, N. R., Phair, J., Ho, D. D., and Koup, R. A. (1996). The role of a mutant CCR5 allele in HIV-1 transmission and disease progression. *Nature Medicine* **2:**1240–1243.
21. Liu, R., Paxton, W. A., Choe, S., Ceradini, D., Martin, S. R., Horuk, R., MacDonald, M. I., Stuhlmann, H., Koup, R. A., and Landau, N. R. (1996). Homozygous defect in HIV-1 coreceptor accounts for resistance of some multiply-exposed individuals to HIV-1 infection. *Cell* **86:**367–377.
22. Paxton, W. A., Martin, S. R., Tse, D., O'Brien, T. R., Skurnick, J., VanDevanter, N. L., Padian, N., Braun, J. F.,

Kotler, D. P., Wolinsky, S. M., and Koup, R. A. (1996). Relative resistance to HIV-1 infection of CD4 lymphocytes from persons who remain uninfected despite multiple high-risk sexual exposures. *Nature Medicine* 2:412–417.

23. Samson, M., Libert, F., Doranz, B. J., Rucker, J., Liesnard, C., Farber, C. M., Saragosti, S., Lapoumerouilie, C., Cognaux, J., Forceille, C., Muyldermans, G., Verhofstede, C., Collman, R. G., Doms, R. W., Vassart, G., and Parmentier, M. (1996). Resistance to HIV-1 infection in Caucasian individuals bearing mutant alleles to the CCR-5 chemokine receptor gene. *Nature* 382:722–725.

24. Cook, D. N., Beck, M. A., Coffman, T. M., Kirby, S. L., Sheridan, J. F., Pragnell, I. B., and Smithies, O. (1995). Requirement of MIP-1α for an inflammatory response to viral infection. *Science* 269:1583–1585.

25. Knudsen, T. B., Kristiansen, T. B., Katsenstein, T. L., and Eugen-Olsen, J. (2001). Adverse effect of the CCR5 promoter -2459A allele on HIV-1 disease progression. *J. Med. Virol.* 65:441.

26. Neubig, R. R., and Siderovski, D. P. (2002). Regulators of G-protein signaling as new central nervous system drug targets. *Nature Rev. Drug Disc.* 1:187–196.

27. Mestas, J., and Hughes, C. C. (2004). Of mice and not men: Differences between mouse and human immunology. *J. Immunol.* 172:2731–2738.

28. Hill, R. (2000). NK1 (substance P) receptor antagonists: Why are they not analgesic in humans? *Trends Pharmacol. Sci.* 21:244–246.

29. Strizki, J. M., Xu, S., Wagner, N. E., Wojcik, L., Liu, J., Hou, Y., Endres, M., Palani, A., Shapiro, S., Clader, J. W., et al. (2001). SCH-C (SCH 351125), an orally bioavailable, small molecule antagonist of the chemokine receptor CCR5, is a potent inhibitor of HIV-1 infection in vitro and in vivo. *Proc. Natl. Acad Sci. USA* 98:12718–12723.

30. Brodde, O.-E., and Leineweber, K. (2005). β₂-adrenoceptor gene polymorphisms. *Pharmacogenet. Genom.* 15:267–275.

31. Kost, T. A., and Condreay, J. P. (2002). Recombinant baculoviruses as mammalian cell gene-delivery vectors. *Trends Biotechnol.* 20:173–180.

32. Douglas, S. A., Ohlstein, E. H., and Johns, D. G. (2004). Techniques: Cardiovascular pharmacology and drug discovery in the 21st century. *Trends Pharmacol. Sci.* 25:225–233.

33. Heldin, C. H. (1995). Dimerization of cell surface receptors in signal transduction. *Cell* 80:213–223.

34. George, S. R., O'Dowd, B. F., and Lee, S. P. (2002). G-protein-coupled receptor oligomerization and its potential for drug discovery. *Nature Rev. Drug Disc.* 1:808–820.

35. Abd'Alla, S., Lother, H., and Quitterer, U. (2000). At1-receptor heterodimers show enhanced G-protein activation and altered receptor sequestration. *Nature* 407:94–98.

36. AbdAlla, S., Lother, H., el Massiery, A., and Quitterer, U. (2001). Increased AT(1) receptor dimers in preeclampsia mediate enhanced angiotensin II responsiveness. *Nature Med.* 7:1003–1009.

37. Mellado, M., Rodríguez-Frade, J. M., Vila-Coro, A. J., Fernández, S., Martín de Ana, A., Jones, D. R., Torán, J. L., and Martínez-Aet, C. (2001). Chemokine receptor homo- or heterodimerization activates distinct signaling pathways. *EMBO J.* 20:2497–2507.

38. Wildoer, M., Fong, J., Jones, R. M., Lunzer, M. M., Sharma, S. K., Kostensis, E., Portoghese, P. S., and Whistler, J. L. (2005). A heterodimer-selective agonist shows *in vivo* relevance of G-protein coupled receptor dimmers. *Proc Natl. Acad. Sci. USA* 102:9050–9055.

39. Milligan, G., Ramsay, D., Pascal, G., and Carrillo, J. J. (2003). GPCR dimerization. *Life Sci.* 74:181–188.

40. Kubinyi, H. (2003). Drug research: Myths, hype, and reality. *Nature Rev. Drug Disc.* 2:665–668.

41. Armour, S. L., Foord, S., Kenakin, T., and Chen, W.-J. (1999). Pharmacological characterization of receptor-activity-modifying proteins (RAMPs) and the human calcitonin receptor. *J. Pharmacol. Toxicol. Meth.* 42:217–224.

42. Foord, S. M., and Marshall, F. H. (1999). RAMPS: Accessory proteins for seven transmembrane domain receptors. *Trends Pharmacol. Sci.* 20:184–187.

43. Fraser, N. J., Wise, A., Brown, J., McLatchie, L. M., Main, M. J., and Foord, S. M. (1999). The amino terminus of receptor activity modifying proteins is a critical determinant of glycosylation state and ligand binding of calcitonin-like receptor. *Mol. Pharmacol.* 55:1054–1059.

44. Kunkel, E. J., Dea, M., Ebens, A., Hytopoulos, E., Melrose, J., Nguyen, D., Ota, K. S., Plavec, I., Wang, Y., Watson, S. R., Butcher, E. C., and Berg, E. L. (2004). An integrative biology approach for analysis of drug action in models of human vascular inflammation. *FASEB J.* 18:1279–1301.

45. Kunkel, E. J., Plavec, I., Nguyen, D., Melrose, J., Rosler, E. S., Kao, L. T., Wang, Y., Hytopoulos, E., Bishop, A. C., Bateman, R., et al. (2004). Rapid structure-activity and selectivity analysis of kinase inhibitors by BioMap analysis in complex human primary cell-based models. *ASSAY Drug Dev. Technol.* 2:431–441.

46. Siegel, P. M., and Massague, J. (2003). Cytostatic and apoptotic actions of TGF-β in homeostasis and cancer. *Nature Rev. Cancer* 3:807–821.

47. Dage, R. C., Roebel, L. E., Hsieh, C. P., Weiner, D. L., and Woodward, J. K. (1982). The effecs of MDL 17,043 on cardiac inotropy in the anaesthetized dog. *J. Cardiovasc. Pharmacol.* 4:500–512.

48. Kenakin, T. P., and Scott, D. L. (1987). A method to assess concomitant cardiac phosphodiesterase inhibition and positive inotropy. *J. Cardiovasc. Pharmacol.* 10:658–666.

49. Kenakin, T. P., and Pike, N. B. (1987). An in vivo analysis of purine-mediated renal vasoconstriction in rat isolated kidney. *Br. J. Pharmacol.* 90:373–381.

50. Kenakin, T. P. (1985). Drug and organ selectivity: Similarities and differences. In: *Advances in drug research*, Vol. 15, edited by B. Test, pp. 71–109. Academic Press, New York.

51. Littman, B. H., and Williams, S. A. (2005). The ultimate model organism: Progress in experimental medicine. *Nature Rev. Drug Disc.* 4:631–638.

52. Polmar, S., Diaz-Gonzalez, F., Dougados, M., Ortiz, P., and del-Miguel, G. (2004). Limited clinical efficacy of a leukotriene B4 receptor (LTB₄) antagonist in patients with active rheumatoid arthritis (RA). *Arthritis Rheum.* 50:S239.

53. Tarazi, F. I., Zhang, K., and Baldessarini, R. J. (2004). Review: Dopamine D4 receptors—beyond schizophrenia. *J. Recept. Sig. Transduct. Res.* 24:131–147.

54. Milne, G. M. (2003). Pharmaceutical productivity—the imperative for new paradigms. *Annu. Rep. Med. Chem.* 38:383–396.

55. Sikora, K. (2002). Surrogate endpoints in cancer drug development. *Drug Disc. Today* 7:951–956.

56. Baczko, I., El-Reyani, N. E., Farkas, A., Virág, L., Iost, N., Leprán, I., Mátyus, P., Varró, A., and Papp, J. G. (2000). Antiarrhythmic and electrophysiological effects of GYK-16638, a novel N-(phenoxyalkyl)-N-phenylalkylamine, in rabbits. *Eur. J. Pharmacol.* 404:181–190.

57. Morphy, R., and Rankovic, Z. (2005). Designed multiple ligands: An emerging drug discovery paradigm. *J. Med. Chem.* **48:**6523–6543.

58. Perez, M., Pauwels, P. J., Fourrier, C., Chopin, P., Valentin, J.-P., John, G. W., Marien, M., and Halazy, S. (1998). Dimerization of sumatriptan as an efficienct way to design a potent, centrally and orally active 5-HT1B agonist. *Bioorg. Med. Chem. Lett.* **8:**675–680.

59. Daniels, D. J., Kulkarni, A., Xie, Z., and Bhushan, R. G. (2005). A bivalent ligand (KDAN-18) containing δ-antagonist and κ-agonist pharmacophores bridges $δ_2$

and $κ_1$ opioid receptor phenotypes. *J. Med. Chem.* **48:**1713–1716.

60. Zhang, M., van de Stolpe, A., Zuiderveld, O., and Timmermans, H. (1997). Combined antagonism of leukotrienes and histamine produces predominant inhibition of allergen-induced early and late phase airway obstruction in asthmatics. *Eur. J. Med. Chem.* **32:**95–102.

61. Kenakin, T. P., and Johnson, S. F. (1985). The importance of α-adrenoceptor agonist activity of dobutamine to inotropic selectivity in the anaesthetized cat. *Eur. J. Pharmacol.* **111:**347–354.

10

"Hit" to Drug: Lead Optimization

It's all a game...sometimes you're cool...sometimes you're lame...
— **GEORGE HARRISON (1943–2001)**
Success is the ability to go from one failure to another with no loss of enthusiasm.
— **SIR WINSTON CHURCHILL (1874–1965)**

10.1 Tracking SARs and Determining Mechanism of Action: Data-driven Drug-based Pharmacology

There are pharmacological tools and techniques designed to determine system-independent measures of the potency and efficacy of drugs. However, to apply them effectively the molecular mechanism of the drug must be known beforehand. In new drug discovery, this is seldom the case, and in fact the observed profile of the molecules must be used to discern the molecular mechanism. In this setting, it is not always possible to apply the correct technique or model for quantification of drug activity and the tool chosen for analysis is based on initial observation of drug activity (i.e., the process is data driven). In practical terms, a wide range of potential drug behaviors can be described by a limited number of molecular models, and it is useful to describe these and their application in the drug discovery process. In general, drugs can be divided into two general initial types: those that do and those that do not initiate pharmacological response in the preparation. As a preface to specific discussion of the use of data-driven analyses, it is useful to consider the application of surrogate parameters.

Ideally, pharmacological data should directly be fit to specific models and parameters derived from the direct fit. However, there are cases where the specific models predict surrogate parameters that can be derived without fitting data to the specific model. This can be an advantage. For example, the equation for simple competitive antagonism of receptors (see Section 6.3) is

$$\text{Response} = \frac{([A]^n \tau^n) E_{max}}{[A]^n \tau^n + ([A] + K_A(1 + [B]/K_B))^n}, \quad (10.1)$$

where n is a fitting parameter for the slopes of the concentration response curves, E_{max} is the maximal response capability of the system, [A] and [B] the agonist and antagonist (respectively), τ the efficacy of agonist, and K_A and K_B the respective equilibrium dissociation constants of the agonist and antagonist receptor complexes. It will be seen that fitting sets of concentration-response curves in the absence ([B] = 0) and presence of a range of concentrations of antagonist can yield a value of K_B. However, this requires fitting to five parameters, some of which (for example, K_A) cannot be independently estimated without separate experiments. Alternatively, it is known that equiactive dose ratios (DRs) from parallel concentration-response curves shifted to the right by the antagonist can be used in Schild analysis. Therefore, DR values can be used as surrogates for the analysis of antagonism without need to fit to the explicit model. Under these circumstances, the data can be fit to a generic sigmoidal curve of the form

$$\text{Response} = \text{Basal} + \frac{\text{Max} - \text{Basal}}{1 + 10^{(\text{LogEC}_{50} - \text{Log}[A])^n}} \quad (10.2)$$

and the shift in EC_{50} values used to calculate DR estimates for Schild analysis (see Section 6.3.1). There are certain instances in data-driven pharmacological analysis where it is useful to use such surrogate parameters.

10.2 Drug Initiation of Response: Agonism

The first observable effect of a drug in a biological preparation is the initiation of some pharmacological effect (referred to as response). If this is seen, then it must be determined that it is specific for the biological target of interest (i.e., not a general nonspecific stimulation of the cell) and that a concentration-response relationship can be determined. Once activity for a given molecule has been confirmed by retest at a single concentration, a dose-response curve for the effect must be determined. The biological effect must be related to the concentration in a predictive manner. Figure 10.1 shows some possible outcomes of determining a possible dose-response curve for an activity determined at one concentration. It can be

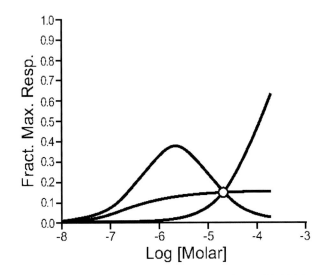

FIGURE 10.1 Possible dose-response curves that could yield the ordinate value shown at a concentration of 20 µM compound.

seen that not all outcomes represent true or useful dose-response activity.

A frequently asked question at this point is, does the array of responses for given concentrations represent a true dose-response relationship or just random noise around a given mean value? It is useful to demonstrate approaches to this question with an example. Assume that a compound is tested in dose-response mode and 11 "responses" are obtained for 11 concentrations of compound giving a maximal ordinal response of 7.45%. On one hand, it might not be expected that noise could present a sigmoid pattern indicative of a concentration-response curve (although such patterns might be associated with location on plates or counters). However, a maximal ordinate response of 7.45% is also extremely low. A useful rule of thumb is to set the criterion of $>3\sigma$ (where σ is the standard error of the mean) of basal noise responses as the definition of a real effect. In this case, the signal from 1,325 wells (for the experiment run that same day; historical data should not be used) obtained in the presence of the lowest concentration of compound (10 pM, assumed to be equivalent to basal response) yielded a mean percentage response of -0.151%, with a standard deviation of 1.86%. Under these circumstances, $3\sigma = 5.58\%$. With this criterion, the response to the agonist would qualify as a signal above noise levels.

A pharmacological method to determine if a very low level of response constitutes a real dose-response curve is to use a maximal concentration of the "very weak partial agonist" to block responses to a standard full agonist. The basis for this method is the premise that the EC_{50} of a weak partial agonist closely approximates its affinity for the receptor (see Chapter 5). For example, assume that a fit to the data points shows a partial agonist to have a maximal response value of 8% and EC_{50} of 3 µM. Under these circumstances, the dose-response curve to the standard agonist would be shifted tenfold to the right by 30 µM of

the weak partial agonist (Figure 10.2). This could indicate that the 8% represents a true response to the compound. Also, it could furnish a lead antagonist series for the screening program. However, this method requires considerable follow-up work for each compound.

Another method of detecting a dose-response relationship is to fit the data to various models for dose-response curves. This method statistically determines whether or not a dose-response model (such as a logistic function) fits the data points more accurately than simply the mean of the values. This method is described fully in Chapter 11. The most simple model would be to assume no dose-response relationship and to calculate the mean of the ordinate data as the response for each concentration of ligand (horizontal straight line parallel to the abscissal axis). A more complex model would be to fit the data to a sigmoidal dose-response function (Equation 10.2). A sum of squares can be calculated for the simple model (response = mean of all response) and then for a fit of the data set refit to the four-parameter logistic shown previously (Equation 10.2). A value for the F statistic is then calculated, which determines whether there is a statistical basis for assuming there is a dose-response relationship. An example of this procedure is given in the next chapter (see Figure 11.13). The remainder of this discussion assumes that it has been determined that the drug in question produces a selective pharmacological response in a biological preparation that can be defined by a concentration-response curve (i.e., it is an agonist).

The next step is to compare the maximal response to the agonist to the maximal response capability of the biological preparation. If there is no statistical difference between the maximal response of the agonist to the maximal response of the tissue, then the drug is a full agonist. If the magnitude of the maximal response to the agonist is lower than that of the tissue, then the drug is a partial agonist. There is separate information that can be gained from either of these two categories of agonist.

10.2.1 Analysis of Full Agonism

As discussed previously, the location parameter of a dose-response curve (potency) of a full agonist is a complex amalgam of the affinity and efficacy of the agonist for the receptor and the ability of the system to process receptor stimulus and return tissue response. This latter complication can be circumvented by comparing the agonists in the same functional receptor system (null methods). Under these circumstances, the receptor density and efficiency of receptor coupling effects cancel each other since they are common for all of the agonists. The resulting relative potency ratios of the full agonists (provided the concentrations are taken at the same response level for each agonist) are system independent measures of the molecular properties of the agonists; namely, their affinity and efficacy for the receptor. This is shown, in terms of both classical receptor theory and the operational model in Section 10.6.1. Such potency ratios for full agonists are sometimes referred to as EMRs (equimolar potency ratios) or EPMRs

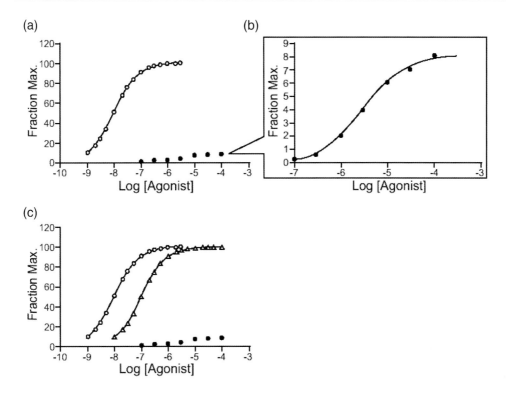

FIGURE 10.2 Dose-response curves for agonists of very low intrinsic activities. (a) A set of responses for a range of concentrations of an unknown molecule is observed. In comparison to a full agonist for the assay, the maximal ordinate value is low (8% of maximal response). (b) An expanded scale shows that the response pattern follows a sigmoidal shape consistent with a true weak agonism for the receptor. (c) Addition of 30 μM of the unknown compound would be predicted to cause a tenfold shift to the right of the agonist dose-response curve if the weak activity truly reflects partial agonist of the unknown at the receptor with an EC_{50} of 3 μM.

(equipotent molar potency ratios) and are a standard method of comparing full agonists across different systems.

There are two major prerequisites for the use of this tool in SARs determination. The first is that the agonists must truly all be full agonists. If one is a partial agonist, then the system independence of the potency ratio measurement is lost. This is because of the different effects that variation in receptor density, efficiency of coupling, and measurement variation have on the location parameters of dose-response curves to partial versus full agonists. For example, Figure 10.3 shows the effect of an increase in receptor number on a high-efficacy agonist ($\tau = 500$) and low-efficacy agonist ($\tau = 5$). It can be seen from this figure that the curve for the high-efficacy agonist shifts to the left directly across the concentration axis, whereas the curve for the lower-efficacy agonist rises upward along the ordinal axis with little concomitant displacement along the concentration axis (i.e., the potency of the full agonist changes, whereas the potency of the partial agonist does not). This is because potency is dependent on efficacy and affinity to different extents for full and partial agonists. Therefore, it is inconsistent to track SARs changes for full and partial agonists with the same tool; in this case, potency ratios.

Figure 10.4 shows the potency ratios of two agonists as a function of the receptor density of the receptor systems in which they are measured (expressed as the operational τ value for the lower-efficacy agonist). Also shown is the maximal response of the higher-efficacy agonist. Of note in this figure is the fact that the potency ratios are not constant until the maximal responses of the two agonists are equal (i.e., until they are both full agonists, region II of the figure). When one of the agonists is a partial agonist, the potency ratio for the two agonists varies with system parameters; in this case, receptor density. Since SARs must be conducted with system-independent measures of drug activity, potency ratios of full and partial agonists are not useful in this region.

The other prerequisite for the use of potency ratios for agonist SARs is that the ratio be independent of the level of response at which it is measured. Figure 10.5 shows dose-response curves to two full agonists. It can be seen that a rigorous fit to the data points results in two curves that are not parallel. Under these circumstances, the potency ratio of these agonists varies depending on which level of response the ratio is measured (see Figure 10.5a). In this situation, the measure of drug activity is system dependent and not useful for SARs. However, the nonparallelism of

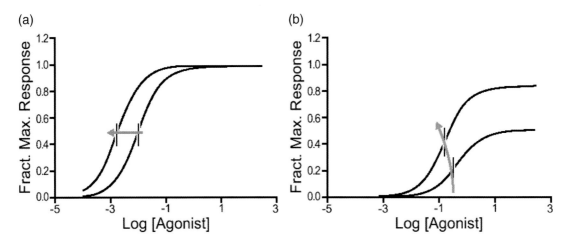

FIGURE 10.3 Comparative potencies of two agonists in two receptor systems containing the same receptor at different receptor densities. (a) Relative potency in system with high receptor density ($\tau_1 = 500$, $\tau_2 = 100$). The potency ratio $= 5$. (b) Dose-response curves for same two agonists in receptor system with $1/100$ the receptor density. Potency ratio $= 1.3$.

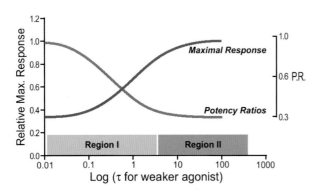

FIGURE 10.4 Changes in the relative maximal response and potency ratio of two agonists tested in a series of preparations with a wide range of sensitivities (due to either differences in receptor densities or efficiencies of receptor coupling or both). Abscissae: Logarithms of the τ value in the system of the weaker agonist. The efficacy of the more efficacious agonist is 3x that of the weaker agonist. Left ordinates: Potency ratios as measured by the ratio of the concentrations of each required to produce 50% of the maximal response to that agonist and relative maxima of the two agonists. Right ordinates: Potency ratios of the agonists. Region I: Both agonists are partial agonists. Region II: Both agonists are full agonists.

data cannot be described by parallel curves (see Chapter 11 for a detailed description of the application of this test). Therefore, the potency ratio can be derived from parallel curves with the result that system-independent data for SARs can be generated.

10.2.2 Error and Agonist Dose-Response Curves

In the course of SARs determination, dose-response curves are replicated to confirm activity. It is important to note that different regions of these curves, depending on the level of agonist activity, will be more subject to random error than others. Specifically, the maximal responses of weak partial agonists will be more sensitive (than that to full agonists) to random variation than their potency (EC_{50} value). (See Figure 10.6a.) The potency of full agonists will be more prone to random variation than the maxima (see Figure 10.6b). Figure 10.6 also shows the effect of random noise on τ values (variation in levels of receptor density and/or efficiency of receptor coupling) on the maxima of agonists (Figure 10.6c) and potency of agonists (Figure 10.6d) as a function of τ. It can be seen from this figure that variation in τ has different effects on intrinsic activity and potency throughout the range of possible τ values. In the SARs process, this should be noted and expected as agonists are modified to produce higher activity.

10.2.3 Quantifying Full Agonist Potency Ratios

The scheme for comparing two full agonists according to the operational model is shown in Figure 10.7. In this case, a surrogate reading of EPMR values from curves fit to a generic sigmoidal function (i.e., Equation 10.2) yields a

these curves may be the result of random variation in response measurement and not a true reflection of the agonist activity. A statistical test can be done to determine whether these curves are from a single population of curves with the same slope (i.e., if the data can be described by parallel curves with the result that the potency ratio will not be system dependent). Application of this test to the curves shown in Figure 10.5a yields the parallel curves shown in Figure 10.5b. In this case, there is no statistical reason the

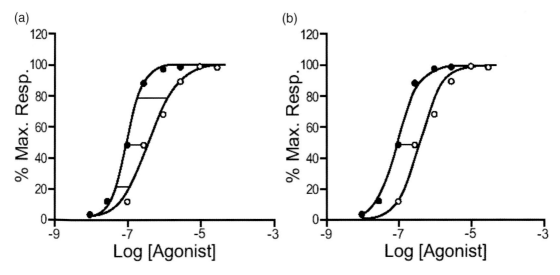

FIGURE 10.5 Full agonist potency ratios. (a) Data fit to individual three-parameter logistic functions. Potency ratios are not independent of level of response. At 20%, PR = 2.4; at 50%, PR = 4.1; and at 80%, PR = 6.9. (b) Curves refit to logistic with common maximum asymptote and slope. PR = 4.1. The fit to common slope and maximum is not statistically significant from individual fit.

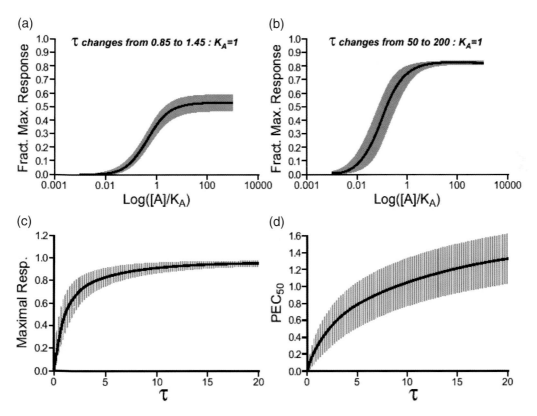

FIGURE 10.6 Effect of 100% variation in values on the maximal response to an agonist (a) and the pEC50 of the agonist (b). Panel c shows the variation (as standard errors) of the maximal response as a function of the efficacy of agonism and ability of the system to produce response (t values for agonists). It can be seen that the magnitude of the error is not uniform as a function of τ but rather is highest at partial agonism (maximal response < 1). Panel d shows the errors associated with the pEC50 as a function of agonism. It can be seen that these increase as the agonist tends toward full agonism.

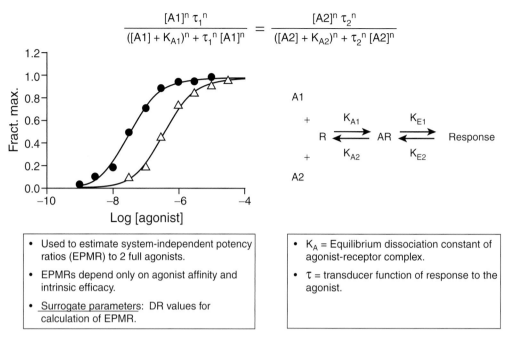

$$\frac{[A1]^n \tau_1^n}{([A1] + K_{A1})^n + \tau_1^n [A1]^n} = \frac{[A2]^n \tau_2^n}{([A2] + K_{A2})^n + \tau_2^n [A2]^n}$$

- Used to estimate system-independent potency ratios (EPMR) to 2 full agonists.
- EPMRs depend only on agonist affinity and intrinsic efficacy.
- <u>Surrogate parameters</u>: DR values for calculation of EPMR.

- K_A = Equilibrium dissociation constant of agonist-receptor complex.
- τ = transducer function of response to the agonist.

FIGURE 10.7 Figure illustrating the comparison of concentration-response curves to two full agonists. Equations describe response in terms of the operational model (variable slope version equation; see Section 10.6.1). Schematic indicates the interacting species; in this case, two full agonists A1 and A2 activating a common receptor R to produce response. Boxes show the relevant measurements (EPMRs) and definitions of the parameters of the model used in the equation.

useful parameter dependent only on the molecular properties of the full agonists (see Section 10.6.1):

$$EPMR = \frac{K_A(1 + \tau')}{K_A'(1 + \tau)}. \qquad (10.3)$$

For full agonists, $\tau, \tau' \gg 1$, allowing the estimate $EC_{50} = K_A/\tau$. Substituting $\tau = [R_t]/K_E$, the potency ratio of two full agonists is

$$EPMR = \frac{EC_{50}}{EC_{50}'} = \frac{K_A \cdot K_E}{K_A' \cdot K_E'}, \qquad (10.4)$$

where K_E is the Micahelis-Menten constant for the activation of the cell by the agonist-bound active receptor complex (a parameter unique to the agonist).

10.2.4 Analysis of Partial Agonism

If the agonist does not produce the full system maximal response, then it is a partial agonist and more information can be gained about its molecular properties. Specifically, the location parameter of the partial agonist concentration-response curve (EC_{50}) is a relatively close estimate of the affinity (K_A) while changes in maximal response are good indicators of changes in efficacy. The scheme for fitting concentration response curves to a full and partial agonist (or to two partial agonists) is shown in Figure 10.8. Unlike

the analysis for full agonists, certain experimentally derived starting points for the fit are evident for partial agonist. For example, E_{max} can be determined separately with full agonists and the K_A for the partial agonist can be approximated by the EC_{50}. Figure 10.9 shows the analysis of the full agonist isoproterenol and partial agonist prenalterol. It can be seen that once the relative efficacy values are determined in one tissue the ratio is predictive in other tissues as well. This advantage can be extrapolated to the situation whereby the relative efficacy and affinity of agonists can be determined in a test system and the activity of the agonist then predicted in the therapeutic system.

10.2.5 Affinity-dependent versus Efficacy-dependent Agonist Potency

In the early stages of lead optimization, agonism is usually detectable but at a relatively low level (i.e., the lead probably will be a partial agonist). Partial agonists are the optimal molecule for pharmacological characterization. This is because there are assays that can estimate the two system-independent properties of drugs; namely, affinity and efficacy (for partial agonists). Under these circumstances, medicinal chemists have two scales of biological activity that they can use for lead optimization. As discussed in Chapter 5, the EC_{50} of a partial agonist is a reasonable approximation of its affinity (see Section 5.9.1). The observed EC_{50} for weak agonists in SARs studies can

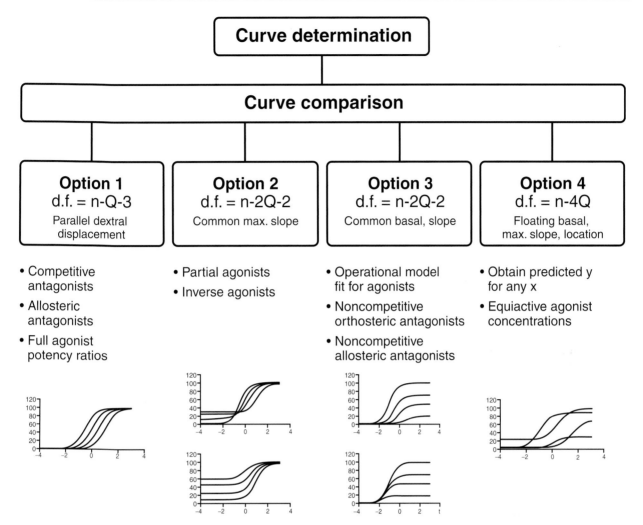

FIGURE 10.13 Data-driven analysis of concentration-response data. Once it is determined that the data points represent concentration-response curves, then comparison of a set of curves is initiated. Generally four characteristic sets of behavior are encountered. The most statistically simple (option 1) utilizes common basal, maximum, and slope values varying only the location parameters. For multiple agonists, this allows calculation of potency ratios. For antagonism, this allows Schild analysis for orthosteric competitive antagonists and allosteric modulators. Option 2 utilizes a common maximum and slope with varying location and basal. This is used for analysis of inverse or partial agonists. Option 3 describes insurmountable antagonism (common basal and slope with varying location and maxima) for orthosteric or allosteric antagonists. Option 4 simply fits the data to separate four parameter logistic functions to enable calculation of equiactive concentrations.

hypothetical data set fit to the orthosteric model in Figure 10.24a and the allosteric model in Figure 10.24b. The circled data points were changed very slightly to cause an F-test to prefer either model for each respective model, illustrating the fallacy of relying on computer fitting of data and statistical tests to determine molecular mechanism. As discussed in Chapter 7, what is required to delineate orthosteric versus allosteric mechanism is the conscious testing of predictions of each mechanism through experiment. Thus, the blockade of a range of agonists through a large range of antagonist concentrations should be carried out to detect possible saturation of effect and probe dependence (see Section 7.6 for further discussion).

10.4 "Short-form" Measures of Antagonism for Following Structure-activity Relationship

The foregoing discussion centers on the in-depth analysis of antagonism through comparison of full curve data to models. While this is important in late-stage drug discovery to determine molecular mechanism of action, it is impractical during the lead optimization phase of drug discovery where potentially large numbers of molecules need to be assessed for antagonist potency. Shorter and less labor-intensive procedures for determining antagonist potency can be used for this purpose.

There are two observable phases to the process of antagonism. The first is characterized by a threshold

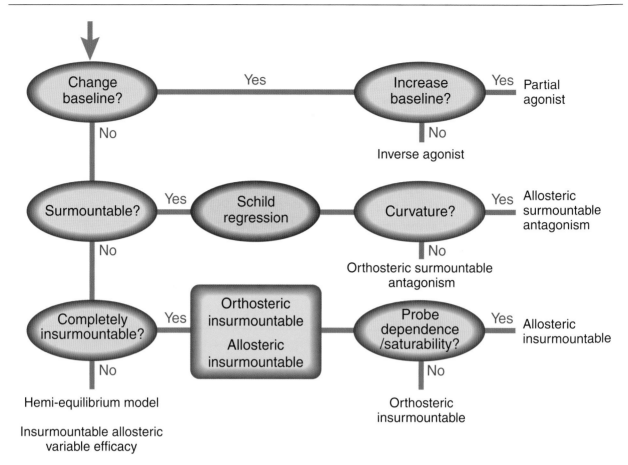

FIGURE 10.14 Schematic diagram of steps involved in analyzing pharmacological antagonism. Key questions to be answered are in blue, beginning with assessments of changes in baseline followed by assessment of whether or not the antagonism is surmountable, followed by assessment of possible probe dependence and/or saturability.

concentration of antagonist that begins to inhibit agonist-induced response. The second is characterized by the pattern of antagonism of agonist response produced by increasing concentrations of antagonist. The first phase can be used to quantify the potency of the antagonist for the receptor. The second can be used to discern the molecular mechanism of the antagonism. In the drug discovery process, often the mechanism is secondary to determining the structure-activity relationships (SARs) involved in optimizing antagonist potency. Defining the threshold for antagonism is sufficient to guide SARs until potency is optimized and the detailed mechanism of action becomes relevant. The threshold for antagonism can readily be measured by determining the concentration dependence of the antagonism of the response to a fixed concentration of agonist (much like a displacement binding curve). The means to do this is through the use of antagonist inhibition curves.

Antagonists can produce varying combinations of dextral displacement and depression of maxima of agonist dose-response curves. The concentration-related effect of an antagonist on the system response to a single concentration of agonist constitutes what will be referred to as an inhibition curve. One of the most straightforward examples

to illustrate this is with a simple competitive antagonist. Figure 10.25a shows the characteristic dextral displacement with no diminution of maxima effects of a simple competitive antagonist on agonist dose-response curves. The filled circles indicate the effect of the antagonist blocking the effects of a concentration of agonist producing a 50% maximal response (EC_{50}). The regression on this response on the concentration of antagonist producing the effect is shown in Figure 10.25b. This is the inhibition curve for that antagonist in this system at the EC_{50} level. The characteristic parameter for this curve is the IC_{50}; namely, the concentration of antagonist that produces 50% antagonism of the agonist response. It should be noted that the location of this inhibition curve along the antagonist concentration axis is determined by the concentration of agonist used for the initial response. For example, if the concentration of agonist producing 80% response is used (EC_{80}; see open circles in Figure 10.25a) then the inhibition curve is shifted to the right (see Figure 10.25b). That is, the more powerful the agonist stimulus the more antagonist is needed to block the effect.

The important relationship, from the point of view of drug development, is the one between the actual molecular

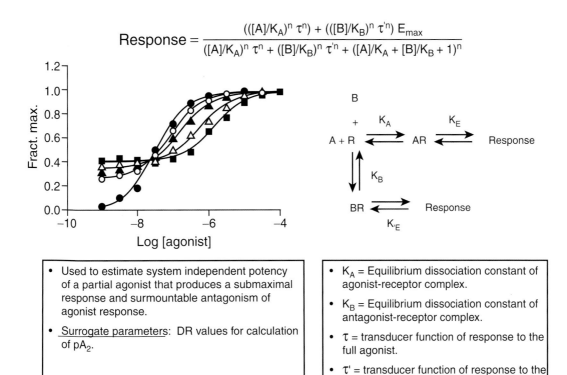

$$\text{Response} = \frac{(([A]/K_A)^n\ \tau^n) + (([B]/K_B)^n\ \tau'^n)\ E_{max}}{([A]/K_A)^n\ \tau^n + ([B]/K_B)^n\ \tau'^n + ([A]/K_A + [B]/K_B + 1)^n}$$

- Used to estimate system independent potency of a partial agonist that produces a submaximal response and surmountable antagonism of agonist response.

- <u>Surrogate parameters</u>: DR values for calculation of pA_2.

- K_A = Equilibrium dissociation constant of agonist-receptor complex.

- K_B = Equilibrium dissociation constant of antagonist-receptor complex.

- τ = transducer function of response to the full agonist.

- τ' = transducer function of response to the partial agonist.

FIGURE 10.15 Figure illustrating the effects of a partial agonist on concentration-response curves to a full agonist. Equations describe response in terms of the operational model (variable slope version equation derived in Section 10.6.2). Schematic indicates the interacting species, in this case, a full agonist A and partial agonist B activating a common receptor R to produce response. Boxes show the relevant measurements and definitions of the parameters of the model used in the equation.

affinity of the competitive antagonist (K_B) and the observed location of the inhibition curve (IC_{50}). As seen in Figure 10.25c, the greater the concentration of agonist that must be antagonized the greater amount of antagonist is needed and this is reflected in an increased value for the IC_{50}. It can be shown that the relationship between the IC_{50} and the K_B in functional experiments is given by Leff and Dougall [3] and derived in Section 10.6.9:

$$K_B = IC_{50}/\big((2 + ([A]/EC_{50})^n)^{1/n} - 1\big), \qquad (10.5)$$

where the concentration of agonist is [A], the concentration of agonist producing 50% maximal response is EC_{50}, and n is the Hill coefficient of the agonist dose-response curve. From Equation 10.5 it can be seen that the K_B, which is a system-independent estimate of antagonist potency, can be made from an estimate of the IC_{50} that is corrected for the level of agonism. However, this is true only for a competitive antagonist. According to Equation 10.5, if the IC_{50} for the blockade of a series of increasing concentrations of agonist is linear this is presumptive evidence that the antagonism is of the simple competitive type. An example of the measurement of antagonist pIC_{50} values through antagonist inhibition curves is given in Section 12.2.12.

10.4.1 Erroneous Application of Correction Factors to Noncompetitive Antagonists

A drawback of relying solely on inhibition curves to determine antagonist potency is the fact that the mechanism of action of the antagonist cannot be discerned from the curve, and this in turn may affect whether or not the correction described by Equation 10.5 should be applied to the IC_{50}. Specifically, if the antagonism is noncompetitive and if there is little receptor reserve for the agonist, then the antagonist will depress the dose-response curve to the agonist (as shown in Figure 10.26a). Under these circumstances, the IC_{50} is not affected by the magnitude of the concentration of agonist used to invoke the initial response and the $IC_{50} = KB$ (see Section 4.6.2). The most simple case of noncompetitive blockade is to assume that the antagonist precludes agonist activation of the receptor with no concomitant effect on agonist affinity. Under these circumstances, the IC_{50} does not depend on the agonist concentration and is equal to the K_B (see Figure 10.26b).

It can be seen that if it is assumed that the antagonism is competitive and a correction is applied according to Equation 10.5, then an error in the calculated K_B, equal to the factor $((2 + ([A]/EC_{50})^n)^{1/n} - 1)$, will be applied to the estimate.

$$\text{Response} = \frac{((\alpha L[A]/K_A\tau)^n + (\beta L[B]/K_B\tau)^n + (L\tau)^n) \, E_{max}}{(\alpha L[A]/K_A\tau)^n + (\beta L[B]/K_B\tau)^n + (L\tau)^n + ([A]/K_A(1+\alpha L) + [B]/K_B(1+\beta L) + L+1)^n}$$

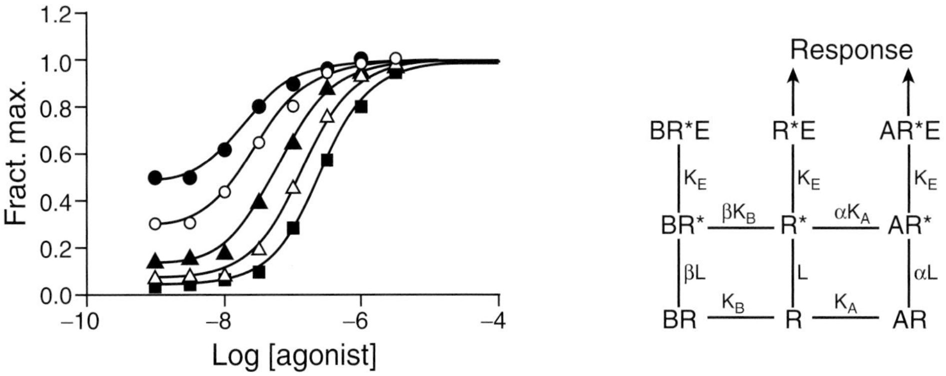

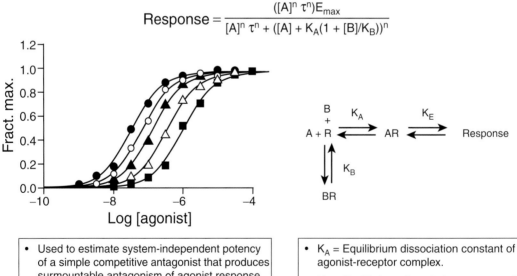

- Used to estimate system independent potency of an inverse agonist in a constitutively active receptor system.
- <u>Surrogate parameters</u>: DR values for calculation of pA_2.

- K_A = Equilibrium dissociation constant of agonist-receptor complex.
- K_B = Equilibrium dissociation constant of antagonist-receptor complex.
- τ = transducer function for response to the full agonist and constitutively active receptor state.
- α = ratio of affinities of agonist for active vs. inactive receptor state.
- β = ratio of affinities of inverse agonist for active vs. inactive receptor state.

FIGURE 10.16 Figure illustrating the effects of an inverse agonist on concentration-response curves to a full agonist. Equations describe response in terms of the operational model (variable slope version equation derived in Section 10.6.3). Schematic indicates the interacting species; in this case, a full agonist A and inverse agonist B activating a common receptor R to produce response. Boxes show the relevant measurements and definitions of the parameters of the model used in the equation.

$$\text{Response} = \frac{([A]^n \, \tau^n) E_{max}}{[A]^n \tau^n + ([A] + K_A(1 + [B]/K_B))^n}$$

- Used to estimate system-independent potency of a simple competitive antagonist that produces surmountable antagonism of agonist response.
- <u>Surrogate parameters</u>: DR values for Schild Analysis.

- K_A = Equilibrium dissociation constant of agonist-receptor complex.
- K_B = Equilibrium dissociation constant of antagonist-receptor complex.
- τ = transducer function of response to the agonist.

FIGURE 10.17 Figure illustrating the effects of an orthosteric competitive antagonist on concentrations-response curves to a full agonist. Equations describe response in terms of the operational model. Schematic indicates the interacting species; in this case, a full agonist A activating the receptor and an antagonist B competing for the receptor but producing no response. Boxes show the relevant measurements and definitions of the parameters of the model used in the equation.

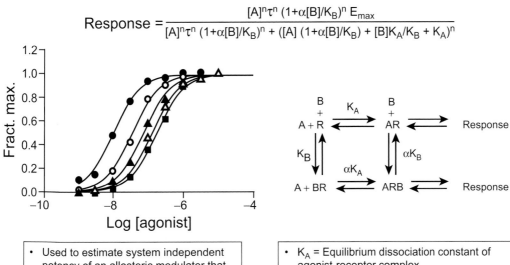

$$\text{Response} = \frac{[A]^n \tau^n (1+\alpha[B]/K_B)^n E_{max}}{[A]^n \tau^n (1+\alpha[B]/K_B)^n + ([A] (1+\alpha[B]/K_B) + [B]K_A/K_B + K_A)^n}$$

- Used to estimate system independent potency of an allosteric modulator that affects the affinity of the agonist for the receptor but does not affect receptor signaling capability.

- Surrogate parameters: DR values for Schild analysis.

- K_A = Equilibrium dissociation constant of agonist-receptor complex.

- K_B = Equilibrium dissociation constant of antagonist-receptor complex.

- τ = transducer function for response to the full agonist and constitutively active receptor state.

- α = ratio of affinities of each ligand when the other ligand is bound to the receptor (cooperativity constant).

FIGURE 10.18 Figure illustrating the effects of an allosteric modulator that alters the affinity of the receptor for agonists but does not interfere with agonist activation of the receptor on concentration-response curves to a full agonist. Equations describe response in terms of the operational model (variable slope version equation derived in Section 10.6.4); Schematic indicates the interacting species (in this case, a full agonist A and allosteric modulator B) that can bind to the receptor simultaneously (species ARB). The affinity of the receptor for each ligand is altered by a factor (α) when one of the ligands is already bound.

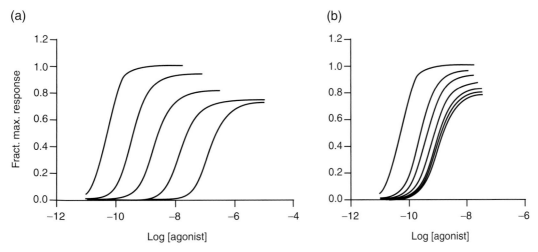

FIGURE 10.19 Patterns of insurmountable blockade of receptors under conditions of (a) hemi-equilibria and (b) allosteric modulation by a modulator that only partially reduces receptor signaling. (a) Concentration-response curves to the full agonist are shifted to the right in a concentration-dependent manner. The maximal response is partially depressed and may attain a plateau level. (b) Curves are shifted to a limiting value characteristic of saturable allosteric modulation. In addition, the maximal response is depressed to a new asymptote. Note that the maximal response is not blocked to basal levels indicative of $\xi > 0$ (see Section 7.8.2 and Equation 7.3).

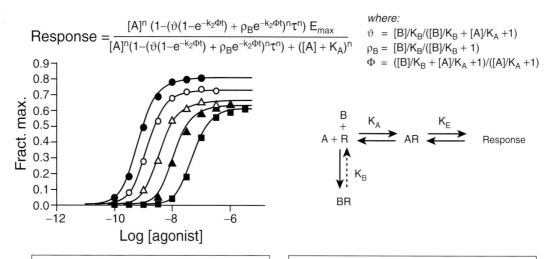

$$\text{Response} = \frac{[A]^n (1-(\vartheta(1-e^{-k_2\Phi t}) + \rho_B e^{-k_2\Phi t})^n \tau^n) E_{max}}{[A]^n(1-(\vartheta(1-e^{-k_2\Phi t}) + \rho_B e^{-k_2\Phi t})^n\tau^n) + ([A] + K_A)^n}$$

where:
$$\vartheta = [B]/K_B/([B]/K_B + [A]/K_A + 1)$$
$$\rho_B = [B]/K_B/([B]/K_B + 1)$$
$$\Phi = ([B]/K_B + [A]/K_A + 1)/([A]/K_A + 1)$$

- Used to estimate system-independent potency of an orthosteric antagonist with a slow rate of offset. Under these conditions, a portion of the receptor population is irreversibly inactivated leading to a depression of agonist maximal response.
- Surrogate parameters: DR values for calculation of pA_2.

- K_A = Equilibrium dissociation constant of agonist-receptor complex.
- K_B = Equilibrium dissociation constant of antagonist-receptor complex.
- τ = transducer function for response to the full agonist and constitutively active receptor state.
- k_2 = rate of offset ($msec^{-1}$) of antagonist from the receptor.

FIGURE 10.20 Figure illustrating the effects of an orthosteric slow-offset antagonist on concentration-response curves to a full agonist in a system demonstrating hemi-equilibrium conditions (see Section 6.5). Equations describe response in terms of the operational model (variable slope version equation derived in Section 10.6.5). Schematic indicates the interacting species; in this case, a full agonist A activating the receptor and an antagonist B competing for the receptor but producing no response. In this case, the rate of offset of the antagonist (k_2) once bound is very slow compared to the time available to measure an equilibrium response to the agonist (note dotted line in schematic). Boxes show the relevant measurements and definitions of the parameters of the model used in the equation.

10.4.2 Noncompetitive Antagonism in Systems with Receptor Reserve

Immediate depression of the maximal responses will be observed for inhibition of responses to partial agonists. However, if the agonist has high efficacy and/or if the receptor system is very efficiently coupled (or the system has a high receptor density) there will be a receptor reserve for the agonist (i.e., the maximal response will be obtained with submaximal receptor activation). Under these circumstances, noncompetitive blockade of receptors will not necessarily depress the maximal response. Maximal responses will be depressed only after the proportion of receptors blocked exceeds the magnitude of the reserve for the agonist. Therefore, a noncompetitive antagonist will produce an initial shift to the right of the agonist dose-response curve with no diminution of maximal response. Higher doses of the antagonist will then produce depression of the maximal response (see Figure 10.27a). The initial dextral displacement of the dose-response curves yields a direct relationship between the IC_{50} needed for antagonism and the K_B according to Equation 10.5 (an apparent linear Cheng-Prusoff relationship). However, once depression of

the maxima occurs, then the IC_{50} ceases to shift to the right and it becomes independent of the concentration of agonist used to produce the initial response. Thus, the inhibition curves for a noncompetitive antagonist in a system with receptor reserve resemble an amalgam of those for competitive (i.e., Figure 10.25c) and noncompetitive antagonism (i.e., Figure 10.26b) (as shown in Figure 10.27b).

10.4.3 Inhibition Curves for Allosteric Modulators

As discussed in Chapter 7, antagonists may bind to a separate loci on the receptor and thereby allosterically modify the affinity of the receptor for the agonist. The maximal change in the agonist affinity is denoted by a term α. It is useful to distinguish allosteric effects in terms of whether the modulator affects signaling and affinity or just affinity. In the latter situation, the modulator produces parallel shifts to the right of the concentration response curve up to a maximal point. This is discussed in Section 7.4.1 ($\xi = 1$). Under these circumstances, the curve for

$$\text{Response} = \frac{[A]^n \tau^n (1 + (\alpha\xi[B]/K_B)^n) E_{max}}{[A]^n \tau^n (1 + (\alpha\xi[B]/K_B)^n) + ([A](1 + \alpha[B]/K_B) + K_A[B]/K_B + K_A)^n}$$

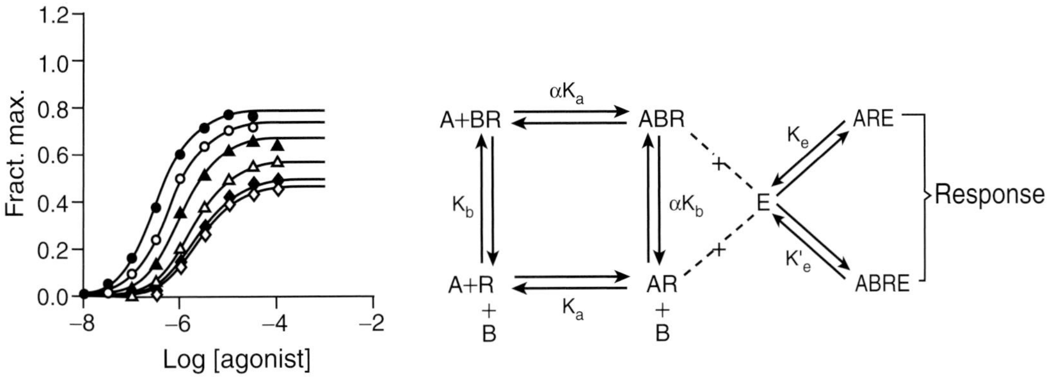

- Used to estimate system-independent potency of an allosteric modulator that affects the affinity (α) and possibly the efficacy (ξ) of the agonist for the receptor.

- K_A = Equilibrium dissociation constant of agonist-receptor complex.

- K_B = Equilibrium dissociation constant of antagonist-receptor complex.

- τ = transducer function for response to the full agonist and constitutively active receptor state.

- α = ratio of affinities of each ligand when the other ligand is bound to the receptor (cooperativity constant).

- ξ = ratio of τ values for the activation of the receptor by agonist alone and agonist through allosterically modulated receptor.

FIGURE 10.21 Figure illustrating the effects of an allosteric modulator that both alters the affinity of the receptor for the agonist and the signaling capability of the agonist on concentration-response curves to a full agonist. Equations describe response in terms of the operational model (variable slope version equation derived in Section 10.6.6). Schematic indicates the interacting species; in this case, a full agonist A activating the receptor and an allosteric modulator B that binds to the receptor to alter agonist affinity (by the factor α) and the signaling capability of the agonist on the receptor (K_e changes to K'_e for agonist response production upon binding of the modulator). Thus, the ratio of the efficacy of the agonist on the receptor in the presence and absence of modulator is given by ξ. Boxes show the relevant measurements and definitions of the parameters of the model used in the equation.

inhibition of agonism by the allosteric modulator is given by

$$\text{Response} = \frac{(1 + \alpha[B]/K_B)([A]/K_A(1+\tau)+1)}{[B]/K_B(1+\alpha[A]/K_A(1+\tau))+[A]/K_A(1+\tau)+1}.$$
(10.6)

A feature of Equation 10.6 is the fact that there are conditions where the binding will not diminish to basal levels in the presence of maximal concentrations of allosteric modulator. This effect is exacerbated by high levels of initial stimulation (as $[A]/K_A \to \infty$). The maximal inhibition by an allosteric modulator for Equation 10.6 is

$$\text{Max. Inhib.} = \frac{\alpha([A]/K_A(1+\tau))}{(1+\alpha[A]/K_A(1+\tau))},$$
(10.7)

which can approach zero for modulators with a very small cooperativity constant ($\alpha \ll 1$) and/or low levels of stimulation ($[A]/K_A \ll 1$). However, it can be seen that as the level of stimulation increases (increased values of $[A]/K_A$) the maximal inhibiton will be greater than zero. In general, the

inhibition curves may not reach zero asymptote values, at some agonist concentrations, if the value of α is greater than 0.1 to 0.05. Figure 10.28 illustrates the effect of the interplay of a values and the level of initial stimulation for inhibition curves for allosteric modulators. It can be seen that the inability to completely inhibit stimulation results from the limited effect modulators may have on the affinity of the receptor for the receptor probe (agonist). This is discussed in more detail in terms of binding in Chapter 4 (see Section 4.2.2).

This effect is not observed with allosteric modulators that block signaling. Under these circumstances, the equation for inhibition of a given agonist response is given by

$$\text{Response} = \frac{([A]/K_A(1+\tau)+1)}{[B]/K_B(1+\alpha[A]/K_A(1+\tau))+[A]/K_A(1+\tau)+1}.$$
(10.8)

The maximal inhibition in this case approaches complete (maximal response to the agonist $\to 0$) as $[B]/K_B \to \infty$ (i.e., there is no condition where the maximal inhibition will fall

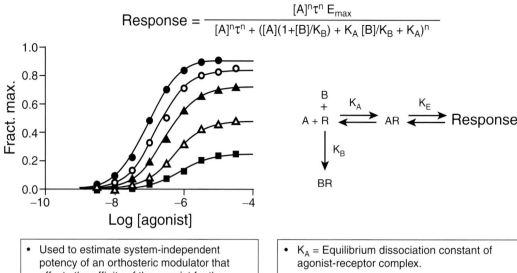

$$Response = \frac{[A]^n \tau^n E_{max}}{[A]^n \tau^n + ([A](1+[B]/K_B) + K_A [B]/K_B + K_A)^n}$$

- Used to estimate system-independent potency of an orthosteric modulator that affects the affinity of the agonist for the receptor and does not allow the receptor to respond to the agonist (no receptor signaling capability).
- Surrogate parameter: DR values for calculation of pA_2.

- K_A = Equilibrium dissociation constant of agonist-receptor complex.
- K_B = Equilibrium dissociation constant of antagonist-receptor complex.
- τ = transducer function for response to the full agonist.

FIGURE 10.22 Figure illustrating the effects of an orthosteric noncompetitive antagonist on concentration-response curves to a full agonist. Equations describe response in terms of the operational model (variable slope version equation derived in Section 10.6.7). Schematic indicates the interacting species; in this case, a full agonist A activating the receptor and an antagonist B binding to the receptor and precluding receptor occupancy and activation. Boxes show the relevant measurements and definitions of the parameters of the model used in the equation.

short of basal levels). The location parameter of the inhibition (defining the IC_{50}) is given by

$$IC_{50} = \frac{K_B([A]/K_A(1+\tau)+1)}{(1+\alpha[A]/K_A(1+\tau))}, \qquad (10.9)$$

where it can be seen that as for orthosteric noncompetitive antagonists if there is a considerable receptor reserve (high τ values) or high levels of receptor stimulation, or both, the IC_{50} is $> K_B$ (i.e., the IC_{50} underestimates the potency of the antagonist just as with orthosteric antagonists in systems with receptor reserve). However, unlike all orthosteric noncompetitive antagonists in cases where the allosteric modulator increases the affinity of the receptor for the agonist while blocking receptor signaling ($\alpha > 1$) the IC_{50} can actually be $< K_B$ (i.e., the IC_{50} may overestimate the potency of the antagonist). This is described in more detail in Section 7.5. With no independent knowledge of the value of an allosteric modulator, it is not possible to calculate a general correction factor for the conversion of IC_{50} values to the K_B for allosteric modulators. However, some guidelines can be discussed for all types of antagonists in general.

10.4.4 Practical Application

Under ideal conditions, inhibition curves should be obtained by blockade of an EC_{50} concentration of agonist.

This yields a correction between the IC_{50} and K_B of 2. However, in practice a large window is required for robust inhibition curves and EC_{80} agonist windows may be preferable. When an EC_{80} is used, a correction factor is required to obtain a system-independent estimate of the antagonist potency. In the case of simple competitive antagonists, the correction factor (Equation 10.5) is clearly indicated since it would make antagonist estimates of antagonism more uniform and system independent. However, for noncompetitive antagonists in systems where the agonist has no receptor reserve application of the factor will introduce an error; namely, an overestimation of antagonist potency. In practical terms, the magnitude of such an error, provided that the slopes of the dose-response curves lie between values of 1 to 2 and assuming that the agonist window is limited by the EC_{80} response (80% maximal response to the agonist), will be (for n = 1, $EC_{80} = [A]/K_A = 5$) on the order of sixfold. For shallow dose-response curves (i.e., 0.5), the correction factor can be quite substantial (for n = 0.5, $EC_{80} = [A]/K_A = 16$, correction = 35). However, standard agonists usually have high efficacy (natural agonist or potent surrogate). Therefore, there probably will be an effective receptor reserve for the agonist and some shift in the agonist dose-response curve upon noncompetitive blockade of a portion of the antagonist population will occur. Under these circumstances, there will be a partial correction factor

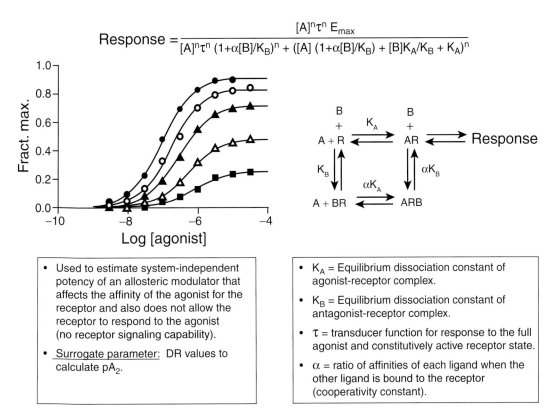

FIGURE 10.23 Figure illustrating the effects of an allosteric noncompetitive modulator on concentrations-response curves to a full agonist. Equations describe response in terms of the operational model (variable slope version equation derived in Section 10.6.8). Schematic indicates the interacting species; in this case, a full agonist A activating the receptor and an allosteric modulator binding to the receptor and precluding receptor activation. Boxes show the relevant measurements and definitions of the parameters of the model used in the equation.

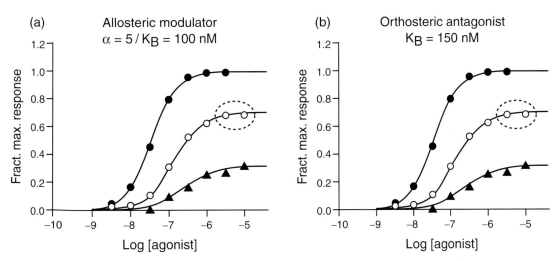

FIGURE 10.24 Simulation data set fit to an allosteric model (Equation 7.6; panel a) and to an orthosteric model (Equation 6.31; panel b). The data points circled with the dotted line were altered very slightly to cause the sum of squares for computer fit of the points to the model to favor either the allosteric or orthosteric model. It can be seen that very small differences can support either model even though they describe completely different molecular mechanisms of action.

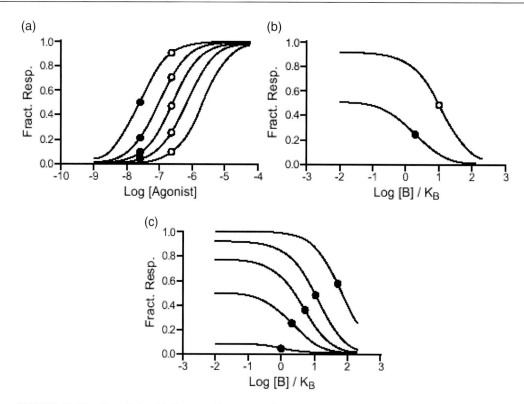

FIGURE 10.25 The relationship between IC_{50} and K_B for simple competitive antagonists. (a) Surmountable antagonism of agonist response. Control dose-response curve (furthest to the left) and repeat dose-response curves in the presence of increasing concentrations of competitive antagonist at 3.3-fold increment increases in concentration from $[B]/K_B = 3$ to 100. Filled circles indicate the EC_{50} for agonism and the effect of the antagonist on the response to this agonist concentration. The open circles represent the same for the EC_{80}. (b) Inhibition curves of the response to the EC_{50} (filled circle indicates the IC_{50}) and the EC_{80} (open circle indicates the IC_{50}). Ordinates: percent maximal response to the agonist. Abscissae: logarithm of the molar concentration of the antagonist. (c) A range of inhibition curves for blockade of responses to the EC_{10}, EC_{50}, EC_{76}, EC_{90}, and EC_{99}. Filled circles indicate the increasing values for the experimentally observed IC_{50} for the antagonist.

applicable and the error for noncompetitive antagonists could be considerably less than 6.

There are two practical reasons to err toward the correction of the IC_{50} of an antagonist to estimate the K_B. The first is that an overestimation of antagonist potency will only result in a readjustment of values upon rigorous measurement of antagonism in subsequent analysis. However, more importantly, if the correction is not applied then there is a risk of not detecting weak but still useful antagonism due to an underestimation of potency (due to nonapplication of the correction factor). These reasons support the application of the correction in all cases as a default.

10.5 Summary and Conclusions

- When dealing with large numbers of investigational compounds to be tested for agonist or antagonist activity, the methods used to determine system-independent measures of activity must be identified from the initial profile of activity (data-driven analysis).
- Short-form measures of activity—potency ratios for agonists, apparent K_B values (pA$_2$, IC_{50})—for

antagonists can adequately drive structure activity relationships if appropriate corrections for system effects are made.
- Surmountable antagonism can be quantified by pA$_2$ values. Insurmountable antagonism through IC_{50} values that in some cases can be corrected for the strength of stimulation in the system.
- In all cases, the molar concentration at which blockade of an agonist response is first encountered is a reasonable indication of the molecular potency of the antagonist, with the possible exception being allosteric modulators that block receptor signaling but increase the affinity of the receptor for the agonist.

10.6 Derivations

- System independence of full agonist potency ratios: classical and operational models
- Operational model for partial agonist interaction with agonist: variable slope
- Operational model for inverse agonist interaction with agonist: variable slope

(a)

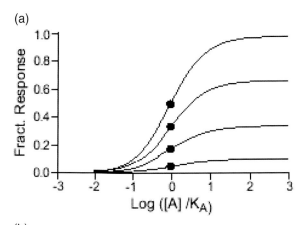

(b)

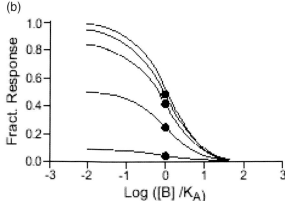

FIGURE 10.26 The effects of noncompetitive antagonism on agonist dose-response curves. (a) A control dose-response curve to an agonist (curve with the highest value for the ordinate asymptote) and the curve to the same agonist in the presence of increasing concentrations of a noncompetitive antagonist ([B]/K_B = 1, 3, and 10). In this panel, the tissue response is directly proportional to receptor occupancy. (b) Inhibition curves for the noncompetitive antagonist producing inhibition of the EC_{10}, EC_{50}, EC_{85}, EC_{95}, and EC_{99}. Note that the IC_{50} for antagonist does not change with increasing agonist concentration.

- Surmountable allosteric antagonism: variable slope
- Functional model for hemi-equilibrium effects: variable slope
- Allosteric antagonism with changes in efficacy: variable slope
- Orthosteric insurmountable antagonism: operational model with variable slope
- Allosteric insurmountable antagonism: operational model with variable slope
- IC_{50} correction factors: competitive antagonists

10.6.1 System Independence of Full Agonist Potency Ratios: Classical and Operational Models

The response to an agonist [A] in terms of the classical model is given as a function of stimulus:

$$\text{Stimulus} = \frac{[A] \cdot e}{[A] + K_A}. \quad (10.10)$$

(a)

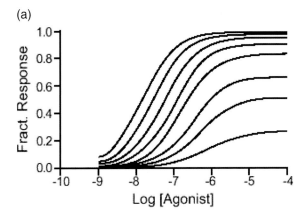

(b)

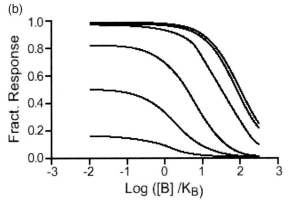

FIGURE 10.27 Noncompetitive antagonism in a system that has a receptor reserve for that agonist (i.e., where maximal response to the agonist can be obtained with submaximal receptor occupancy). (a) Control dose-response curve to the agonist (furthest to the left) and dose-response curves to the agonist in the presence of increasing concentrations of noncompetitive antagonist (3.33-fold increments of concentration beginning with [B]/K_B = 1). (b) Inhibition curves for the noncompetitive antagonist in this system for inhibition of increasing concentrations of agonist.

Assuming that a hyperbola of the form Response = Stimulus/(Stimulus + β) translates stimulus to response. Response is given as

$$\text{Response} = \frac{[A]/K_A \cdot e}{[A]/K_A(e + \beta) + \beta}. \quad (10.11)$$

From Equation 10.11 the observed EC_{50} is given as

$$EC_{50} = \frac{K_A \cdot \beta}{(e + \beta)}. \quad (10.12)$$

The potency ratio of two agonists (ratio denoted as EC'_{50}/EC_{50}) is

$$\text{Potency Ratio} = \frac{K_A \cdot (e' + \beta)}{K'_A \cdot (e + \beta)}. \quad (10.13)$$

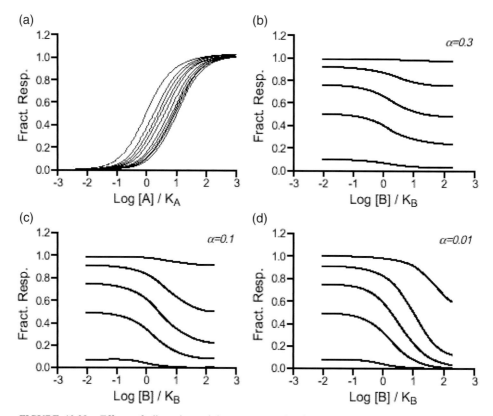

FIGURE 10.28 Effects of allosteric modulators on agonist dose-response curves and the resulting inhibition curves for the modulators. (a) Effects of various concentrations of an allosteric modulator with $\alpha = 0.1$ (producing a tenfold decrease in the affinity of the receptor for the agonist). A limiting value of a tenfold shift to the right of the curves is observed. (b) Inhibition curves for the blockade of various concentrations of agonist for an allosteric modulator $\alpha = 0.3$, $\alpha = 0.1$ (panel c), and $\alpha = 0.01$ (panel d). Note how ordinate curves for the higher concentrations of agonist do not diminish to zero.

In well-coupled systems where both agonists are full agonists, $\beta \rightarrow 0$. Therefore, the potency ratio approximates:

$$\text{Potency Ratio} = \frac{K_A \cdot e'}{K_A' \cdot e}. \tag{10.14}$$

These are system-independent constants relating only to the agonists. The same analysis can be done with the operational model. The response to an agonist [A] in terms of the operational model is given as

$$\text{Response} = \frac{E_{max} \cdot [A] \cdot \tau}{[A](1 + \tau) + K_A}, \tag{10.15}$$

where E_{max} is the maximal response of the system, τ is a factor quantifying the ability both the agonist (in terms of the agonist efficacy) and the system (in terms of the receptor density $[R_t]$ and the efficiency of stimulus-response coupling K_E, $\tau = [R_t]/K_E$). From Equation 10.15, the EC_{50} for a full agonist is

$$EC_{50} = \frac{K_A}{1 + \tau}, \tag{10.16}$$

where K_A is the equilibrium dissociation constant of the agonist-receptor complex. For full agonists $\tau \gg 1$.

Therefore, the $EC_{50} = K_A/\tau$. Substituting $\tau = [R_t]/K_E$, the potency ratio of two full agonists is

$$\text{Potency Ratio} = \frac{EC_{50}'}{EC_{50}} = \frac{K_A' \cdot K_E'}{K_A \cdot K_E}. \tag{10.17}$$

It can be seen that the potency ratio of two full agonists, as defined by Equation 10.17, is comprised of factors unique to the agonists and not the system, assuming that the stimulus-response coupling components of K_E (being common for both agonists) cancel.

10.6.2 Operational Model for Partial Agonist Interaction with Agonist: Variable Slope

The response-producing species for a partial agonist that competes for the agonist is given by Equation 6.85 (rewritten here):

$$\text{Response} = \frac{\rho_A[R_t]/K_E + \rho_B[R_t]/K_E'}{\rho_A[R_t]/K_E + \rho_B[R_t]/K_E' + 1}. \tag{10.18}$$

Defining $[R_t]/K_E$ as τ, $[R_t]/K_E'$ as τ' employing the operational forcing function for variable slope (see Section 3.13.4) yields

Resp.

$$= \frac{((([A]/K_A)^n \tau^n + ([B]/K_B)^n \tau'^n) E_{max}}{([A]/K_A)^n \tau^n + ([B]/K_B)^n \tau'^n + ([A]/K_A + [B]/K_B + 1)^n}. \tag{10.19}$$

Total Receptor species

$$= [A]/K_A(1 + \alpha[B]/K_B) + [B]/K_B + 1. \tag{10.27}$$

The operational forcing function for variable slope (see Section 3.13.4) yields

$$\text{Response} = \frac{([A]/K_A \tau)^n((1 + \alpha[B]/K_B)^n E_{max}}{([A]/K_A \tau)^n(1 + \alpha[B]/K_B)^n + ([A]/K_A(1 + \alpha[B]/K_B) + [B]/K_B + 1)^n}. \tag{10.28}$$

10.6.3 Operational Model for Inverse Agonist Interaction with Agonist: Variable Slope

From Equation 6.64, the expressions for the response producing species can be identified as

$$[AR^*] = \alpha L[A]/K_A, \tag{10.20}$$

$$[BR^*] = \beta L[B]/K_B, \quad \text{and} \tag{10.21}$$

$$[R^*] = L, \tag{10.22}$$

and the total receptors as $= [A]/K_A(1 + \alpha L) + [B]/K_B(1 + \beta L) + [B]/K_B + 1$. The operational forcing function for variable slope (see Section 3.13.4) yields

10.6.5 Functional Model for Hemi-equilibrium Effects: Variable Slope

The agonist receptor occupancy according to the hemi-equilibrium model of orthosteric antagonism (see Section 6.5) is given by Equation 6.2. The response species is

$$[AR] = ([A]/K_A)(1 - (\vartheta(1 - e^{-k_2 \Phi_t}) + \rho_B e^{-k_2 \Phi_t})), \tag{10.29}$$

and the total receptor species is given by $([A]/K_A + 1)$. The operational forcing function for variable slope

$$\text{Resp.} = \frac{(\alpha L[A]/K_A \tau)^n + (\beta L[B]/K_B \tau)^n + (L\tau)^n E_{max}}{(\alpha L[A]/K_A \tau)^n + (\beta L[B]/K_B \tau)^n + (L\tau)^n + ([A]/K_A(1 + \alpha L) + [B]/K_B(1 + \beta L) + [B]/K_B + 1)^n}. \tag{10.23}$$

10.6.4 Surmountable Allosteric Antagonism: Variable Slope

The fraction of response-producing species for a modulator that affects the affinity of the receptor for the agonist but does not alter signaling is given by

$$\rho_{AR} = \frac{[A]/K_A(1 + \alpha[B]/K_B)}{[A]/K_A(1 + \alpha[B]/K_B) + [B]/K_B + 1}, \tag{10.24}$$

leading to the response species:

$$[AR] = [A]/K_A, \tag{10.25}$$

$$[ABR] = \alpha[A]/K_A[B]/K_B, \quad \text{and} \tag{10.26}$$

(see Section 3.13.4) yields

Response

$$= \frac{([A])^n(1 - (\vartheta(1 - e^{-k_2 \Phi_t}) + \rho_B e^{-k_2 \Phi_t}))^n \tau^n E_{max}}{([A])^n(1 - (\vartheta(1 - e^{-k_2 \Phi_t}) + \rho_B e^{-k_2 \Phi_t}))^n \tau^n([A] + K_A)^n}. \tag{10.30}$$

10.6.6 Allosteric Antagonism with Changes in Efficacy: Variable Slope

In this case the modulator may alter both the affinity (through α) and efficacy (through ξ) of the agonist effect on the receptor (see Section 7.4). The fractional receptor occupancy by the agonist is given by Equation 7.3, leading to the response species:

$$[AR] = [A]/K_A, \qquad (10.31)$$

$$[ABR] = \alpha[A]/K_A[B]/K_B, \quad \text{and} \qquad (10.32)$$

$$\text{Total Receptor} = [A]/K_A(1 + \alpha[B]/K_B) + [B]/K_B + 1. \qquad (10.33)$$

The operational forcing function for variable slope (Equation 3.14.4) yields

term α) and prevents receptor activation of the receptor by the agonist. It is assumed that the only response-producing species is [AR]:

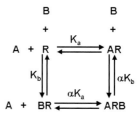

$$\text{Resp.} = \frac{([A]/K_A\tau)^n + (\alpha[A]/K_A[B]/K_B\tau')^n E_{max}}{([A]/K_A\tau)^n + (\alpha[A]/K_A[B]/K_B\tau')^n + ([A]/K_A(1 + \alpha[B]/K_B) + [B]/K_B + 1)^n}. \qquad (10.34)$$

Defining ξ as τ'/τ and rearranging, Equation 10.34 becomes

The resulting equilibrium equations are

$$K_a = [AR]/[A][R], \qquad (10.38)$$

$$\text{Resp.} = \frac{[A]^n\tau^n(1 + \alpha\xi[B]/K_B)^n E_{max}}{[A]^n\tau^n(1 + \alpha\xi[B]/K_B)^n + ([A](1 + \alpha[B]/K_B) + K_A[B]/K_B + K_A)^n}. \qquad (10.35)$$

10.6.7 Orthosteric Insurmountable Antagonism: Operational Model with Variable Slope

The antagonist blocks the receptor and does not allow reequilibration with the agonist according to mass action. The receptor occupancy equation for the agonist is given by Equation 6.8, leading to the response species

$$[AR] = [A]/K_A \qquad (10.36)$$

and total receptor species given by $[A]/K_A$ $(1 + [B]/K_B) + [B]/K_B + 1$. The operational forcing function for variable slope (Equation 3.14.4) yields

$$\text{Response} = \frac{[A]^n\tau^n E_{max}}{[A]^n\tau^n + ([A](1 + [B]/K_B) + K_A[B]/K_B + K_A)^n}. \qquad (10.37)$$

10.6.8 Allosteric Insurmountable Antagonism: Operational Model with Variable Slope

In this case the antagonist blinds to its own site on the receptor to affect the affinity of the agonist (through the

$$K_b = [BR]/[B][R], \qquad (10.39)$$

$$\alpha K_a = [ARB]/[BR][A], \quad \text{and} \qquad (10.40)$$

$$\alpha K_b = [ARB]/[AR][B]. \qquad (10.41)$$

Solving for the agonist-bound response-producing receptor species [AR] as a function of the total receptor species ($[R_{tot}] = [R] + [AR] + [BR] + [ARB]$) yields

$$\frac{[AR]}{[R_{tot}]} = \frac{((1/\alpha[B]K_b) + 1)}{((1/\alpha[B]K_b) + (1/\alpha K_a) + (1/\alpha[A]K_aK_b) + 1)}. \qquad (10.42)$$

Simplifying and changing association to dissociation constants (i.e., $K_A = 1/K_a$) yields

$$\rho_{AR} = \frac{[A]/K_A}{[A]/K_A(1 + \alpha[B]/K_B) + [B]/K_B + 1}. \qquad (10.43)$$

The operational forcing function for variable slope (see Section 3.13.4) for Equation 10.43 yields

$$\text{Response} = \frac{[A]^n\tau^n E_{max}}{[A]^n\tau^n + ([A](1 + \alpha[B]/K_B) + K_A[B]/K_B + K_A)^n}. \qquad (10.44)$$

10.6.9 *IC$_{50}$ Correction Factors: Competitive Antagonists*

The relationship between the concentration of antagonist that produces a 50% inhibition of a response to an agonist (antagonist concentration referred to as the IC$_{50}$) and the equilibrium dissociation constant of the antagonist-receptor complex (K$_B$) can be derived from the mass action equations describing the agonist receptor response in the presence and absence of the antagonist. The response in the absence of antagonist can be fit to a logistic curve of the form

$$\text{Response} = \frac{E_{max}[A]^n}{[A]^n + [EC_{50}]^n},\qquad(10.45)$$

where the concentration of agonist is [A], E_{max} is the maximal response to the agonist, n is the Hill coefficient of the dose-response curve, and [EC$_{50}$] is the molar concentration of agonist producing 50% maximal response to the agonist.

In the presence of a competitive antagonist, the EC$_{50}$ of the agonist dose-response curve will be shifted to the right by a factor equal to the dose ratio. This is given by the Schild equation as [B]/K$_B$ + 1, where the concentration of the antagonist is [B] and K$_B$ is the equilibrium dissociation constant of the antagonist-receptor complex:

$$\text{Response} = \frac{E_{max}[A]^n}{[A]^n + ([EC_{50}](1 + [B]/K_B))^n}.\qquad(10.46)$$

The concentration of antagonist producing a 50% diminution of the agonist response to concentration [A] is defined as the IC$_{50}$ for the antagonist. Therefore:

$$\frac{0.5E_{max}[A]^n}{[A]^n + [EC_{50}]^n} = \frac{E_{max}[A']^n}{[A']^n + ([EC_{50}](1 + [IC_{50}]/K_B))^n}.\qquad(10.47)$$

After rearrangement [3]:

$$K_B = \frac{[IC_{50}]}{(2 + ([A]/[EC_{50}])^{n1/n}) - 1}.\qquad(10.48)$$

References

1. Stephenson, R. P. (1956). A modification of receptor theory. *Br. J. Pharmacol.* **11**:379–393.
2. Kenakin, T. P., and Beek, D. (1980). Is prenalterol (H 133/80) really a selective beta-1 adrenoceptor agonist? Tissue selectivity resulting selective beta-1 adrenoceptor agonist? Tissue selectivity resulting from difference in stimulus-response relationships. *J. Pharmacol. Exp. Ther.* **213**:406–413.
3. Leff, P., and Dougall, I. G. (1993). Further concerns over Cheng-Prusoff analysis. *Trends Pharmacol. Sci.* **14**:110–112.

11

Statistics and Experimental Design

To call in the statistician after the experiment is done may be no more than asking him to perform a postmortem examination: he may be able to say what the experiment died of.

— RONALD FISHER

11.1 Structure of This Chapter

This chapter is divided into three main sections. The first is devoted to methods, ideas, and techniques aimed at determining whether a set of pharmacological data is internally consistent (i.e., to what extent a given value obtained in the experiment will be obtained again if the experiment is repeated). The second section is devoted to methods and techniques aimed at determining to what extent the experimentally observed value is externally consistent with literature, other experimental data sets, or values predicted by models. This second section is divided into two subsections. The first deals with comparing experimental data to models that predict values for the entire population (i.e., curve fitting, and so on) and the second subsection is concerned with differences between experimentally determined data (or an experimentally determined data set) and values from the literature. Finally, some ideas on experimental design are discussed in the context of improving experimental techniques.

11.2 Introduction

Statistics in general is a discipline dealing with ideas on description of data, implications of data (relation to general pharmacological models), and questions such as what effects are real and what effects are different? Biological systems are variable. Moreover, often they are living. What this means is that they are collections of biochemical reactions going on in synchrony. Such systems will have an intrinsic variation in their output due to the variances in the rates and set points of the reactions taking place during the natural progression of their function. In general, this will be referred to as biological "noise" or variation. For example, a given cell line kept under culture conditions will have a certain variance in the ambient amount of cellular cyclic AMP present at any instant. Pharmacological experiments strive to determine whether or not a given chemical can change the ambient physiological condition of a system and thus demonstrate pharmacological activity. The relevant elements in this quest are the level of the noise and the level of change in response of system imparted by the chemical (i.e., the signal-to-noise ratio).

11.3 Descriptive Statistics: Comparing Sample Data

In general, when a pharmacological constant or parameter is measured it should be done so repeatedly to give a measure of confidence in the value obtained (i.e., the likelihood that if the measurement were repeated it would yield the same value). There are various statistical tools available to determine this. An important tool and concept in this regard is the Gaussian distribution.

11.3.1 Gaussian Distribution

When an experimental value is obtained numerous times, the individual values will symmetrically cluster around the mean value with a scatter that depends on the number of replications made. If a very large number of replications are made (i.e., $>2,000$), the distribution of the values will take on the form of a Gaussian curve. It is useful to examine some of the features of this curve since it forms the basis of a large portion of the statistical tools used in this chapter. The Gaussian curve for a particular population of N values (denoted x_i) will be centered along the abscissal axis on the mean value where the mean (η) is given by

$$\eta = \frac{\sum_i x_i}{N}. \tag{11.1}$$

The measure of variation in this population is given by the standard deviation of the population (σ):

$$\sigma = \sqrt{\frac{\sum (x_i - \eta)^2}{N}}. \tag{11.2}$$

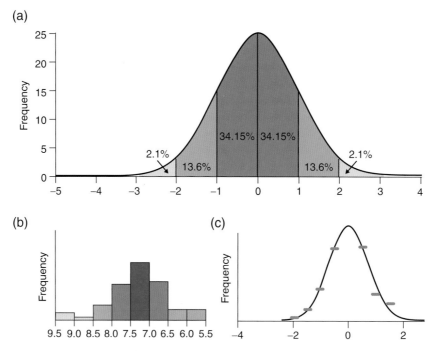

FIGURE 11.1 Normal distributions. (a) Gaussian distribution showing the frequency of values in a population expressed as a function of distance away the value is from the mean of the population. Percentage values represent areas in the strips of curve (i.e., between 0 and 1 represents the area within 1 standard deviation unit from the mean). (b) Histogram showing the pK_B of an antagonist (TAK 779, an antagonist of HIV infection) divided into bins consisting of 1 SEM unit away from the mean value. (c) The histogram is an approximation of a Gaussian normal distribution shown in panel b.

The ordinates of a Gaussian curve are the relative frequency that the particular values on the abscissae are encountered. The frequency of finding these values for a particular value diminishes the farther away it is from the mean. The resulting curve is shown in Figure 11.1a. The abscissal axis is divided into multiples of σ values. Thus, $+1$ or -1 refers to values that are within one standard deviation either greater than or less than the mean. It is useful to consider the area under the curve at particular points along the abscissae since this gives a measure of the probability of finding a particular value within the standard deviation limits chosen. For example, for a standard Gaussian curve 68.3% of all the values reside within 1 standard deviation unit of the mean. Similarly, 95.5% of all the values lie within 2σ units, and 99.7% of the values within 3σ units (see Figure 11.1a). Most statistical tests used in pharmacology are parametric (i.e., require the assumption that the distribution of the values being compared are from a normal distribution). If enough replicates are obtained, a normal distribution of values will be obtained. For example, Figure 11.1b shows a collection of 58 replicate estimates of the pK_B of a CCR5 antagonist TAK 779 as an inhibitor of HIV infection. It can be seen that the histograms form a relatively symmetrical array around the mean value. As more values are added to such

collections, they take on the smoother appearance of a Gaussian distribution (Figure 11.1c). It should be noted that the requirements of normal distribution are paramount for the statistical tests that are to be described in this chapter. As discussed in Chapter 1, while pK_I, pEC_{50}, and pK_B estimates are normally distributed because they are derived from logarithmic axes on curves the corresponding IC_{50}, EC_{50}, and K_B values are not (see Figure 1.16) and thus cannot be used in parametric statistical tests.

11.3.2 Populations and Samples

Populations are very large collections of values. In practice, experimental pharmacology deals with samples (much smaller collections) from a population. The statistical tools used to deal with samples differ somewhat from those used to deal with populations. When an experimental sample is obtained, the investigator often wants to know about two features of the sample: central tendency and variability. The central tendency refers to the most representative estimate of the value, while the variability defines the confidence that the estimate is a true reflection of that value. Central tendency estimates can be the median (value that divides the sample into two equal halves) or the mode (most commonly occurring value). These values (especially the median) are not affected by extreme values (outliers). However, the most common estimate of central tendency in experimental work is the mean (x_m) defined for a set of n values as

$$x_m = \frac{\sum_i x_i}{n}. \tag{11.3}$$

The estimate of variability for a sample mean is the standard error of the mean:

$$s_x = \sqrt{\frac{\sum (x_i - x_m)^2}{(n-1)}}. \tag{11.4}$$

Alternatively, this frequently used quantity can be calculated as

$$s_x = \sqrt{\frac{n \sum x^2 - \left(\sum x\right)^2}{n(n-1)}}. \tag{11.5}$$

There are instances where deviations, as measured by the standard error, are scaled to the magnitude of the mean to yield the coefficient of variation. This is calculated by

$$C.V. = 100 \times standard\ deviation/mean. \tag{11.6}$$

A frequently asked question is: Are two experimentally derived means equal? In fact, this question really should be stated: Do the two experimentally derived samples come from the same population? Hypothesis testing is used to answer this question. This process is designed to disprove what is referred to as the *null hypothesis* (i.e., condition of no difference). Thus, the null hypothesis states that there is no difference between the two samples (i.e., that they both come from the same population). It is important to note that experiments are designed to disprove the null hypothesis, not prove the hypothesis correct. Theoretically speaking, a hypothesis can never be proven correct since failure to disprove the hypothesis may only mean that the experiment designed to do so is not designed adequately. There could always be an as yet to be designed experiment capable of disproving the null hypothesis. Thus, it is a Sysyphean task to prove it "correct." However, the danger of overinterpreting failure to disprove the null hypothesis cannot be overemphasized. As put by the statistician Finney (1955), "Failure to disprove that certain observations no not disprove a hypothesis does not amount to proof of a hypothesis."

This concept is illustrated by the example shown in Table 11.1. Shown are three replicate pEC_{50} values for the agonist human calcitonin obtained from two types of cells: wild-type HEK 293 cells and HEK 293 cell enriched with $G_{\alpha s}$-protein. The respective pEC_{50} values are 7.47 ± 0.15 and 8.18 ± 0.21. The question is: Do these two estimates come from the same population? That is, is there a statistically significant difference between the sensitivity of cells enriched and not enriched with $G_{\alpha s}$-protein to human calcitonin? To go further toward answering this question

TABLE 11.1

T-test for differences between experimental means.

pEC_{50} values for human calcitonin in wild-type HEK 293 cells (x_2) and HEK 293 cells enriched with $G_{\alpha s}$-protein (x_1).

x_1	x_2
7.9	7.5
8.2	7.3
8.3	7.6
$\sum x_1 = 24.4$	$\sum x_2 = 22.4$
$\sum x_1^2 = 198.54$	$\sum x_2^2 = 167.3$
$x_{m1} = 8.13$	$x_{m2} = 7.47$
$s_{x1} = 0.21$	$s_{x2} = 0.15$
$s_p^2 = 0.033$	
difference $= 0.67$	
$SE_{(difference)} = 0.149$	
$t = 4.47$	
df $= 4$	

Data from [9].

requires discussion of the concepts of probability and the t-distribution.

Statistical tests do not declare anything with certainty; they only assess the probability that the result is true. Thus, values have a "level of confidence" associated with them. Within the realm of hypothesis testing, where the verisimilitude of a data set to predictions made by two hypotheses is examined, a probability is obtained. As discussed previously, the approach taken is that the data must disprove the null hypothesis (stating that there is no difference). For example, when testing whether a set of data is consistent with or disproves the null hypothesis a level of confidence of 95% states that the given hypothesis is disproved but that there is 5% chance that this result occurred randomly. This means that there is a small (5%) chance that the data supported the hypothesis but that the experiment was unable to discern the effect. This type of error is termed a type I error (rejection of a true hypothesis erroneously) and is often given the symbol α. Experimenters preset this level before the experiment (i.e., $\alpha = 0.05$ states that the investigator is prepared to accept a 5% chance of being incorrect). Statistical significance is then reported as $p < 0.05$, meaning that there is less than a 5% probability that the experiment led to a type I error. Another type of error (termed type II error) occurs when a hypothesis is erroneously accepted (i.e., the data appears to be consistent with the null hypothesis) but in fact the samples do come from separate populations and are indeed different.

So how does one infer that two samples come from different populations when only small samples are available? The key is the discovery of the t-distribution by Gosset in 1908 (publishing under the pseudonym of Student) and development of the concept by Fisher in 1926. This revolutionary concept enables the estimation of σ (standard deviation of the population) from values of standard errors of the mean and thus to estimate population means from sample means. The value t is given by

$$t = (x_m - \eta)/SE_x, \tag{11.7}$$

where SE_x is the standard deviation and η is the mean of the population. Deviation of the estimated mean from the population mean in SE_x units yields values that can then be used to calculate the confidence that given sample means come from the same population. Returning to the data in Table 11.1 (two sample means x_{m1} and x_{m2} of size n_1 and n_2, respectively), the difference between the two means is $(x_{m1} - x_{m2}) = 0.67$ log units. A standard error of this difference can be calculated by

$$S.E._{\text{difference}} = s_p^2 (1/n_1 + 1/n_2)^{1/2}, \tag{11.8}$$

where s_p^2 is the pooled variance given as

$$s_p^2 = \frac{\sum x_1^2 + \sum x_2^2 - (\sum x_1/n_1) - (\sum x_2/n_2)}{n_1 + n_2 - 2}. \tag{11.9}$$

For the example shown in Table 11.1, $S.E._{\text{difference}} = 0.15$. The value of t is given by

$$t = (x_{m1} - x_{m2})/S.E._{\text{difference}}. \tag{11.10}$$

For the example shown in Table 11.1, the calculated t is 4.47. This value is associated with a value for the number of degrees of freedom in the analysis. For this test, the degrees of freedom (df) is $n_1 + n_2 - 2 = 4$. This value can be compared to a table of t values (Appendix A) to assess significance. There are t values for given levels of confidence. Referring to Appendix A, it can be seen that for df = 4 the value for t at a level of significance of 95% is 2.776. This means that if the calculated value of t is less than 2.776 there is a greater than a 5% chance that the two samples came from the same population (i.e., they are not different). However, as can be seen from Table 11.1 the calculated value of t is > 2.776, indicating that there is less than a 5% chance ($p < 0.05$) that the samples came from the same population.

In fact, a measure of the degree of confidence can be gained from the t calculation. Shown in Appendix A are columns for greater degrees of confidence. The value for df = 4 for a 98% confidence level is 3.747 and it can be seen that the experimentally calculated value is also greater than this value. Therefore, the level of confidence that these samples came from different populations is raised to 98%. However, the level of confidence in believing that these two samples came from separate populations does not extend to 99% (t = 4.604). Therefore, at the 98% confidence level this analysis indicates that the potency of human calcitonin is effectively increased by enrichment of $G_{\alpha s}$-protein in the cell.

A measure of variability of the estimate can be gained from the standard error but it can be seen from Equations 11.4 and 11.5 that the magnitude of the standard error is inversely proportional to n (i.e., the larger the sample size the smaller will be the standard error). Therefore, without prior knowledge of the sample size a reported standard error cannot be evaluated. A standard error value of 0.2 indicates a great deal more variability in the estimate if n = 100 than if n = 3. One way around this shortcoming is to report n for every estimate of mean ± standard error. Another, and better, method is to report confidence intervals of the mean.

11.3.3 Confidence Intervals

The confidence interval for a given sample mean indicates the range of values within which the true population value can be expected to be found and the probability that this will occur. For example, the 95% confidence limits for a given mean are given by

$$c.l._{95} = x_m + s_x(t_{95}), \tag{11.11}$$

where s_x is the standard error and the subscripts refer to the level of confidence (in the case previously cited, 95%). Values of t increase with increasing levels of confidence. Therefore, the higher the level of confidence required for defining an interval containing the true value from a sample mean the wider the confidence interval. This is intuitive since it would be expected that there would be a greater probability of finding the true value within a wider range. The confidence limits of the mean pEC_{50} value for human calcitonin in wild-type and $G_{\alpha s}$-protein-enriched HEK 293 cells are shown in Table 11.2. A useful general rule (but not always explicitly accurate especially for small samples, see Section 11.6.1) is to note that if the mean values are included in the 95% confidence limits of the other mean (if $p < 0.05$ is the predefined level of significance in the experiment) then the means probably are from the same population. In general, reporting variability as confidence limits eliminates ambiguity with respect to the sample size since the limits are calculated with a t value which itself is dependent on degrees of freedom (the sample size).

While statistical tests are helpful in discerning differences in data, the final responsibility in determining difference remains with the researcher. While a given statistical test may indicate a difference, it will always do so as a

TABLE 11.2
Confidence intervals for the means in Table 11.1.

	x_{m1}			x_{m2}		
	Lower c.l.	Mean	Greater c.l.	Lower c.l.	Mean	Greater c.l.
95%	7.55	8.13 to 8.13	8.71	7.05	7.47 to 7.47	7.89
98%	7.34	8.13 to 8.13	8.92	6.91	7.47 to 7.47	8.03
99%	7.16	8.13 to 8.13	9.10	6.78	7.47 to 7.47	8.16
99.5%	6.95	8.13 to 8.13	9.31	6.63	7.47 to 7.47	8.31

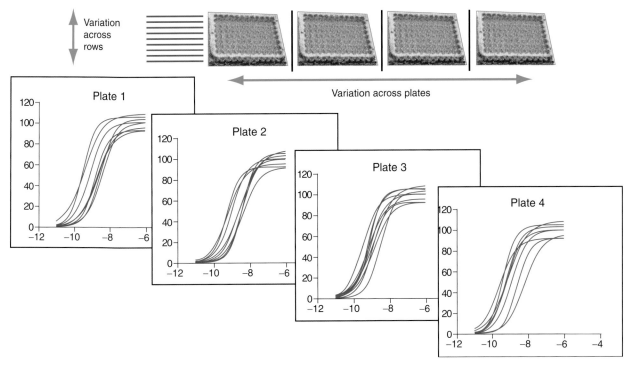

FIGURE 11.4 Two-way analysis of variance. Arrangement of data in rows and columns such that each row of the cell culture plate (shown at the top of the figure) defines a single dose-response curve to the agonist. Also, data is arranged by plate in that each plate defines eight dose-response curves and the total data set is comprised of 32 dose-response curves. The possible effect of location with respect to row on the plate and/or which plate (order of plate analysis) can be tested with the two-way analysis of variance.

models describing dose-response curves. The aim is to take a selected sample of data and predict the behavior of the system, generating that data over the complete concentration range of the drug (i.e., predict the population of responses). Nonlinear curve fitting is the technique used to do this.

The process of curve fitting utilizes the sum of least squares (denoted SSq) as the means of assessing "goodness of fit" of data points to the model. Specifically, SSq is the sum of the differences between the real data values (y_d) and the value calculated by the model (y_c) squared to cancel the effects of arithmetic sign:

$$SSq = \sum (y_d - y_c)^2. \tag{11.21}$$

There are two approaches to curve fitting. The first uses empirical models that may yield a function that closely fits the data points but has no biological meaning. An example of this was given in Chapter 3 (see Figure 3.1). A danger in utilizing empirical models is that nuances in the data points that may be due to random variation may be unduly emphasized as true reflections of the system. The second approach uses parameters rooted in biology (i.e., the constants have biological meaning). In these cases, the

TABLE 11.6

Results of the two-way analysis of variance for data shown in Figure 11.4 and Table 11.5.

	SSq	df	MSq	F
Between rows	0.66	7	0.09	0.56
Between columns	0.51	3	0.17	1.02
Residual	3.52	21	0.17	
Total	4.69	31		

model may not fit the data quite as well. However, this latter strategy is preferable since the resulting fit can be used to make predictions about drug effect that can be experimentally tested.

It is worth considering hypothesis testing in general from the standpoint of the choice of models one has available to fit data. On the surface, it is clear that the more complex a model is (more fitting parameters) the greater the verisimilitude of the data to the calculated line (i.e., the smaller will be the differences between the real and predicted values). Therefore, the more complex the model the more likely it will accurately fit the data. However, there are other factors that must be considered. One is the physiological relevance

(a)

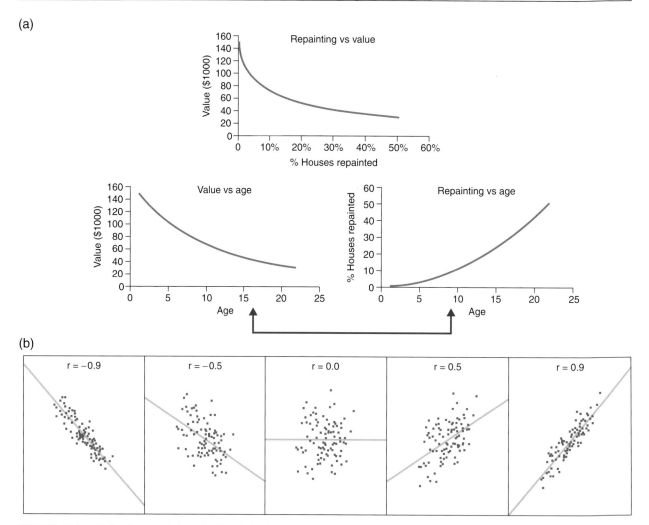

(b)

FIGURE 11.5 Misleading correlations. (a) Correlation between percentage of houses that are repainted and house value. It can be seen that the relationship is inverse (i.e., painting a house will decrease its value). This correlation comes from two other correlations showing that the value of a house decreases as it ages and the fact that as a house ages there is greater probability it will need to be repainted. (b) Some correlations. A very good negative correlation ($r = -0.9$), a weak negative correlation ($r = -0.5$), no correlation ($r = 0$), a weak positive correlation ($r = 0.5$), and strong positive correlation ($r = 0.9$).

of the mathematical function used to fit the data. For example, Figure 11.7 shows a collection of responses to an agonist. A physiologically relevant model to fit this data is a variant of the Langmuir adsorption isotherm (i.e., it is likely that these responses emanate from a binding reaction such as that described by the isotherm followed by a series of Michaelis-Menten type biochemical reactions that also resemble the adsorption isotherm). Therefore, a model such as that described by an equation rooted in biology would seem to be pharmacologically relevant. The fit to such a model is shown in Figure 11.7a. However, a better mathematical fit can be obtained by a complex mathematical function of the form

$$\text{Response} = \sum_{n=4}^{n=1} a^n e^{-([A]-b_n)/c_n}. \qquad (11.22)$$

While better from a mathematical standpoint, the physiological relevance of Equation 11.22 is unknown. Also, the more complex is a fitting function is the greater the chance that problems in computer curve fitting will ensue. Fitting software generally use a method of least squares to iteratively come to a best fit (i.e., each parameter is changed stepwise and the differences between the fit function and real data calculated). The best fit is concluded when a "minimum" in the calculated sum of those differences is found. The different fitting parameters often have different weights of importance in terms of the overall effect produced when they are changed. Therefore, there can occur "local minima," where further changes in parameters do not appear to produce further changes in the sum of the differences. However, these minima may still fall short of the overall minimum value that could be

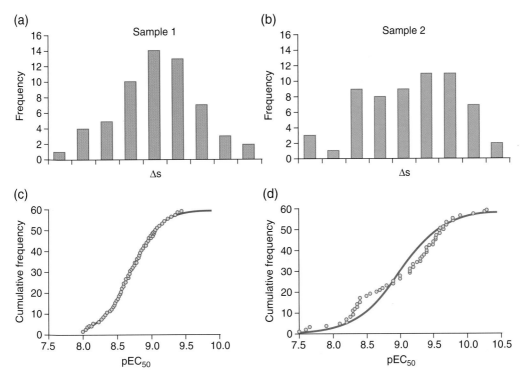

FIGURE 11.6 Distibution of 59 pEC$_{50}$ values. (a) Frequency of pEC$_{50}$ values displayed as a function of binning by increments of 0.5 x standard error (mean pEC$_{50}$ = 8.7 ± 0.36). (b) Another data set with an equivocal (with respect to single or bimodal) distribution (mean pEC$_{50}$ = 9.0 ± 0.67). (c) Cumulative distribution curve for the data set shown in panel a. The data is best fit by a single-phase curve. (d) Cumulative distribution curve for the data set shown in panel b. In this case, a single-phase curve clearly deviates from the data that indicates bimodality.

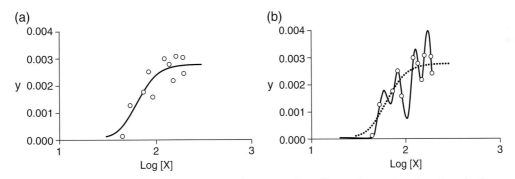

FIGURE 11.7 Fitting dose-response data. (a) Data points fit to Langmuir adsorption isotherm with E$_{max}$ = 0.00276, n = 4.36 and EC$_{50}$ = 65. (b) Data fit to empirical model of the form $y = (600^{-1})e^{-((x-60)2/80)} + (400^{-1})e^{-((x-85)2/400)} + (320^{-1})e^{-((x-130)2/300)} + (280^{-1})e^{-((x-180)2/800)}$.

attained if further iteration were allowed. The likelihood of encountering such local minima (which in turn leads to incorrect fitting of functions to data) increases as the model used to fit the data is more complex (has many fitting parameters). Therefore, complex models with many fitting parameters can lead to practical problems in computer fitting of data. A sampling of mathematical fitting functions is given in Appendix A for application to fitting data to empirical functions.

Local minima will rarely be observed if the data has little scatter, if an appropriate equation has been chosen, and if the data is collected over an appropriate range of x values. A way to check whether or not a local minimum has been encountered in curve fitting is to observe the effect

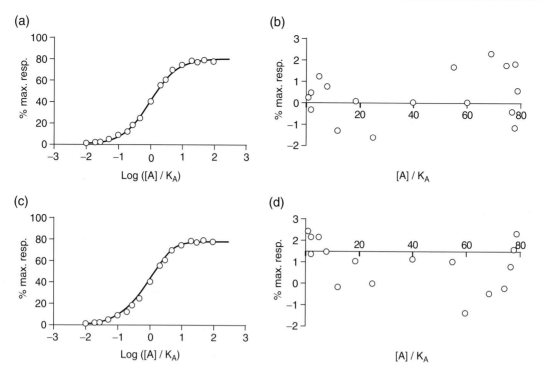

FIGURE 11.8 Residual distribution as a test for goodness of fit. (a) Data points fit to a physiologically appropriate model (Langmuir isotherm). (b) Residuals (sum of squares of real data points minus calculated data points) expressed as a function of the x value on the curve. It can be seen that the residuals are relatively symmetrically centered around the mean of the residual value over the course of the data set. (c) Same data fit to a general mathematical function (Equation 3.1). (d) The residuals in this case group below the mean of the residuals, indicating a nonsymmetrical fitting of the values.

of making large changes in one of the variables on the sum of squares. If there is a correspondingly large change in the sum of squares, it is possible that a local minimum is operative. Ideally, the sum of squares should converge to the same value with any changes in the values of parameters.

Another criterion for goodness of fit is to assess the residual distribution (i.e., how well the model predicts values throughout the complete pattern of the data set). Some models may fit some portions of the data well but not other portions, and thus the residuals (differences between the calculated and real values) will not be uniformly distributed over the data set. Figure 11.8 shows a set of data fit to an empirical model (Equation 3.1) and the Langmuir adsorption isotherm. Inspection of the fit dose-response curves does not indicate a great difference in the goodness of fit. However, an examination of the residuals, expressed as a function of the concentration, indicates that while the adsorption isotherm yields a uniform distribution along the course of the data set (uniform distribution of values greater than and less than zero) the empirical fit shows a skewed distribution of errors (values at each end positive and values in the middle negative). A uniform distribution of the residual errors is desired, and models that yield such balanced residuals statistically are preferred.

Finally, complex models may be inferior for fitting data purely in statistical terms. The price of low sums of differences between predicted and real values obtained with a complex model is the loss of degrees of freedom. This results in a greater $(df_s - df_c)$ value for the numerator of the F-test calculation and a greater denominator values since this is SSq_c divided by df_c (see following section on hypothesis testing). Therefore, it is actually possible to decrease values of F (leading to a preference for the more simple model) by choosing a more complex model (*vide infra*).

11.4.2 Curve Fitting: Good Practice

There are practical guidelines that can be useful for fitting pharmacological data to curves.

1. All regions of the function should be defined with real data. In cases of sigmoidal curves it is especially important to have data define the baseline, maximal asymptote, and mid-region of the curve.
2. In usual cases (slope of curve is unity), the ratio of the maximum to the minimum concentrations should be on the order of 3,200 (approximately 3.5 log units).

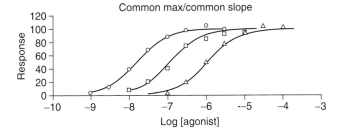

	SSq	AIC	n	K
Individual fits	102	66.1	23	9
Common maxima	211	73.0	23	7
Common slope/maxima	119	65.4	23	5

FIGURE 11.15 Aikake's information criteria (AIC) calculations. Lower panel shows three dose-response curves that can alternately be fit to a three-parameter logistic such that each curve is fit to its own particular value of maximum and slope (individual fits), with common maxima but individual slopes (common maxima), or with common maxima and slope. The IAC values (Equation 11.30) for the fits are shown in the table above the figure. It can be seen that the lowest value corresponds to the fit with common maxima and slope. Therefore, this fit is preferred.

SSq_s (df_s) is number of data points minus the common max, common slope, and four fitted values for EC_{50}. Thus, $df_s = 24 - 6 = 18$. The value for F for comparison of the simple model (common maximum and slope) to the complex model (individual maxima and slopes) for the data shown in Figure 11.14 is $F = 2.4$. To be significant at the 95% level of confidence (5% chance that this F actually is not significant), the value of F for $df = 12, 18$ needs to be > 2.6. Therefore, since F is less than this value there is no statistical validation for usage of the most complex model. The data should then be fit to a family of curves of common maximum and slope and the individual EC_{50} values used to calculate values of DR.

The same conclusion can be drawn from another statistical test for model comparison; namely, through the use of Aikake's information criteria (AIC) calculations. This is often preferred, especially for automated data fitting, since it is more simple than F tests and can be used with a wider variety of models. In this test, the data is fit to the various models and the SSq determined. The AIC value is then calculated with the following formula

$$AIC = n \cdot \ln\left[\frac{SSq}{n}\right] + 2 \cdot K + \left[\frac{2 \cdot K \cdot (K+1)}{(n-K-1)}\right], \quad (11.31)$$

where n is the number of total data points and K is the number of parameters used to fit the models. The fit to the model with the lowest AIC value is preferred. A set of dose-response curves is shown in Figure 11.15. As with the previous example, the question is asked: Can these data points be fit to a model of dose-response curves with common maximum and slope? The AIC values for the various models for the data are given in the table shown in Figure 11.15. It can be seen that the model of common slope and maximum has the lowest AIC value. Therefore, this model is preferred.

11.4.6 One Curve or Two? Detection of Differences in Curves

There are instances where it is important to know the concentration of a drug, such as a receptor antagonist, that first produces a change in the response to an agonist. For example, a competitive antagonist will produce a twofold shift to the right of an agonist dose-response curve when it is present in the receptor compartment at a concentration equal to the K_B. A tenfold greater concentration will provide a tenfold shift to the right. With an antagonist of unknown potency, a range of concentrations is usually tested and there can be ambiguity about small differences in the dose-response curves at low antagonist concentrations. Hypothesis testing can be useful here. Figure 11.16 shows what could be two dose-response curves: one control curve and one possibly shifted slightly to the right by an antagonist. An alternative interpretation of these data is that the antagonist did nothing at this concentration and what is being observed is random noise around a second measurement of the control dose-response curve. To resolve this, the data is fit to the most simple model (a single dose-response curve with one max, slope, and location parameter EC_{50} for all 12 data points) and then refit to a more complex model of two dose-response curves with a common maximum and slope but different location parameters EC_{50}. Calculation of F can then be

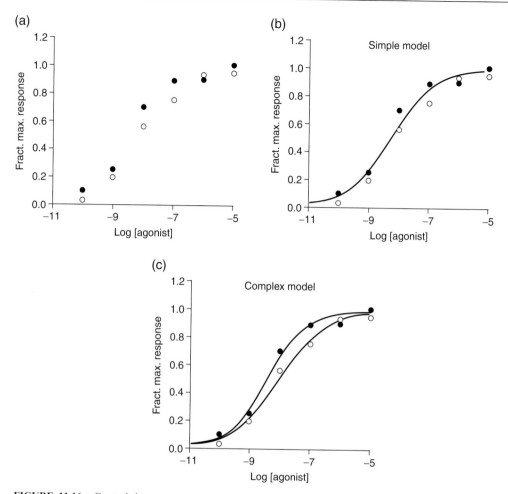

FIGURE 11.16 Control dose-response curve and curve obtained in the presence of a low concentration of antagonist. Panel a: data points. Panel b: data fit to a single dose-response curve. $SSq_s = 0.0377$. Panel c: data fit to two parallel dose-response curves of common maximum. $SSq_c = 0.0172$. Calculation of F indicates that a statistically significant improvement in the fit was obtained by using the complex model (two curves; $F = 4.17$, $df = 7, 9$). Therefore, the data indicate that the antagonist had an effect at this concentration.

used to resolve whether the data is better fit to a single curve (indicating noise around the control curve and no antagonism) or to two separate curves (antagonist produces a low level of receptor blockade). For the data shown in Figure 11.16, the value for F indicated that a statistically significant improvement in the fit was obtained with two dose-response curves as opposed to one. This indicates, in turn, that the antagonist had an effect at this concentration.

11.4.7 Asymmetrical Dose-Response Curves

As noted in Chapter 1, the most simple and theoretically sound model for drug-receptor interaction is the Langmuir adsorption isotherm. Other models, based on receptor behavior (see Chapter 3), are available. One feature of all of these models (with the exception of some instances of the operational model) is that they predict symmetrical curves. A symmetrical curve is one where the half maximal abscissal point (EC_{50}, concentration of x that yields 50% of the maximal value of y) and the inflection point of the curve (where the slope is zero) are the same (see Figure 11.17a). However, many experimentally derived dose-response curves are not symmetrical because of biological factors in the system. Thus, there can be curves where the EC_{50} does not correspond to the point at which the slope of the curve is zero (see Figure 11.17b). Attempting to fit such data with symmetrical functions leads to a lack of fit on either end of the data set. For example, Figure 11.18a shows an asymmetrical data set fit to a symmetrical Langmuir isotherm. The values $n = 0.65$ and $EC_{50} = 2.2$ fit the upper end of the curve, whereas a function $n = 1$ and $EC_{50} = 2$ fit the lower end. No single symmetrical function fits the entire data set. There are a number of options, in terms of empirical models, for fitting asymmetrical data sets. For example,

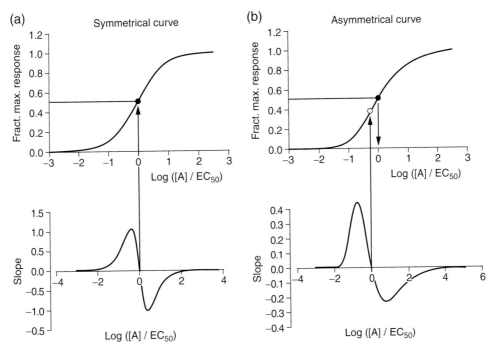

FIGURE 11.17 Symmetrical and asymmetrical dose-response curves. (a) Symmetrical Hill equation with n = 1 and EC_{50} = 1.0. Filled circle indicates the EC_{50} (where the abscissa yields a half maximal value for the ordinate). Below this curve is the second derivative of the function (slope). The zero ordinate of this curve indicates the point at which the slope is zero (inflection point of the curve). It can be seen that the true EC_{50} and the inflection match for a symmetrical curve. (b) Asymmetrical curve (Gompertz function with m = 0.55 and EC_{50} = 1.9). The true EC_{50} is 1.9, while the point of inflection is 0.36.

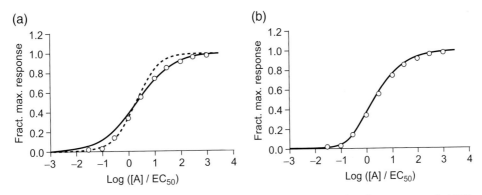

FIGURE 11.18 Asymmetrical dose-response curves. (a) Dose-response data fit to a symmetrical Hill equation with n = 0.65 and EC_{50} = 2.2 (solid line) or n = 1, EC_{50} = 2 (dotted line). It can be seen that neither symmetrical curve fits the data adequately. (b) Data fit to the Gompertz function with m = 0.55 and EC_{50} = 1.9.

the Richards function can be used [7]:

$$y = \frac{E_{max}}{1 + 10^{m(Log[A]+pEC_{50})^s}}. \qquad (11.32)$$

In this model, the factor s introduces the asymmetry. Alternatively, a modified Hill equation can be used [1]:

$$y = \frac{E_{max}}{1 + 10^{(Log[A]+pEC_{50})^p}}. \qquad (11.33)$$

The introduction of the p factor yields asymmetry. Finally, the Gompertz function can be used [5]:

$$y = \frac{E_{max}}{e^{10 \cdot m(Log[A]+pEC_{50})}}. \qquad (11.34)$$

For this model, the factor m introduces asymmetry. The asymmetrical data set shown in Figure 11.18a is fit well with the Gompertz model.

In general, these models are able to fit asymmetrical data sets but require the use of added parameters (thereby reducing degrees of freedom). Also, some of the parameters can be seriously correlated (see discussion in [2, 3, 8]). Most importantly, these are empirical models with no correspondence to biology.

11.4.8 Comparison of Data to Linear Models

There are instances where data are compared to models that predict linear relationships between ordinates and abscissae. Before the widespread availability of computer programs allowing nonlinear fitting techniques, linearizing data was a common practice because it yielded simple algebraic functions and calculations. However, as noted in discussions of Scatchard analysis (Chapter 4) and double reciprocal analysis (Chapter 5), such procedures produce compression of data points, abnormal emphasis on certain data points, and other unwanted aberrations of data. For these reasons, nonlinear curve fitting is preferable. However, in cases where the pharmacological model predicts a linear relationship (such as Schild regressions, see Chapter 6), there are repeated questions asked in the process: (1) Is the relationship linear?, (2) Do two data sets form one, two, or more lines? It is worth discussing these questions with an example of each.

11.4.9 Is a Given Regression Linear?

There are instances where it is important to know if a given regression line is linear. For example, simple competitive antagonism should yield a linear Schild regression (see Chapter 6). A statistical method used to assess whether or not a regression is linear utilizes analysis of covariance. A prerequisite to this approach is that there

must be multiple ordinates for each value of the abscissae. An example of this method is shown in Figure 11.19, where a Schild regression for the α-adrenoceptor antagonist phentolamine is shown for blockade of norepinephrine responses in rat anococcygeus muscle. Saturation of neuronal catecholamine uptake is known to produce curvature of Schild regressions and resulting aberrations in pK_B estimates. Therefore, this method can be used to determine whether the regression is linear (with a slope less than unity) or curved. The conclusions regarding the relationship between the intercept and the pK_B differ for these two outcomes. The data for this example are given in Table 11.9. The calculations for this procedure are detailed in Table 11.10a. As can be seen in Table 11.10b, the value for F_2 is significant at the 1% level of confidence, indicating that the regression is curved ($p < 0.05$).

Curvature in a straight line can be a useful tool to detect departures from model behavior. Specifically, it is easier for the eye to detect deviations from straight lines than from curves (i.e., note the detection of excess protein in the binding curve in Figure 4.4 by linearization of the binding curve). An example of this is detection of cooperativity in binding. Specifically, a biomolecular interaction between a ligand and a receptor predicts a sigmoidal binding curve (according to the Langmuir adsorption isotherm) with a slope of unity if there is no cooperativity in the binding. This means that the binding of one ligand to the receptor population does not affect the binding of another ligand to the population. If there is cooperativity in the binding (as, for example, the binding of oxygen to the protein hemoglobin), then the slope of the binding curve will deviate from unity. Figure 11.20 shows a series of binding curves with varying degrees of cooperativity ($n = 0.8$ to 2). While there are differences between the curves, they must be compared to each other to detect them. In contrast, if the binding curves are linearized (as, for example, through the Scatchard transformation see Chapter 4), then the

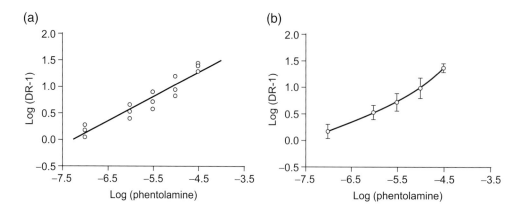

FIGURE 11.19 Test for linearity. Schild regressions for phentolamine antagonism of norepinephrine responses in rat anococcygeus muscle. Ordinates: log (dose ratio − 1). Abscissae: logarithms of molar concentrations of phentolamine. (a) Individual Log (DR-1) values plotted and a best fit straight line passed through the points. (b) Joining the means of the data points (shown with SEM) suggests curvature. The statistical analysis of these data is shown in Table 11.3. Data redrawn from [6].

TABLE 11.9

Schild regression data for phentolamine blockade of norepinephrine responses in rat anococcygeus muscle (data shown in Figure 11.19).

Log[phent]	Log[DR-1]	T
−7	0.25	0.4
−7	−0.05	
−7	0.2	
−6	0.53	1.53
−6	0.3	
−6	0.7	
−5.5	0.71	2.19
−5.5	0.57	
−5.5	0.91	
−5	1	3.12
−5	0.82	
−5	1.3	
−4.5	1.7	4.25
−4.5	1.1	
−4.5	1.45	
$\sum x = -84.00$	$\sum y = 11.49$	$\sum T_i^2/n_i = 11.70$
$\sum x^2 = 481.50$	$\sum y^2 = 12.19$	
$n = 20$	$\sum xy = -58.75$	
$k = 5$		

T = sum of each replicate Log (DR-1) value.

Data from [6].

TABLE 11.10

Test for linearity.

(a) Procedure

	SSq	df	Mean Sq.	Var. Ratio
Due to regression	A	1	s_1^2	
Deviation of means	D	$k - 2$	s_2^2	$F_1 = s_1^2/s_3^2$
Within-assay residual	B	$n - k$	s_3^2	$F_2 = s_2^2/s_3^2$
Total	C	$n - 1$		

$$A = \frac{\left[\sum xy - \left(\sum x \sum y\right)/n\right]}{\sum x^2 \left(\left(\sum x\right)/n\right)^2} \qquad B = \sum y^2 - \sum T_i^2/n_i$$

$$C = \sum y^2 - \frac{\sum y^2}{n} \qquad D = C - A - B$$

(b) Calculations

	SSq	df	Mean Sq.	Var. Ratio
Due to regression	0.86	1	0.855	
Deviation of means	4.24	3	1.414	26.19
Within-assay residual	0.49	15	0.033	43.29
Total	5.59	19		

deviations can readily be seen. This is because the eye is accustomed to identifying linear plots (no cooperativity, n=1) and therefore can identify nonlinear regressions with no required comparison.

11.4.10 One or More Regression Lines? Analysis of Covariance

There are methods available to test whether or not two or more regression lines statistically differ from each other in the two major properties of lines in Euclidean space; namely position (or elevation) and slope. This can be very useful in pharmacology. An example is given later in the chapter for the comparison of Schild regressions (see Chapter 6).

A Schild regression for an antagonist in a given receptor preparation is equivalent to a fingerprint for that receptor and antagonist combination. If the receptor population is uniform (i.e., only one receptor is interacting with the agonist) and the antagonist is of the simple competitive type, then it should be immaterial which agonist is used to produce the receptor stimulation. Under these circumstances, all Schild regressions for a given antagonist in a given uniform preparation should be equivalent for blockade of all agonists for that receptor. However, if there is receptor heterogeneity and the antagonist does not have equal affinity for the receptor types unless the agonists used to elicit response all have identical efficacy for the receptor types there will be differences in the Schild regressions for the antagonist when different agonists are used. Before the advent of recombinant systems, natural cells and/or tissues were the only available test systems available and often these contained mixtures of receptor subtypes. Therefore, a test of possible receptor heterogeneity is to use a number of agonists to elicit response and block these with a single antagonist. This is a common practice for identifying mixtures of receptor populations. Conformity of Schild regressions suggests no receptor heterogeneity. A useful way to compare Schild regressions is with analysis of covariance of regression lines.

Figure 11.21 shows three sets of Log (DR-1) values for the β_1-adrenoceptor antagonist atenolol in guinea pig tracheae. The data points were obtained by blocking the effects of the agonists norepinephrine, isoproterenol, and salbutamol. These were chosen because they have differing efficacy for β_1- versus β_2-adrenoceptors (see Figure 11.21). If a mixture of two receptors mediates responses in this tissue, then responses to the selective agonists should be differentially sensitive to the β_1-adrenoceptor selective antagonist. In the example shown in Figure 11.21, it is not immediately evident if the scatter around the abscissal values is due to random variation or there is indeed some dependence of the values on the type of agonist used. The data for Figure 11.21 are shown in Table 11.11. The procedure for determining possible differences in slope of the regressions is given in Table 11.12a; for the data set in Figure 11.21, the values are given in Table 11.12b. The resulting F value indicates that there is no statistical difference in the slopes of the Schild regressions obtained with each agonist.

The procedure for determining possible differences in position of regression lines is given in Table 11.13a. In contrast to the analysis for the slopes, these data indicate

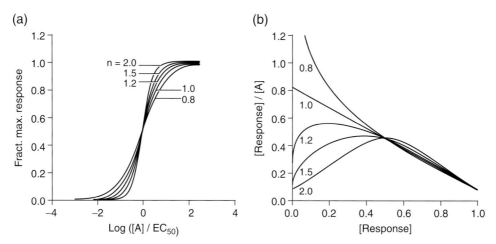

FIGURE 11.20 Use of linear transformation to detect deviation from model behavior. (a) A series of binding curves with various levels of cooperativity in the binding. Numbers next to the curves show the value of the slope of the binding curve according to the equation $[AR] = B_{max} [A]^n/([A]^n + K^n)$. (b) Scatchard tranformation of the curves shown in panel A according to the equation $[AR]/[A] = (B_{max}/K) - ([AR]/K)$ (Equation 4.5). Numbers are the value of the slope of the binding curves. Cooperativity in binding occurs when $n \neq 1$.

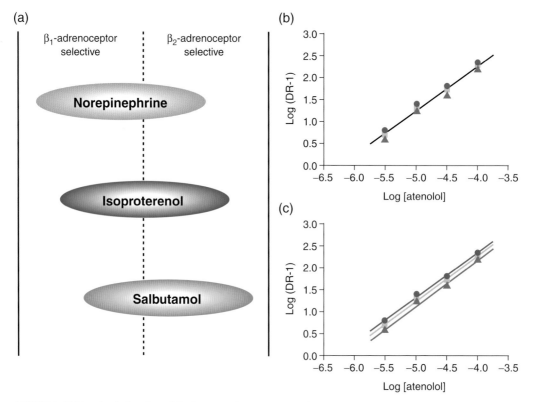

FIGURE 11.21 Analysis of straight lines to detect receptor heterogeneity. (a) Schematic diagram depicts the relative affinity and efficacy of three agonists for two subtypes of β-adrenoceptors. Norepinephrine is relatively β₁-adrenoceptor selective while salbutamol is relatively β₂-adrenoceptor selective. (b) Schild regression for blockade of β-adenoceptor mediated relaxation of guinea pig tracheae using the three agonists: salbutamol (filled circles), isoproterenol (open circles), and norepinephrine (open triangles). Data points fit to a single regression. (c) Regression for each agonist fit to a separate regression. Analysis for these data given in Tables 11.12b and 11.13b. In this case, the data best fits three separate regressions, indicating that there is a difference in the antagonism produced by atenolol and therefore probably a heterogeneous receptor population mediating the responses.

that there is a statistical difference in the elevation of these regressions ($F = 9.31$, df $= 2$, 8; see Table 11.13b). This indicates that the potency of the antagonist varies with the type of agonist used in the analysis. This, in turn, indicates that the responses mediated by the agonists are not due to activation of a homogeneous receptor population.

TABLE 11.11

Analysis of covariance of regression lines.

Data for Figure 11.15. Schild analysis for atenolol in guinea pig trachea.

Log[phentol]	Norepi	Iso.	Salb.
−5.5	0.8	0.7	0.6
−5	1.4	1.29	1.25
−4.5	1.8	1.75	1.6
−4	2.35	2.3	2.2
$\sum x_i = -19$	$\sum y_i = 6.35$	6.04	5.65
$\sum x_i^2 = 91.15$	$\sum y_i^2 = 11.36$	10.51	9.32
	$\sum xy_i = -28.90$	−27.38	−25.55

Norepi = norepinephrine; Iso = isoproterenol; salb. = salbutamol; phentol = phentolamine.

11.5 Comparison of Samples to "Standard Values"

In the course of pharmacological experiments, a frequent question is: Does the experimental system return expected (standard) values for drugs? With the obvious caveat that "standard" values are only a sample of the population that have been repeatedly attained under a variety of circumstances (different systems, different laboratories, different investigators), there is a useful statistical test that can provide a value of probability that a set of values agree or do not agree with an accepted standard value. Assume that four replicate estimates of an antagonist affinity are made (pK_B values) to yield a mean value (see Table 11.14). A value of t can be calculated that can give the estimate probability that the mean value differs from a known value with the formula

$$t_{calculated} = \frac{|known\ value - x_m|}{s}\sqrt{n}, \qquad (11.35)$$

where x_m is the mean of the values. For the example shown in Table 11.14, $t = 2.36$ (df $= 3$). Comparison of this value to the table in Appendix A indicates that there is 95% confidence that the mean value obtained in the experimental system is not different from the accepted standard

TABLE 11.12

Regression line and slope covariance.

(a) Analysis of Covariance of Regression Lines (Comparison of Slopes)			
	SSq	df	Mean Sq
Due to common slope	A	1.00	
Differences between slopes	B	k − 1	C
Residual	D	n − 2k	E

$$F = C/E, df = (k-1), (n-2k).$$

$$s_x^2 = \sum x_i^2 - \frac{\left(\sum x\right)^2}{n_i} \qquad s_y^2 = \sum y_i^2 - \frac{\left(\sum y\right)^2}{n_i}$$

$$s_{xy} = \sum xy_i - \frac{\left(\sum x_i\right)\left(\sum y_i\right)}{n_i} \qquad A = \left[\sum_{i=1}^{k}(s_{xy})_i\right]^2 \Big/ \sum_{i=1}^{k}(s_x^2)_i$$

$$B = \sum_{i=1}^{k}\left[\frac{(s_{xy})_i^2}{(s_x^2)_i}\right] - A \qquad C = \frac{B}{(k-1)}$$

$$D = \sum_{i=1}^{k}(s_y^2)_i - \sum^{k}\frac{(s_{xy}^2)}{s_x^2 (s_x^2)_i} \qquad E = \frac{D}{(n-2k)}$$

(b) Calculations: Analysis of Data in Figure 11.15 (Analysis of Covariance of Slopes)			
	SSq	df	Mean Sq
Due to common slope	3.98	1	
Difference between slopes	0.001	2	0.0006
Residual	0.03	6	0.00042
		F = 0.13	
		df = 2.6	

TABLE 11.13

Covariance of position.

(a) Analysis of Covariance of Regression Lines (Comparison of Position)

	s_x^2	s_{xy}	s_y^2	SSq	df	Mean Sq
Within groups	A	B	C	D	$n - k - 1$	E
Total	F	G	H	I		
Between groups				J	$k - 1$	K

$$F = K/E, df = (k-1), (n-k-1).$$

$$A = \left(\sum x^2\right)_{total} = \sum_{i=1}^{k} \frac{\left(\sum x\right)_i^2}{n_i} \qquad B = \left(\sum x^2\right)_{total} = \sum_{i=1}^{k} \frac{\left(\sum x\right)_i \left(\sum y\right)_i}{n_i}$$

$$D = C - \frac{(B)^2}{A}$$

$$C = \left(\sum y^2\right)_{total} = \sum_{i=1}^{k} \frac{\left(\sum y\right)_i^2}{n_i} \qquad E = \frac{D}{n-k-1}$$

$$F = \left(\sum x^2\right)_{total} = \frac{\left(\sum x\right)_{total}^2}{n_{total}} \qquad G = \left(\sum xy\right)_{total} = \frac{\left(\left(\sum x\right)_{total}\left(\sum y\right)_{total}\right)}{n_{total}}$$

$$I = C - \frac{(G)^2}{F}$$

$$J = |D - I|$$

$$H = \left(\sum y^2\right)_{total} = \frac{\left(\sum y\right)_{total}^2}{n_{total}} \qquad K = \frac{J}{K-1}$$

(b) Analysis of Covariance of Position (Calculations for Data Shown in Figure 11.15)

	s_x^2	s_{xy}	s_y^2	SSq	df	Mean Sq
Within groups	3.75	3.86	4.01	0.03	8.00	0.003
Total	3.75	3.86	4.07	0.09		
Between groups				0.06	2.00	0.03
			F = 9.31			
			df = 2, 8			

value of $pK_B = 7.4$. Therefore, there is a 95% level of certainty that the experimental value falls within the accepted normal standard for this particular antagonist in the experimental system.

11.5.1 Comparison of Means by Two Methods or in Two Systems

Another frequent question asked considers whether the mean of a value measured by two separate methods differs significantly. For example, does the mean pK_B value of an antagonist measured in a binding experiment differ significantly from its affinity as an antagonist of agonist function? The value of t for the comparison of the mean values x_{m1} and x_{m2} can be calculated with the following equation:

$$t_{calculated} = \frac{(x_{m1} - x_{m2})}{s_{pooled}} \sqrt{\frac{n_1 n_2}{n_1 + n_2}}, \qquad (11.36)$$

TABLE 11.14

Experimental estimates of antagonist affinity: comparison to standard value.

pK_B	Standard
7.6	Value = 7.4
7.9	
8.1	
7.5	t = 2.36
	df = 3
mean = 7.775	
s = 0.28	

where:

$$s_{pooled} = \sqrt{\frac{s_1^2(n_1 - 1) + s_2^2(n_2 - 1)}{n_1 + n_2 - 2}} \qquad (11.37)$$

and s_1^2 and s_2^2 are given by Equation 11.4.

Table 11.15 shows the mean of four estimates of the affinity of an antagonist measured with radioligand binding

TABLE 11.15

Comparing two mean values to evaluate method/assays.

Binding pK_I	Function pK_B
8.1	7.6
8.3	7.7
7.9	7.9
7.75	7.5
mean $= 8.01$	7.68
$s = 0.24$	0.17
$s_{pooled} = 0.21$	
$t = 2.29$	
$df = 6$	

TABLE 11.16

Multiple values to compare methods.

pK_B values for human α_{1B}-adrenoceptor antagonists obtained in binding studies with SPA and filter binding.[1]

	pK_I	pK_I	Difference
Prazosin	10.34	10.27	0.0049
5-CH3 urapidil	7.05	7.32	0.0729
Yohimbine	6.1	6.31	0.0441
BMY7378	7.03	7.06	0.0009
Phentolamine	7.77	7.91	0.0196
		mean $= 0.03$	
		$s = 0.03$	
		$t = 2.12$	
		$df = 5$	

[1]Data from [4].

and also in a functional assay. Equation 11.34 yields a value for t of 2.29. For $n_1 + n_2 - 2$ degrees of freedom, this value of t is lower than the t for confidence at the 95% level (2.447, see Appendix A table of t values). This indicates that the estimate of antagonist potency by these two different assay methods does not differ at the 95% confidence level. It should be noted that the previous calculation for pooled standard deviation assumes that the standard deviation for both populations is equal. If this is not the case, then the degrees of freedom are calculated by

degrees of freedom

$$= \left[\frac{s_1^2/n_1 + s_2^2/n_2}{((s_1^2/n_1)^2/(n_1 + 1)) + ((s_2^2/n_2)^2/(n_2 + 1))} \right] - 2. \quad (11.38)$$

11.5.2 Comparing Assays/Methods with a Range of Ligands

One way to compare receptor assays is to measure a range of agonist and antagonist activities in each. The following example demonstrates a statistical method by which two pharmacological assays can be compared. Table 11.16 shows the pK_B values for a range of receptor antagonists for human α_{1B} adrenoceptors carried out with a filter binding assay and with a scintillation proximity assay (SPA). The question asked is: Does the method of measurement affect the measured affinities of the antagonists? The relevant measurement is the difference between the estimates made in the two systems (defined as $x_{1i} - x_{2i} = d$):

$$t_{calculated} = \frac{d_m}{s_d} \sqrt{n}, \quad (11.39)$$

where d_m is the mean difference and s_d is given by

$$s_d = \frac{\sqrt{\sum (d_i - d)^2}}{n - 1}. \quad (11.40)$$

As seen in Table 11.16, the values for α_{1B}-adrenoceptor antagonists obtained by filter binding and SPA do not differ significantly at the $p < 0.05$ level. This suggests that

there is no difference between the two methods of measurement.

11.6 Experimental Design and Quality Control

11.6.1 Detection of Difference in Samples

In a data set it may be desirable to ask the question: Is any one value significantly different from the others in the sample? A t statistic (for $n - 1$ degrees of freedom where the sample size is n) can be calculated that takes into account the difference of the magnitude of that one value (x_i) and the mean of the sample (x_m):

$$t_{n-1} = \frac{(x_m - x_i)}{s\sqrt{((1/n) + 1)}}, \quad (11.41)$$

where s is the standard error of the means. This can be used in screening procedures where different compounds are tested at one concentration and there is a desire is to detect a compound that gives a response significantly greater than basal noise. As samples get large, it can be seen that the square root term in the denominator of Equation 11.39 approaches unity and the value of t is the deviation divided by the standard error. In fact, this leads to the the standard rule where values are different if they exceed $t \cdot s$ limits (i.e., for t_{95} these would be the 95% confidence limits; see Section 11.3.3). This notion leads to the concept of control charts (visual representation of confidence intervals for the distribution), whereby the scatter and mean of a sample are tracked consecutively to detect possible trends of deviation. For example, in a drug activity screen a standard agonist is tested routinely for quality control and the pEC_{50} noted chronologically throughout the screen. If on a given day the pEC_{50} of the control is outside of the 95% c.l. of the sample means collected throughout the course of the screen, then the data collected on that day is suspect and the experiment may need to be repeated. Figure 11.22a shows such a chart where the definitions of a warning limit are the values that exceed 95.5% (> 2 s units) of the confidence limits of the

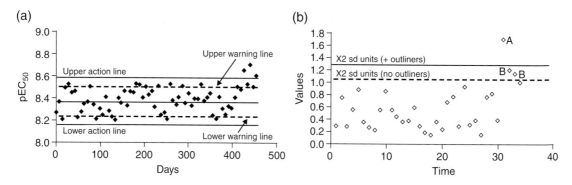

FIGURE 11.22 Control charts and outliers. (a) pEC_{50} values (ordinates) run as a quality control for a drug screen over the days on which the screen is run (abscissae). Dotted lines are the 95% c.l. and the solid lines the 99.7% c.l. Data points that drift beyond the action lines indicate significant concern over the quality of the data obtained from the screen on those days. (b) The effect of significant outliers on the criteria for rejection. For the data set shown, the inclusion of points A and B lead to a c.l. for 95% confidence that includes point B. Removal of point A causes the 95% limits to fall below points B, causing them to be suspect as well. Thus, the presence of the data to be possibly rejected affects the criteria for rejection of other data.

TABLE 11.17

Q-test for rejection of data points.

Table A		Table B	
pK_B Values	Gap	Q@90%	n
7.5		0.76	4
7.6	0.1	0.64	5
7.6	0	0.56	6
7.7	0.1	0.51	7
7.8	0.1	0.47	8
7.8	0	0.44	9
8.1	0.3	0.41	10
	Range = 0.6		
	Q = 0.5		
	n = 7		

mean and action (removal of the data) applies to values >99.7% (3 s units) c.l. of the mean. Caution should be included in this practice since the presence of outliers themselves alters the outcome of the criteria for the test (in this case, the standard mean and standard error of that mean). Figure 11.22b shows a collection of data where inclusion of the outlier significantly alters the mean and standard error to the extent that the decision to include or exclude two other data points are affected. This effect is more serious with smaller samples and loses importance as sample size increases.

Another method that may be employed to test whether single data points should be included in a sample mean is the Q-test. This simple test determines the confidence with which a data point can or cannot be considered part of the data set. The test calculates a ratio of the gap between the data point and its nearest neighbor and the range of the complete data set:

$$Q_{calculated} = gap/range. \qquad (11.42)$$

If Q is greater than values from a table yielding Q values for 90% probability of difference, then the value may be removed from the data set (p < 0.10). An example of how this test is used is given in Table 11.17a. In this case, the pK_B value of 8.1 appears to be an outlier with respect to the other estimates made. The calculated Q is compared to a table of Q values for 90% confidence (Table 11.17b) to determine the confidence with which this value can be accepted into the data set. In the case shown in Table 8.17, Q < 0.51. Therefore, there is <90% probability that the value is different. If this level of probability is acceptable to the experimenter, then the value should remain in the set.

Scientifically, the question of outliers is a difficult one. On one hand, they could be due to high random biological and/or measurement variation and therefore legitimately rejected. On the other hand, they might be the most interesting data in the set and indicative of a rare but important effect. For instance, in a psychological cognition test outliers may represent a rare but real cognitive problem leading to a fractal change in the test score. As with hypothesis testing, the ultimate responsibility lies with the investigator.

11.6.2 Power Analysis

There is an increasing appreciation of the importance of power analysis in the drug discovery process. This method enables decisions regarding the size of the experimental sample needed to make accurate and reliable judgments and to estimate the likelihood that the statistical tests will find differences of a given magnitude. The size of the sample is important since too small a sample will be useless (the result will be too imprecise for definitive conclusions to be drawn) and too large a sample leads to diminishing returns and wasted resources. These ideas can be dealt with in sampling theory and power analysis.

Essentially, the decision regarding the sample size involves the question: How large does a sample need to be to accurately reflect the characteristics of the population? For example, the question could be stated: Is the potency of a given agonist in a recombinant assay equal to the known potency of the same agonist in a secondary therapeutic assay? The true value of the potency in the recombinant system, denoted λ, is estimated by choosing a sample of n values from the population. The mean observed potency of the agonist in this sample is denoted ψ. Unless the sample size is nearly infinite, the value λ will not equal ψ since λ was obtained by random sampling. The magnitude of the difference is referred to as the sampling error. The larger the value of n, the lower is the sampling error. Computer calculation of power curves can yield guidelines for the sample needed to find a defined difference between the population and the sample (if there is one), the probability that this difference is real, and the likelihood that the defined sample size will be successful in doing so (i.e., find the minimal sampling error).

Statistical power can be illustrated with a graphical example. There are three principal components to power analysis: (1) define the magnitude of the difference δ that one wishes to detect, (2) quantify the error in measuring the values, and (3) choose the power (make the experimental choice of defining the probability that the experiment will reject the null hypothesis). Assume that the aim of a study is to find values that are greater than 95% of a given population ($p < 0.05$ for difference). The sample of data we obtain will be represented by a normal distribution. The difference we wish to find is denoted δ (see Figure 11.23). We want to know what proportion of the sample distribution is greater than the 95th percentile of the population distribution (shaded area of the sample distribution in Figure 11.23). The proportion of the

sample distribution that lies in the defined region (in this case, >95th percentile) is defined as the power to be able to detect the sample value that is greater than the 95th percentile. The sample for a given experiment will yield a distribution of values. In Figure 11.23a, the percentage of the sample distribution greater than the 95th percentile of the population is 67%. Therefore, that is the power of the analysis as shown. This means that with the experiment designed in the present manner there will be a 67% chance that the defined difference δ will be detected with >95% probability. One way to increase the chances of detecting the defined difference δ is to produce a sampling distribution that has a larger area lying to the right of the 95th percentile. Figure 8.23b shows a distribution with 97% of the area >95th percentile of the population. This second situation has a much greater probability of finding a value >δ (i.e., has a higher statistical power). It can be seen that this is because the distribution is more narrow. One way of getting from the situation shown in Figure 11.23a to the one in Figure 11.23b (more narrow sampling distribution) is to reduce the standard error. This can be done by increasing the number of samples (n, see Equation 11.4). Therefore, the power and n are interrelated, allowing researchers to let power define the value of n (sample size) needed to determine a given difference δ with a defined probability. The number of samples given by power analysis to define a difference δ, the measurement of which has a standard deviation s, is given by

$$n \geq \frac{2(t_i + t_p)^2 s^2}{\delta^2}, \qquad (11.43)$$

where t_i is the t value for significance level desired (in the example in Figure 11.23, this was 95%) and t_p is the level of power (67% for Figure 11.23a and 97% for Figure 11.23b).

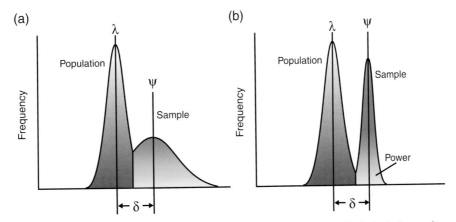

FIGURE 11.23 Power analysis. The desired difference is >2 standard deviation units ($\lambda - \psi = \delta$). The sample distribution in panel a is wide and only 67% of the distribution values are >δ. Therefore, with an experimental design that yields the sample distribution shown in panel a will have a power of 67% to attain the desired endpoint. In contrast, the sample distribution shown in panel b is much less broad and 97% of the area under the distribution curve is >δ. Therefore, an experimental design yielding the sample distribution shown in panel B will gave a much higher power (97%) to attain the desired end point. One way to decrease the broadness of sample distributions is to increase the sample size.

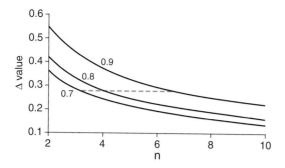

FIGURE 11.24 Power curves. Abscissae is the sample size required to determine a difference between means shown on the ordinate. Numbers next to the curves refer to the power of finding that difference. For example, the gray lines show that a sample size of n = 3 will find a difference of 0.28 with a power of 0.7 (70% of the time) but that the sample size would need to be increased to 7 to find that same difference 90% of the time. The difference of 0.28 has previously been defined as being 95% significantly different.

This latter value (t_p) is given by power analysis software and can be obtained as a power curve. Figure 11.24 shows a series of power curves giving the samples sizes required to determine a range of differences. From these curves, for example, it can be seen that a sample size of 3 will be able to detect a difference of 0.28 with a power of 0.7 (70% of time) but that a sample size of 7 would be needed to increase this power to 90%. In general, power analysis software can be used to determine sample sizes for optimal experimental procedures.

11.7 Chapter Summary and Conclusions

- Descriptive statistics quantify central tendency and variance of data sets. The probability of occurrence of a value in a given population can be described in terms of the Gaussian distribution.
- The t distribution allows the use of samples to make inferences about populations
- Statistical tests simply define the probability that a hypothesis can be disproven. The experimenter still must assume the responsibility of accepting the risk that there is a certain probability that the conclusion may be incorrect.
- The most useful description of variance are confidence limits since these take into account the sample size.
- While a t-test can be used to determine if the means of two samples can be considered to come from the same population, paired data sets are more powerful to determine difference.

- Possible significant differences between samples can be estimated by one-way and two-way analysis of variance.
- While correlation can indicate relationships, it does not imply cause and effect.
- There are statistical methods to determine the verisimilitude of experimental data to models. One major procedure to do this is nonlinear curve fitting to dose-response curves predicted by receptor models.
- Choosing models that have parameters that can be related to biology is preferable to generic mathematical functions that may give better fits.
- There are statistical procedures available to choose models (hypothesis testing), assess outliers (or weight them), and deal with partial curves.
- Procedures can also be used to analyze straight lines with respect to slope and position, compare sample values to standard population means, compare methods, and detect differences in small samples.
- Power analysis can be used to optimize experiments for detection of difference with minimal resources.

References

1. Boeynaems, J. M., and Dumont, J. E. (1980). *Outlines of receptor theory.* Elsevier/North-Holland, Amsterdam.
2. Freund, R. J., and Litell, C. R. (eds.) (1991). Non-linear models. In: *SAS system recognition.* Cary, SAS Institute.
3. Giraldo, J., Vivas, N. M., Vila, E., and Badia, A. (2002). Assessing the (a)symmetry of concentration-effect curves: Empirical versus mechanistic models. *Pharmacol. Therap.* **95**:21–45.
4. Gobel, J., Saussy, D. L., and Goetiz, A. S. (1999). Development of scintillation-proximity assays for alpha adrenoceptors. *J. Pharmacol. Toxicol. Meth.* **42**:237.
5. Gompertz, B. (1825). On the nature of the function expressive of the law of human mortality. *Philo. Trans. R. Soc. Lond.* **36**:513–585.
6. Kenakin, T. P., and Beek, D. (1981). The measurement of antagonist potency and the importance of selective inhibition of agonist uptake processes. *J. Pharmacol. Exp. Ther.* **219**:112–120.
7. Richards, F. J. (1959). A flexible growth function for empirical use. *J. Exp. Bot.* **10**:290–300.
8. Van Der Graaf, P. H., and Schoemaker, R. C. (1999). Analysis of asymmetry of agonist concentration-response curves. *J. Pharmacol. Toxicol. Methods* **41**:107–115.
9. Watson, C., Chen, G., Irving, P. E., Way, J., Chen, W.-J., and Kenakin, T. P. (2000). The use of stimulus-biased assay systems to detect agonist-specific receptor active states: Implications for the trafficking of receptor stimulus by agonists. *Mol. Pharmacol.* **58**:1230–1238.

12

Selected Pharmacological Methods

In mathematics you don't understand things. You just get used to them.
— JOHANN VON NEUMANN (1903–1957)

12.1 Binding Experiments

12.1.1 Saturation Binding

Aim: To measure the binding of a radioligand (or ligand that is traceable by other means) to a receptor. The object is to obtain an estimate of the equilibrium dissociation constant of the radioligand-receptor complex (denoted K_d) and the maximal number of binding sites (denoted B_{max}).

General Procedure: The receptor preparation is incubated with a range of concentrations of radioligand (to give a measure of total binding) and again in the presence of a high concentration of nonradioactive receptor-selective ligand (present at a concentration of $100 \times Kd$ for the nonradioactive ligand) to give a measure of nonspecific binding (nsb). After a period of equilibration (30 to 90 minutes), the amount of bound ligand is quantified and the total binding and nsb are fit to simultaneous equations to yield a measure of the ligand specifically bound to receptor.

Procedure:

1. A range of concentrations of radioligand are added to a range of tubes (or wells). An example of such a range of concentrations (in pM) is shown in Table 12.1. A parallel array of tubes is prepared with an added concentration of nonradioactive ligand (to define nsb) at a concentration $100\times$ the K_d for binding to receptor.
2. The membrane (or cell) preparation is added to the tubes to begin the binding reaction. The reagents are equilibrated for 30 to 90 minutes (time required for equilibration must be determined experimentally) and then the amount of bound ligand is quantified (either by separation or reading of scintillation proximity beads). The nsb and total binding are obtained from this experiment as shown (in bound pM).
3. The total binding and nsb are plotted as a function of added radiolabel (as shown in Figure 12.1a), and fit simultaneously with nonlinear curve fitting

techniques. For the example shown in Figure 12.1a, the data are fit to

$$\text{Total Binding} = \frac{[A^*]^n \cdot B_{max}}{[A^*]^n + K_{d^n}} + k \cdot [A^*] \qquad (12.1)$$

and

$$\text{nsb} = k \cdot [A^*]. \qquad (12.2)$$

4. The data for Table 12.1 columns A through C were fit to Equations 12.1 and 12.2 simultaneously to yield $B_{max} = 6.63 \pm 1.5$ pmoles/mg protein, $n = 0.95 \pm 0.2$, and $K_d = 26.8$ pM ($pK_d = 10.57 \pm 0.3$). The fitted curves are shown in Figure 12.1b along with a dotted line to show the calculated specific binding.

12.1.2 Displacement Binding

Aim: To measure the affinity of a ligand by observing the inhibition it produces of a receptor-bound radioligand (or ligand that is traceable by other means). The object is to obtain an estimate of the equilibrium dissociation constant of the nonradioactive ligand receptor complex (alternately denoted K_B or K_i). The pattern of displacement curves can also be used to determine whether or not the antagonism is competitive.

General Procedure: The receptor preparation is incubated with a single concentration of radioligand (this furnishes the B_0 value) in the absence and presence of a range of concentrations of nonradioactive displacing ligand. This is also done in the presence of a high concentration of nonradioactive ligand (present at a concentration of $100 \times Kd$ for the nonradioactive ligand) to give a measure of nonspecific binding (nsb). After a period of equilibration (30 to 90 minutes), the amount of bound ligand is quantified. The nsb value is subtracted from the estimates of total binding to yield a measure of the ligand specifically bound to receptor. The resulting displacement curves are fit to models to yield the equilibrium dissociation constant of the displacing ligand-receptor complex.

Procedure:

1. Choice of radioligand concentration. The optimal concentration is one that is below the K_d for

TABLE 12.1

Data for saturation binding curves.

A [A*]: M	B nsb	C Total Binding
4.29×10^{-12}	0.16	0.97
1.3×10^{-11}	0.45	2.42
2.7×10^{-11}	0.81	3.87
4.0×10^{-11}	1.29	5.16
6.86×10^{-11}	2.10	6.77
1.37×10^{-10}	4.19	10.00
2.2×10^{-10}	6.94	12.58

Binding in pmoles/mg protein.

saturation binding (i.e., $[A*] = 0.1 \, K_d$, $0.3 \, K_d$) such that the IC_{50} of the displacement curves will not be significantly higher than the K_B for the antagonist. This will minimize the extrapolation required for determination of the K_i. However, a higher concentration may be required to achieve a useful window of specific binding and sufficient signal-to-noise ratio. The amount of membrane protein can also be adjusted to increase the signal strength, with the caveat that too much protein will deplete the radioligand and produce error in the measurements (see Section 4.4.1). For this example, four radioligand concentrations are chosen to illustrate the Cheng-Prusoff correction for determination of K_B from the IC_{50} values.

2. The chosen concentration of radioligand is added to a set of tubes (or wells). To a sample of these a concentration of a designated nonradioactive ligand used to define nsb is added at a concentration 100 nM the K_d for binding to receptor. Then, a range of concentrations of the nonradioactive ligand for which the displacement curve will be determined is added to the sample of tubes containing prebound

radioligand. The concentrations for this example are shown in Table 12.2a.

3. The membrane (or cell) preparation is added to the tubes to begin the binding reaction. The reagents are equilibrated for 30 to 90 minutes (see considerations of temporal effects for two ligands coming to equilibrium with a receptor in Section 4.4.2), and then the amount of bound ligand is quantified (either by separation or reading of scintillation proximity beads). The nsb and total binding are obtained from this experiment, as shown in Table 12.2a (in bound pM). For a radioligand concentration of $[A*]/K_d = 0.1$, the total binding is shown in Table 12.2a. For three higher concentrations of radioligand, the data are shown in the columns to the right in Table 12.2a.

4. The nsb for this example was shown to be 15.2 ± 0.2 pM/mg protein. This value is subtracted from the total binding numbers or the total binding fit to displacement curves. Total binding with a representation of nsb is used in this example and is shown in Figure 12.2a.

5. Nonlinear fitting techniques (for example, to Equation 12.3) are used to fit the data points to curves. The IC_{50} values form the fit curves are shown in Table 12.2b.

$$\rho^* = \frac{B_0 - \text{basal}}{1 + (10^{\text{Log}[B]}/10^{\text{Log}[IC50]})} + \text{Basal} \qquad (12.3)$$

Here, B_0 is the initial binding of radioligand in the absence of displacing ligand and basal is the nsb.

6. It can be seen that the IC_{50} increases with increasing values of $[A*]/K_d$ in accordance with simple competitive antagonism. This can be tested by comparison of the data to the Cheng-Prusoff equation (Equation 4.12). The data in Table 12.2b

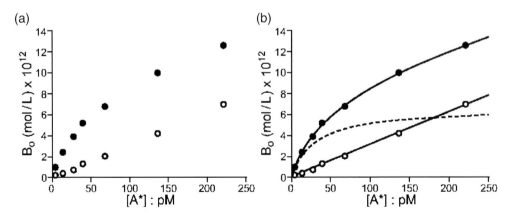

FIGURE 12.1 Human calcitonin receptor binding. Ordinates: pmole ^{125}I-AC512 bound/mg protein. Abscissae: concentration of ^{125}I-AC512 (pM). Total binding (filled circles) and nsb (open circles). Curves fit simultaneously to Equations 4.3 and 4.1. ($B_{max} = 6.63$ pmoles/mg protein, $n = 0.95$, $K_d = 26.8$ pM). Dotted line shows specific binding.

TABLE 12.2
Displacement binding.

	(a) Data for Displacement of Radioligand Binding Curves			
[B]: M	$[A^*]/$ $K_d = 0.1$	$[A^*]/$ $K_d = 0.3$	$[A^*]/$ $K_d = 1.0$	$[A^*]/$ $K_d = 3.0$
10^{-14}	17.7	21.87	29.93	37.44
3×10^{-14}	17.65	21.77	29.78	37.33
10^{-13}	17.5	21.43	29.29	36.95
3×10^{-13}	17.14	20.63	28.04	35.93
10^{-12}	16.43	18.91	25	33
3×10^{-12}	15.73	17.09	21	27.86
10^{-11}	15.27	15.8	17.5	21.43
3×10^{-11}	15.1	15.29	15.94	17.65
10^{-10}	15.03	15.09	15.29	15.87
3×10^{-10}	15.01	15.03	15.1	15.3

Concentration of displacing ligand in pM.

Binding shown as pmoles/mg protein.

nsb = 15 ± 0.2 pmoles/mg protein.

	(b) Fit Parameters to Data Shown in A	
$[A^*]/K_d$	IC_{50} (M)	n
0.9	1.1×10^{-12}	0.95
2.7	1.3×10^{-12}	0.97
9	2×10^{-12}	0.92
27	3.9×10^{-12}	0.95

is fit to

$$IC_{50} = K_B \cdot ([A^*]/K_d + 1). \qquad (12.4)$$

The resulting fit is shown in Figure 12.2c. The regression is linear with a slope not significantly different from unity (slope = 0.95 ± 0.1). The intercept yields the K_B value; in this case, 1 pM.

7. In cases where the plot of $[A^*]/K_d$ vs IC_{50} is not linear, other mechanisms of antagonism may be operative. If there is a nearly vertical relationship, this be due to noncompetitive antagonism in a system with no receptor reserve (see Figure 12.2d). Alternatively, if the plot is linear at low values of $[A^*]/K_d$ and then approaches an asymptotic value the antagonism may be allosteric (the value of α defines the value of the asymptote) or noncompetitive in a system with receptor reserve (competitive shift until the maximal response is depressed, Figure 12.2d).

12.2 Functional Experiments

12.2.1 Determination of Equiactive Concentrations on Dose-response Curves

Aim: Mathematical estimation of concentrations on a dose-response curve that produce the same magnitude of response as those on another dose-response curve. This is a procedure common to many pharmacological methods aimed at estimating dose-response curve parameters.

General Procedure: A function is fit to both sets of data points and a set of responses are chosen that have data points for at least one of the curves within the range of the other curve. A metameter of the fitting function is then used to calculate the concentrations of agonist for the other curve that produce the designated responses from the first curve.

Procedure:

1. Dose-response data are obtained and plotted on a semi-logarithmic axis, as shown in Figure 12.3a (data shown in Table 12.3a).

2. The data points are fit to a function with nonlinear fitting procedures. For this example, Equation 12.5 is used:

$$Response = Basal + \frac{E_{max}[A]^n}{[A]^n + (EC_{50})^n}. \qquad (12.5)$$

The procedure calculates the concentrations from both curves that produce the same level of response. Where possible, one of the concentrations will be defined by real data and not the fit curve (see Figure 12.3b). The fitting parameters for both curves are shown in Table 12.3b. Some alternative fitting equations for dose-response data are shown in Figure 12.4.

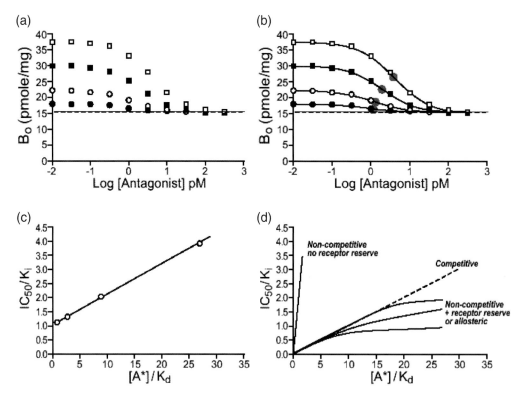

FIGURE 12.2 Displacement of a radioligand by a nonradioactive competitive ligand. Ligand displaces signal to nsb, which in this case is 15 pmoles/mg protein. Ordinates: pmoles/mg protein bound. Abscissae concentration of displacing ligand in pM on a logarithmic scale. (a) Data for displacement curves shown for increasing concentrations of radioligand. Curves shown for $[A^*]/K_d = 0.1$ (filled circles), $[A^*]/K_d = 0.3$ (open circles), $[A^*]/K_d = 1.0$ (filled squares), and $[A^*]/K_d = 3.0$ (open squares). (b) Nonlinear curve fitting according to Equation 4.9. (c) Cheng-Prusoff correction for IC_{50} to K_B values for data shown in panel b. (d) Theoretical Cheng-Prusoff plots for competitive antagonist (dotted line) and noncompetitive and/or allosteric antagonists in different systems.

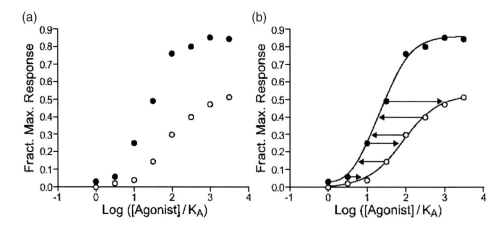

FIGURE 12.3 Determination of equiactive concentrations of agonist. (a) Two dose-response curves. (b) Concentrations of agonist (denoted with filled and open circles) that produce equal responses are joined with arrows that begin from the real data point and end at the calculated curve.

TABLE 12.3

Estimation of equiactive agonist concentrations.

(a) Dose-Response Data for Two Curves

[A]: M	Control Curve 1	Treated Curve 2	
10^{-9}	0.025	0	
3×10^{-8}	0.06	0.02	*
10^{-8}	0.25	0.04	* Designated responses
3×10^{-8}	0.49	0.145	*
10^{-7}	0.755	0.3	
3×10^{-7}	0.8	0.4	
10^{-6}	0.85	0.47	
3×10^{-6}	0.84	0.51	

(b) Parameters for Fit Curves

Curve 1	Curve 2
$E_{max} = 0.86$	$E'_{max} = 0.52$
$EC_{50} = 22$ nM	$EC'_{50} = 79$ nM
$n = 1.2$	$n' = 1$
$Basal = 0$	$Basal' = 0$

(c) Equiactive Agonist Concentrations

Responce	[A$_1$]: M	[A$_2$]: M
0.06	$\mathbf{3 \times 10^{-9}}$	1.03×10^{-8}
0.145	3.38×10^{-9}	$\mathbf{3.0 \times 10^{-8}}$
0.25	$\mathbf{10^{-8}}$	7.3×10^{-8}
0.3	7.8×10^{-9}	$\mathbf{10^{-7}}$
0.4	1.17×10^{-8}	$\mathbf{3.0 \times 10^{-7}}$
0.49	$\mathbf{3.0 \times 10^{-8}}$	1.29×10^{-6}

Real data points in bold font. Calculated from fit curves in normal font.

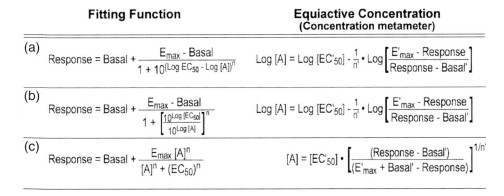

	Fitting Function	Equiactive Concentration (Concentration metameter)
(a)	$Response = Basal + \dfrac{E_{max} - Basal}{1 + 10^{(Log\,EC_{50} - Log\,[A])^n}}$	$Log\,[A] = Log\,[EC'_{50}] - \dfrac{1}{n'} \bullet Log\left[\dfrac{E'_{max} - Response}{Response - Basal'}\right]$
(b)	$Response = Basal + \dfrac{E_{max} - Basal}{1 + \left[\dfrac{10^{Log\,[EC_{50}]}}{10^{Log\,[A]}}\right]^n}$	$Log\,[A] = Log\,[EC'_{50}] - \dfrac{1}{n'} \bullet Log\left[\dfrac{E'_{max} - Response}{Response - Basal'}\right]$
(c)	$Response = Basal + \dfrac{E_{max}\,[A]^n}{[A]^n + (EC_{50})^n}$	$[A] = [EC'_{50}] \bullet \left[\dfrac{(Response - Basal')}{(E'_{max} + Basal' - Response)}\right]^{1/n'}$

FIGURE 12.4   Metameters for determining equiactive concentrations of agonist.

3. A range of responses (corresponding to real data points) are chosen from the dose-response curves. For this example, the responses from concentrations of curve 1 (3, 10, and 30 nM) and responses from curve 2 (30, 100, and 300 nM) are compared. The corresponding responses are 0.06, 0.145, 0.25, 0.3, and 0.145 $(1-(T_r/T_i))$ units (melanophore responses).

4. These responses are used for response in the concentration metameter for the fit for the second curve. For example, the response defined by real data for curve 1 at 3 nM is 0.06. The corresponding equiactive concentration from curve 2 is given by Equation 12.6, with Response = 0.06, basal = 0, and the values of n', E'_{max}, and EC'_{50} derived from the fit

of curve 2 (1, 0.52, and 79 nM, respectively, see Table 12.3b). The calculated equiactive concentration for curve 2 from Equation 12.6 is 10.3 nM.

$$[A] = (EC'_{50}) \cdot \left[\frac{(\text{Response} - \text{Basal}')}{E'_{max} - \text{Response}}\right]^{1/n'} \qquad (12.6)$$

5. The complete set of equiactive concentrations (real data in bold font, calculated data in normal font) is shown in Table 12.3c.

12.2.2 Method of Barlow, Scott, and Stephenson for Measurement of the Affinity of a Partial Agonist

Aim: To measure the affinity of partial agonists.

General Procedure: Full dose-response curves to a full and partial agonist are obtained in the same receptor preparation. It is essential that the same preparation be used as there can be no differences in the receptor density and/or stimulus-response coupling behavior for the receptors for all agonist curves. From these dose-response curves, concentrations are calculated that produce the same response (equiactive concentrations). These are used in linear transformations to yield estimates of the affinity of the partial agonist.

Procedure:

1. A dose-response curve to a full agonist is obtained. Shown for this example (see Table 12.4) are data to the full agonist histamine in guinea pig ileal smooth muscle (responses as a percentage of the maximal response to histamine).

2. After a period of recovery for the preparation (to avoid possible desensitization), a dose-response curve to a partial agonist is obtained. Data are shown in Table 12.4a for the histamine partial agonist E-2-P ((N,N-diethyl-2-(1-pyridyl) ethylamine). Response to E-2-P is expressed as a percentage of the maximal response to histamine.

3. Data points are subjected to nonlinear curve fitting. For these data, Equation 12.5 is used to fit the curve with basal = 0. The fitting parameters for histamine and E-2-P are given in Table 12.4b. The curves are shown in Figure 12.5a.

4. Equiactive concentrations of histamine and E-2P are calculated (see method in Section 12.2.1). For this calculation, responses produced by E-2-P are used since they covered a convenient range to characterize both dose-response curves. The equiactive concentrations are shown in Table 12.4c.

5. A regression of 1/[E-2-P] versus 1/[Histamine] is constructed. This is shown in Figure 12.5b.

TABLE 12.4
Method of Barlow, Scott, and Stephenson for partial agonist affinity.

(a) Data for Dose-Response Curves.

[Histamine]: M	Response	[E-2-P]: M	Response
10^{-8}	0.12	10^{-6}	0.04
3×10^{-8}	0.27	3×10^{-6}	0.12
10^{-7}	0.53	10^{-5}	0.26
3×10^{-7}	0.76	3×10^{-5}	0.42
10^{-6}	0.93	10^{-4}	0.53
3×10^{-6}	1.01	3×10^{-4}	0.58
		10^{-3}	0.61

(b) Parameters for Fit Curves

	Histamine	E-2-P
$E_{max} = 1.05$		0.62
$EC_{50} = 90 \, nM$		12.5 μM
$n = 0.95$		0.95

(c) Equiactive Agonist Concentrations

Response	[Histamine]: M	1/[Hist]	[E-2-P]: M	1/[E-2-P]
0.12	5×10^{-8}	2×10^{7}	3×10^{-6}	3.3×10^{5}
0.26	1.3×10^{-7}	7.7×10^{6}	10^{-5}	10^{5}
0.42	2.8×10^{-7}	3.57×10^{6}	3×10^{-5}	3.3×10^{4}
0.53	4.4×10^{-7}	2.27×10^{6}	10^{-4}	10^{4}
0.58	5.4×10^{-7}	1.85×10^{6}	3×10^{-4}	3.33×10^{3}

TABLE 12.7

Resultant analysis.

		(a) Concentration Scheme for Resultant Analysis		
Ref Antagonist Scopol. (M)	Regression I Test Antag. Atropine	Regression II Test Antag. Atropine (M)	Regression III Test Antag. Atropine (M)	Regression IV Test Antag. Atropine (M)
10^{-9}	0			
3×10^{-9}	0	3×10^{-9}		
10^{-8}	0	3×10^{-9}	10^{-8}	3×10^{-8}
3×10^{-8}	0	3×10^{-9}	10^{-8}	3×10^{-8}
10^{-7}		3×10^{-9}	10^{-8}	3×10^{-8}
3×10^{-7}			10^{-8}	3×10^{-8}

Test antagonist = atropine. Reference antagonist = scopolamine. The Schild regression is obtained to the concentrations of scopolamine shown in the left-hand column in the presence of the concentrations of atropine shown in columns labeled Regression I to IV.

	(b) Data Describing Schild Analyses for Scopolamine (I) and Scopolamine and Atropine (II to IV)		
Regression	Slope	95% c.l.	Intercept
I	1.3	0.9 to 1.5	11.88
II	1.2	0.9 to 1.4	10.34
III	1.06	0.76 to 1.3	8.77
IV	0.95	0.78 to 1.1	7.5

	(c) Parameters for Schild Regressions Fit to Unit Slope and Data for Resultant Regression (Log [Atropine] vs. Log ($\kappa-1$))			
Regression	pK_B from slope = 1	κ	[Atropine]: M	Log ($\kappa-1$)
I	9.4 + 0.1			
II	8.7 + 0.07	5	3.00E−09	0.60
III	8.29 + 0.04	12.9	1.00E−08	1.08
IV	7.9 + 0.02	31.6	3.00E−08	1.49

regressions are plotted on a common antagonist concentration axis and their dextral displacement along the concentration axis used to construct a resultant plot. This plot, if linear with a slope of unity, yields the pK_B of the test antagonist as the intercept.

Procedure:

1. Schild regressions to a reference antagonist are obtained according to standard procedures (see Section 12.2.4) in the absence and presence of a range of concentrations of the test antagonist. In the cases where the test antagonist is present, it is included in the medium for the control dose-response curve as well as the curves obtained in the presence of the reference antagonist. For this example, the scheme for the dose-response curves used for the construction of regressions I to IV is shown in Table 12.7a. The test antagonist is atropine and the reference antagonist is scopolamine.

2. The Schild regressions for scopolamine, obtained in the absence (regression I) and presence of a range of concentrations of atropine (regressions II to IV) are shown in Figure 12.8a. The data describing these regressions is given in Table 12.7b.

3. The displacement, along the antagonist concentration axis, of the Schild regressions is calculated. To obtain a value for [B']/[B] (shift along the concentration axis) that is independent of Log (DR-1) values the Schild regressions must be parallel. The first step is to fit the regressions to a common slope of unity. This can be done if the 95% confidence limits of the slopes of each regression include unity (which is true for this example, see Table 12.7b).

4. The pK_B values for scopolamine from slopes I to IV, each fit to a slope of unity, are given in Table 12.7c.

5. The resultant plot is constructed by calculating the shift to the right of the Schild regressions produced by the addition of atropine (pK_B for unit slope regression for scopolamine regressions II to IV divided by the pK_B for scopolamine found for regression I). (See Table 12.7c.) These yield values of κ for every concentration of atropine added. For example, the κ value for regression II ([Atropine] = 3 nM) is $10^{-8.7}/10^{-9.4} = 5$. These values of κ are used in a resultant plot of Log ($\kappa-1$) versus the concentration of the test antagonist (atropine) used for the regression. The resultant plot is shown in Figure 12.8c.

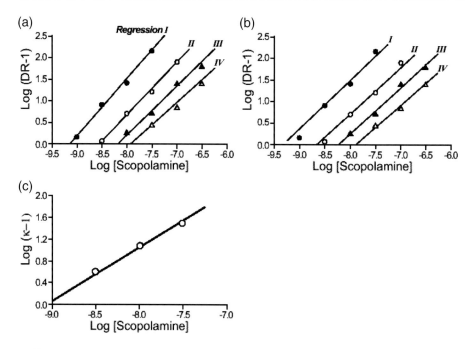

FIGURE 12.8 Resultant analysis. (a) Schild regressions for scopolamine in the absence (I, filled circles) and presence of atropine 3 nM (II, open circles), 10 nM (III, filled triangles), and 30 nM (IV, open triangles). (b) Schild regressions shown in panel a fit to regressions of unit slope. (c) Resultant plot for atropine. Displacements of the Schild regressions shown in panel b furnish values for κ for a regression according to Equation 6.18.

6. The resultant regression is linear and has a slope not significant from unity (slope = 0.9 ± 0.07; 95% c.l. = 0.4 to 1.35). A refit of the data points to a linear plot with unit slope yields a pK_B for atropine of 9.05 ± 0.04 (95% c.l. 8.9 to 9.2).

12.2.6 Method of Stephenson for Measurement of Partial Agonist Affinity

Aim: This procedure measures the affinity of a partial agonist by quantifying the antagonism of responses to a full agonist by the partial agonist.

General Procedure: Dose-response curves to a full agonist are obtained in the absence and presence of a range of concentrations of partial agonist. For a single pair of curves (full agonist alone and in the presence of one concentration of partial agonist), a plot of equiactive concentrations of full agonist yields a linear regression. The K_p for the partial agonist can be calculated from the slope of this regression. An extension of this method utilizes a number of these slopes for a more complete analysis. For this method, the individual slopes are used in a metameter of the equation to yield a single linear regression from which the K_p can be calculated much like in Schild analysis.

Procedure:

1. A dose-response curve to a full agonist is obtained. A concentration of partial agonist is equilibrated with the same preparation (30 to 60 minutes) and then the dose-response curve is repeated in the presence of the partial agonist. The data are fit to curves (for this example, Equation 12.5) to yield a pair of curves like those shown in Figure 12.9a. For this example, the full agonist is isoproterenol, the partial agonist is chloropractolol, and the response emanates from rat atria containing β-adrenoceptors.

2. Equiactive concentrations of isoproterenol, in the absence [A] and presence [A'] of chloropractolol (100 nM), are calculated according to the general procedure described in Section 12.2.1. These are given in Table 12.8a. A plot of these equiactive concentrations yields a linear regression (according to Equation 6.25, see Figure 12.9b). The x values are the concentrations of isoproterenol [A'] in the presence of chloropractolol and the y values are the control concentrations of isoproterenol [A].

3. The slope of this regression is given in Table 12.8a (slope = 0.125). The K_p for the partial agonist is given by Equation 6.26 ($K_p = [P] \cdot Slope/(1 - Slope) \cdot \vartheta$). The term ϑ represents an efficacy term modifying the estimate of affinity $(1 - (\tau_p/\tau_a))$ in terms of the operational model and $(1 - (e_p/e_a))$ in terms of the classical model. For weak partial agonists and highly efficacious full agonists, this factor approaches unity and the method approximates the affinity of the partial agonist.

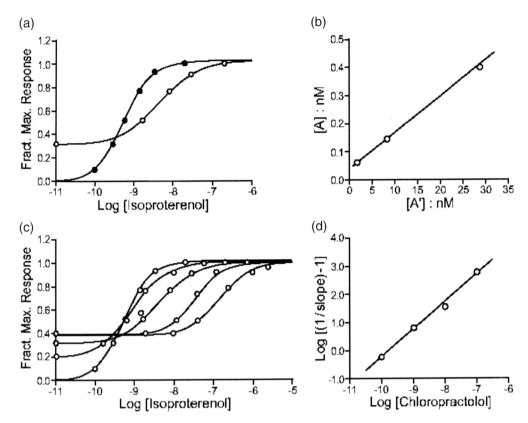

FIGURE 12.9 Method of Stephenson for measurement of partial agonist affinity. (a) Dose-response curves to isoproterenol in the absence (filled circles) and presence of chloropractolol (100 nM, open circles). (b) Regressions of equiactive concentrations of isoproterenol in the absence (ordinates) and presence (abscissae) of chloropractolol (100 nM, data from panel a). Regression is linear with a slope of 0.125. (c) Extension of this method by Kaumann and Marano. Dose-response curves to isoproterenol in the absence and presence of a range of concentrations of chloropractolol. (d) Each shift of the isoproterenol dose-response curve shown in panel c yields a regression such as that shown in panel b. A regression of the respective slopes of these regressions is made upon the 4 concentrations of partial agonist (chloropractolol) according to Equation 6.13. The regression is linear with a slope of 0.96 ± 0.

12.2.7 *Extension of the Stephenson Method: Method of Kaumann and Marano*

1. The previous procedure can be repeated for a number of concentrations of partial agonist (see Figure 12.9c) to provide a wider base of data on which to calculate the partial agonist affinity. Thus, a number of regressions (like that shown in Figure 12.9b) are constructed to yield a number of slopes for a range of partial agonist concentrations. An example is shown in Table 12.8b.

2. The slope values are used in a metameter (Log(1/slope) − 1)) as the y values for the corresponding Log concentrations of the partial agonists (x values) to construct a linear regression according to Equation 6.27. The regression for chloropractolol is shown in Figure 12.9d.

3. This regression is linear, with a slope of 0.96 ± 0.05. This slope is not significantly different from unity. Thus, the data points are refit to a linear regression with a slope of unity. The intercept of this regression

yields an estimate of the pK_p for the partial agonist (as for Schild analysis). For this example, $pK_p = 7.74 \pm 0.05$ (95% c.l. = 7.6 to 7.9).

12.2.8 *Method of Gaddum for Measurement of Noncompetitive Antagonist Affinity*

Aim: This method is designed to measure the affinity of a noncompetitive antagonist.

General Procedure: Dose-response curves to a full agonist are obtained in the absence and presence of the non-competitive antagonist. From these curves, equiactive concentrations of full agonist are compared in a linear regression (see Section 12.2.1). The slope of this regression is used to estimate the K_B for the noncompetitive antagonist.

Procedure:

1. A dose-response curve is obtained to the agonist. Then the same preparation is equilibrated with a

TABLE 12.8
Method of Stephenson for affinity of partial agonists
(+ method of Lemoine and Kaumann).

(a)

Response	[A]	[A′]
0.51	5.9×10^{-10}	1.79×10^{-9}
0.76	1.43×10^{-9}	8.3×10^{-9}
0.90	4.0×10^{-9}	2.89×10^{-8}

Slope $= 0.125$
$K_p = 1.43 \times 10^{-8}$

(b)

[Chloro]: M	Slope(s)
10^{-8}	0.619
10^{-7}	0.127
10^{-6}	0.023
10^{-5}	0.0018

Slope $= 0.96 \pm 0.05$
$pK_p = 7.74 \pm 0.05$
95%c.l. $= 7.6$ to 7.9

known concentration of noncompetitive antagonist (for 30 to 60 minutes, depending on the time needed to reach temporal equilibrium) and a dose-response curve to the agonist repeated in the presence of the antagonist. A hypothetical example is shown in Figure 12.10a. The data are given in Table 12.9a. For this example, the preparation is equilibrated with a 100-nM antagonist.

2. The data points are fit to an appropriate function (Equation 12.5). (See Figure 12.10b.) From the real data points and calculated curves, equiactive concentrations of agonist in the absence and presence of the antagonist are calculated (see Section 12.2.1). For this example, real data points for the blocked curve were used and the control concentrations calculated (control curve $E_{max} = 1.01$, $n = 0.9$, and $EC_{50} = 10 \, \mu M$). The equiactive concentrations are shown in Table 12.9b.

3. A regression of $1/[A]$ where [A] equal the equiactive concentrations for the control curve (no antagonist) upon $1/[A']$ (x values) where [A′] equal the equiactive concentrations in the presence of the antagonist is constructed. For the example, this is shown in Figure 12.10c. This regression is linear, with a slope of 13.4.

4. The K_B for the noncompetitive antagonist is calculated with Equation 6.35 ($K_B = [B]/(slope-1)$). For this example, the calculated K_B for the antagonist is 8.06 nM.

12.2.9 Measurement of the Affinity and Maximal Allosteric Constant for Allosteric Modulators Producing Surmountable Effects

Aim: This procedure measures the affinity and cooperativity constant of an allosteric antagonist. It is used for known allosteric antagonists or molecules that produce a saturable antagonism that does not appear to follow the Gaddum equation for simple competitive antagonism.

General Procedure: Dose-response curves are obtained for an agonist in the absence and presence of a range of concentrations of the antagonist. The dextral displacement of these curves (EC_{50} values) are fit to a hyperbolic equation to yield the potency of the antagonist and the maximal value for the cooperativity constant (α) for the antagonist.

Procedure:

1. Dose-response curves are obtained for an agonist in the absence and presence of a range of concentrations of the antagonist and the data points fit with standard linear fitting techniques (Equation 12.5) to a common maximum asymptote and slope. An example of acetylcholine responses in the presence of a range of concentrations of gallamine are shown in Table 12.10a. The curves are shown in Figure 12.11a.

2. The EC_{50} values for the fit curves (see Table 9.10b) are then fit to a function of the form (variant of Equation 7.2)

$$\frac{EC'_{50}}{EC_{50}} = \frac{(x/B + 1)}{(Cx/B + 1)}, \qquad (12.7)$$

where EC'_{50} and EC_{50} are the location parameters of the dose-response curves in the absence and presence of the allosteric antagonist, respectively, x is the molar concentration of antagonist, and B and C are fitting constants.

3. The data in Table 12.10b are fit to Equation 12.7 to yield estimates of $B = 9.5 \times 10^{-7}$ and $C = 0.011$. (See Figure 12.11b.) These values can be equated to the model for allosteric antagonism (Equation 6.31) to yield a K_B value of 95 nM and a value for α of 0.011.

12.2.10 Measurement of pA_2 for Antagonists

Aim: This method allows estimation of the potency of an unknown antagonist. It does not indicate the mechanism of antagonism (by the pattern of effect on dose-response curves) but does give a starting point for further analysis. The potency of the antagonist is quantified as the pA_2 defined as the molar concentration of antagonist that produces a twofold shift to the right of the agonist dose-response curve.

General Procedure: A dose-response curve to an agonist is obtained and a concentration of agonist that produces

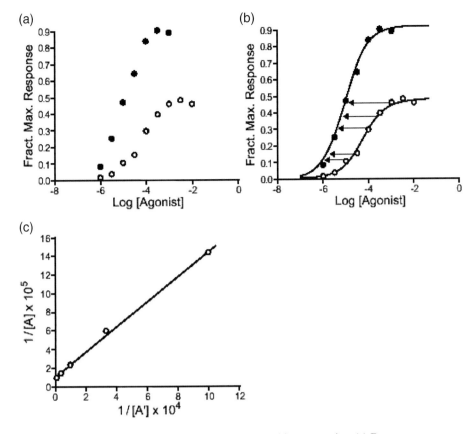

FIGURE 12.10 Measurement of affinity for noncompetitive antagonists. (a) Dose-response curve to an agonist in the absence (filled circles) and presence (open circles) of a noncompetitive antagonist. (b) Data points in panel A fit to dose-response curves. Equiactive concentrations of agonists determined as in Section 12.2.1. Real data points used from curve in the presence of antagonist. Equiactive concentrations of agonist from control curve calculated (see arrows). (c) Double reciprocal plot of equiactive concentrations of agonist in the absence (ordinates) and presence (abscissae) of antagonist. Regression is linear with a slope of 13.4.

between 50 and 80% maximal response chosen for further study. Specifically, the effects of a range of antagonist concentrations on the response produced to the chosen agonist concentration is measured and the IC_{50} (concentration of antagonist that produces a 50% blockade of the initial agonist response) is measured to yield an inhibition curve. This concentration is then corrected to yield an estimate of the antagonist pA_2.

Procedure:

1. A dose-response curve to the agonist is obtained. Ideally, it should be done as near the time for analysis of antagonism as possible to negate possible variances in preparation sensitivity. Dose-response data are shown in Table 12.11a (and Figure 12.12a). The data are fit to a curve. For this example, they are fit to Equation 12.8 with fitting parameters $E_{max} = 96$, $n = 0.7$, and $EC_{50} = 20\,nM$. The curve is shown in Figure 12.12b.

$$Response = \frac{E_{max} \cdot [A]^n}{[A]^n + EC_{50}^n} \qquad (12.8)$$

2. A target agonist concentration is chosen. For this example, a concentration of 0.3 mM agonist was used. This approximates the concentration that produces an 80% maximal response. The antagonist is tested against the response produced by 0.3 μM agonist.

3. A set of responses to the target agonist concentration is measured in the absence and presence of a range of antagonist concentrations. The fit agonist response curve is shown in Figure 12.12c. For this example, the repeat test of the target concentration (0.3 μM agonist) gives a response value of 86. The repeat response to the target agonist concentration is shown as the open circle. The addition of the antagonist to the preparation theoretically produces a shift of the agonist dose-response curve shown as the dotted lines. The arrow on Figure 12.12c indicates the expected response to the target concentration of agonist as increasing concentrations of the antagonist are added.

4. The responses to the target concentration of agonist in the presence of a range of concentrations of the

TABLE 12.9

Gaddum method for measuring the affinity of a noncompetitive antagonist.

(a)

[A]	Control Resp.	Blocked Resp.
10^{-6}	0.08	0.01
3.0×10^{-6}	0.25	0.03
10^{-5}	0.47	0.1
3.0×10^{-5}	0.64	0.15
10^{-4}	0.84	0.29
3.0×10^{-4}	0.9	0.39
10^{-3}	0.89	0.46
3.0×10^{-3}		0.48
10^{-2}		0.46

(b)

Response	[A']	1/[A']	[A]	1/[A]
0.1	10^{-5}	10^5	7.0×10^{-7}	1.4×10^6
0.15	3×10^{-5}	3.33×10^4	1.7×10^{-6}	5.88×10^5
0.29	10^{-4}	10^4	4.3×10^{-6}	2.32×10^5
0.39	3×10^{-4}	3.33×10^3	7.26×10^{-6}	1.37×10^5
0.46	10^{-3}	10^3	1.02×10^{-5}	9.76×10^4

Intercept $= 1.01 \times 10^5$

Slope $= 13.4$

TABLE 12.10

Allosteric antagonism.

(a) Dose-Response Data for Gallamine Blockade of Acetylcholine Responses

[A]: M	Control Resp.	[A]: M	1×10^{-5}M Gallamine	[A]: M	3.0×10^{-5}M Gallamine
10^{-9}	3.1	3×10^{-8}	9.38	3×10^{-7}	29.69
10^{-8}	20.3	10^{-7}	25	5×10^{-7}	41
3×10^{-8}	53.1	3×10^{-7}	45	10^{-6}	56.25
10^{-7}	74	10^{-6}	76.56	2×10^{-6}	67.19
2×10^{-7}	85.9				
3×10^{-7}	92.2				
5×10^{-7}	93.7				

[A]: M	1.00×10^{-4}M Gallamine	[A]: M	3×10^{-4}M Gallamine	[A]: M	5.00E − 04 Gallamine
5×10^{-7}	25	10^{-7}	3.1	10^{-6}	31.2
10^{-6}	40.6	5×10^{-7}	15.6	2×10^{-6}	46.87
3×10^{-6}	71.87	10^{-6}	31.25	5×10^{-6}	65.62
10^{-5}	87.5	2×10^{-6}	46.87	10^{-5}	78.12
		5×10^{-6}	73.44		
		10^{-5}	79.69		
		3×10^{-5}	89.06		

(b) Parameters for Fit Dose-Response Curves for Acetylcholine

Curve	EC_{50}(M)
I	2.94×10^{-8}
II	2.9×10^{-7}
III	7.5×10^{-7}
IV	1.3×10^{-6}
V	2×10^{-6}
VI	2.4×10^{-6}

common $E_{max} = 97.6$

common slope $= 1.09$

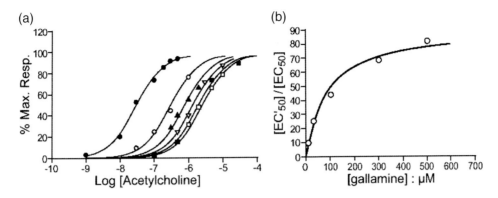

FIGURE 12.11 Measurement of allosteric antagonism. (a) Dose-response curves to acetylcholine in the absence (filled circles) and presence of gallamine 10 μM (open circles), 30 μM (filled triangles), 100 μM (open inverted triangles), 300 μM (filled squares), and 500 μM (open squares). Data points fit to curves with a common maximum and slope. (b) Displacement of dose response curves shown in panel a used to furnish dose ratios for acetylcholine ([EC'_{50} in the presence of gallamine]/[EC_{50} for control curve]) ordinates. Abscissae are concentrations of gallamine. Line is the best fit according to Equation 12.7.

TABLE 12.11

Measurement of antagonist pA_2.

(a) Dose-Response Data for Agonist

[A]	Resp
10^{-9}	10
3×10^{-9}	25
10^{-8}	33
3×10^{-8}	55
10^{-7}	72
3×10^{-7}	80
10^{-6}	90
3×10^{-6}	93

(b) Response to the Test Concentration of Agonist in the Presence of a Range of Concentrations of the Antagonist

[B]	Response
10^{-8}	86
3×10^{-8}	86
5×10^{-8}	84
10^{-7}	80
3×10^{-7}	65
10^{-6}	43
3×10^{-6}	26
10^{-5}	10

antagonist are given in Table 12.11b and shown in Figure 12.12d.

5. The data points are fit to a function. For this example, Equation 12.9 is used:

$$\text{Response} = \text{Basal} - \frac{\text{Resp}_0[B]^n}{[B]^n + IC_{50}^{n}}, \quad (12.9)$$

where Resp_0 refers to the response produced by the target agonist concentration in the absence of

antagonist. For the example, values for the fit curve are $\text{Resp}_0 = 86$, $n = 0.93$, and $IC_{50} = 1 \mu M$. The fit curve is shown in Figure 12.12e.

6. The IC_{50} is used in a version of the Cheng-Prusoff equation for functional assays. Thus, the apparent K_B (apparent equilibrium dissociation constant for the antagonist-receptor complex) is given by

$$\text{Antilog } pA_2 = IC_{50}/((2 + ([A]/EC_{50})^n)^{1/n} - 1),$$
$$(12.10)$$

where the values of n and EC_{50} are the values from the control agonist dose-response curve (n = 0.7, $EC_{50} = 20$ nM, and [A] = 30 nM). Equation 12.10 yields the molar concentration that produces a twofold shift to the right of the agonist dose-response curve. The negative logarithm of this value is the pA_2. For this example ($IC_{50} = 1 \mu M$), the antilog $pA_2 = 48$ nM and the $pA_2 = 7.3$.

12.2.11 Method for Estimating Affinity of Insurmountable Antagonist (Dextral Displacement Observed)

Aim: This method is designed to measure the affinity of an antagonist that produces insurmountable antagonism (depression of maximal response to the agonist) but also shifts the curve to the right by a measurable amount.

General Procedure: Dose-response curves to a full agonist are obtained in the absence and presence of the antagonist. At a level of response approximately 30% of the maximal response of the depressed concentration response curve, an equiactive dose ratio for agonist concentrations is measured. This is used to calculate a pA_2.

Procedure:

1. A dose-response curve is obtained to the agonist. Then the same preparation is equilibrated with a

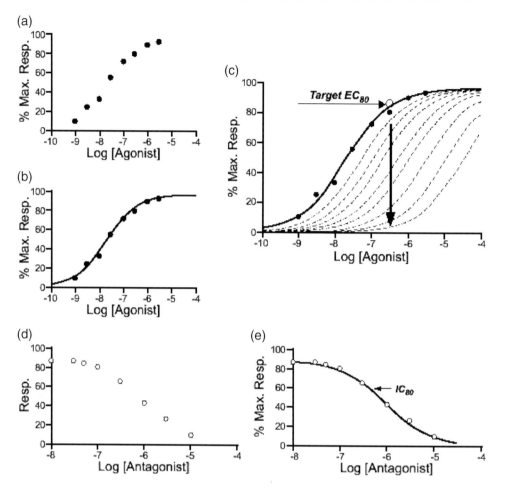

FIGURE 12.12 Measurement of pA$_2$ values for antagonists. (a) Dose-response curve data for an agonist. (b) Curve fit to data points according to Equation 12.8. (c) Open circle represents EC$_{80}$ concentration of agonist chosen to block with a range of concentrations of antagonist. The antagonist, if competitive, will produce shifts to the right of the agonist dose-response curve as shown by dotted lined. The inhibition curve tracks the response to the target concentration of agonist (open circle), as shown by arrow. Note that if the antagonism is noncompetitive the curves will not shift to the right but rather will be depressed. This will still produce diminution of the response to the target agonist concentration and production of an inhibition curve. (d) Inhibition curve produced by a range of antagonist concentrations (abscissae) producing blockade of response to the target concentration of agonist. (e) Data points fit to curve according to Equation 12.9. The IC$_{50}$ is shown. The pA$_2$ of the antagonist is calculated from this value.

known concentration of noncompetitive antagonist (for 30 to 60 minutes, depending on the time needed to reach temporal equilibrium) and a dose-response curve to the agonist repeated in the presence of the antagonist. A hypothetical example is shown in Figure 12.13a. The data are given in Table 12.12. For this example, the preparation is equilibrated with 2 μM antagonist.

2. The data points are fit to an appropriate function (Equation 12.5). (See Figure 12.13b.) At a response level of 0.3, an equiactive dose ratio of agonist is calculated. The respective concentrations of agonist

producing this response are 50 nM (control) and 0.20 μM in the presence of the antagonist. The dose ratio is (DR = 2.0/0.5 = 4).

3. The value for DR is converted to Log (DR-1) value, which in this case equals 0.48. The pA$_2$ is calculated with the equation

$$pA_2 = -Log[B] + Log(DR-1), \qquad (12.11)$$

which in this case is $5.7 + 0.48 = 6.18$. This translates into a K$_B$ value of 0.67 μM.

4. This value should be considered an upper limit for the potency of the antagonist as the pA$_2$ corresponds

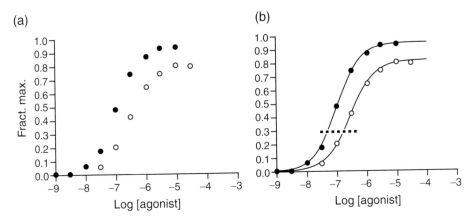

FIGURE 12.13 Calculation of a pA_2 value for an insurmountable antagonist. (a) Concentration-response curve for control (filled circles) and in the presence of 2 μM antagonist (open circles). (b) Data points fit to logistic functions. Dose ratio measured at response value 0.3 (dotted line). In this case, the DR = (200 nM/50 nM = 4).

TABLE 12.12

Responses in the absence and presence of an insurmountable antagonist that causes dextral displacement of the concentration-response curve.

Conc.	Control Response[1]	Modulated Response[1]
1×10^{-8}	0.06	
3×10^{-8}	0.17	0.05
1×10^{-7}	0.47	0.2
3×10^{-7}	0.73	0.42
1×10^{-6}	0.86	0.64
3×10^{-6}	0.92	0.74
1×10^{-5}	0.93	0.8
3×10^{-5}		0.79

[1]Fraction of system maximal response.

to the pK_B according to the equation (6.38)

$$pK_B = pA_2 - Log(1 + 2[A]/K_A) \qquad (12.12)$$

for orthosteric insurmountable antagonists and (Equation 7.8)

$$pK_B = pA_2 - Log(1 + 2\alpha[A]/K_A) \qquad (12.13)$$

for allosteric insurmountable antagonists.

It is worth examining the possible magnitudes of the error with various scenarios. The maximal value for $[A]/K_A$ can be approximated, assuming a system where response is directly proportional to receptor occupancy. Under these circumstances, Response $= 0.3 = [A]/K_A/([A]/K_A + 1)$, which in this case is $[A]/K_A = 0.5$. Therefore, the pA_2 is $pK_B + Log(2)$ (i.e., the pA_2 will overestimate the affinity of the antagonist by a maximal factor of 2). If the insurmountable antagonist is allosteric antagonist that reduces the affinity of the receptor for agonist ($\alpha < 1$), then the error will be <2. However, if the modulator

increases the affinity of the receptor for the agonist, then the error could be as high as 2α, where $\alpha > 1$.

12.2.12 Method for Estimating Affinity of Insurmountable Antagonist (No Dextral Displacement Observed)

Aim: This method is designed to measure the affinity of a noncompetitive antagonist, which produces depression of the maximal response of the agonist concentration-response curve with no dextral displacement.

General Procedure: The response to the agonist is determined in the absence and presence of a range of concentrations of the insurmountable antagonist. The data points may be fit to logistic functions (for observation of trends; this is not necessary for calculation of IC_{50}). A concentration of agonist is chosen, and the response to that concentration (expressed as a fraction of control) is plotted as a function of the concentration of antagonist to form an inhibition curve. This curve is fit to a function and the midpoint (IC_{50}) calculated. This is an estimate of the affinity of the insurmountable antagonist. To detect possible allosteric increase in affinity of the antagonist with agonist concentration, more than one concentration may be chosen for this procedure.

Procedures:

1. Responses to the agonist are obtained in the absence and presence of a range of concentrations of antagonist. A sample set of data is given in Table 12.13 and Figure 12.14a.
2. Data may be fit to an appropriate function (i.e., Equation 12.5), but this is not necessary for the analysis (see Figure 12.14b).
3. Two concentrations of agonist are chosen for further analysis. These should be two concentrations as widely spread as possible along the

TABLE 12.13

Responses in the absence and presence of an insurmountable antagonist that causes no dextral displacement of the concentration-response curve.

(a) Concentration Response Curve Data

Agonist Concentration	Control	1×10^{-7} Antagonist	2×10^{-7} Antagonist	5×10^{-7} Antagonist	1×10^{-6} Antagonist	2×10^{-6} Antagonist
1×10^{-8}	0.02					
3×10^{-9}	0.05	0.03	0.04	0.03	0.02	0.02
1×10^{-7}	0.15	0.13	0.12	0.085	0.05	0.03
3×10^{-7}	0.3	0.25	0.22	0.15	0.1	0.05
1×10^{-6}	0.5	0.38	0.32	0.22	0.13	0.06
3×10^{-6}	0.6	0.45	0.37	0.23	0.15	0.07
1×10^{-5}	0.646	0.48	0.39	0.25	0.13	0.06
1×10^{-5}	0.67	0.49	0.4	0.26	0.16	0.07

(b) Conversion to Inhibition Curves

Concentration Antagonist	Concentration Agonist 1×10^{-7} Response	Percent Response	Concentration Agonist 1×10^{-5} Response	Percent Response
0	0.15	100	0.64	100
1×10^{-7}	0.13	87	0.48	75
2×10^{-7}	0.12	80	0.39	61
5×10^{-7}	0.09	60	0.25	39
1×10^{-6}	0.05	33	0.13	29
2×10^{-6}	0.03	20	0.07	11

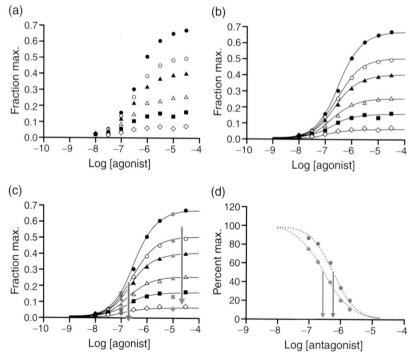

FIGURE 12.14 Measurement of potency of a noncompetitive antagonist that produces little dextral displacement of the agonist concentration-response curve. (a) Data points for control response to agonist (filled circles) and response in the presence of noncompetitive antagonist at concentrations = 0.1 μM (open circles), 0.2 μM (filled triangles), 0.5 μM (open triangles), 1 μM (filled squares), and 2 μM (open diamonds). (b) Logistic function (Figure 12.4) fit to data points (optional). (c) Response to two specific concentrations of agonist identified (10 μM in red and 100 nM in blue). (d) Effects of antagonist on responses to 10 μM (red) and 100 nM (blue) agonist expressed as a percent of the control response plotted as a function of the concentration of antagonist to yield an inhibition curve (data shown in Table 12.13b). Arrows indicate the IC_{50} values for each curve.

concentration axis and with the lowest producing a robust size of response. For this example, responses chosen for agonist concentration were 100 nM (blue, Figure 12.14c) and 10 μM (red, Figure 12.14c).

4. The responses to the respective concentrations of agonist are expressed as a percentage of the initial control response (obtained in the absence of antagonist) as a function of the concentration of antagonist. The data for this step are shown in Table 12.13b, and the resulting inhibition curves (plot on a semi-logarithmic concentration scale) are shown in Figure 12.14d.

5. The inhibition curves are fit to an appropriate function to allow estimation of the half maximal value for blockade (IC_{50}). For example, the data from Table 13.13b were fit to

$$\text{Percent} = 100 - \frac{100[B]^n}{[B]^n + (IC_{50})^n}, \qquad (12.14)$$

where the concentration of antagonist is [B], n a slope fitting parameter, and IC_{50} the half maximal value for blockade. For this example, the IC_{50} values for the two curves are 0.65 μM (n = 1.15) for 100 nM agonist (blue) and 0.3 μM (n = 1.05) for 10 μM agonist (red).

6. It can be seen from this example that the inhibition curve shifts to the left with increasing concentration of agonist, indicating an allosteric mechanism whereby the modulator blocks receptor signaling but increases the affinity of the receptor for the agonist.

Glossary of Pharmacological Terms

Affinity/affinity constant, ligands reside at a point of minimal energy within a binding locus of a protein according to a ratio of the rate that the ligand leaves the surface of the protein (k_{off}) and the rate it approaches the protein surface (k_{on}). This ratio is the equilibrium dissociation constant of the ligand–protein complex (denoted $K_{eq} = k_{off}/k_{on}$) and defines the molar concentration of the ligand in the compartment containing the protein where 50% of the protein has ligand bound to it at any one instant. The "affinity" or attraction of the ligand for the protein is the reciprocal of K_{eq}.

Agonist, a molecule that produces physiological response through activation of a receptor.

Alkylating agent, a reactive chemical that forms a covalent bond with chemical moieties on the biological target (usually a protein). For instance, β-haloalkylamines generate an aziridinium ion in aqueous base that inserts into –SH, –CHOH, or other chemical structures in peptides. Once inserted, the effects of the alkylating agent are irreversible.

Allele, different forms of a gene at a given locus.

Allosterism (allosteric), the imposition of an effect on a protein through interaction of a molecule with a site on the protein distinct from the natural binding locus for the endogenous ligand for that protein. Interactions between the allosteric molecule and the endogenous ligand occur through the protein and not through direct steric interaction.

Allosteric modulators, unlike competitive antagonists that bind to the same domain on the receptor as the agonist, allosteric modulators bind to their own site on the receptor and produce an effect on agonism through a protein conformational change. Allosteric modulators can affect the affinity or the responsiveness of the receptor to the agonist. A hallmark of allosteric interaction is that the effect reaches a maximal asymptote corresponding to saturation of the allosteric sites on the receptor. For example, an allosteric modulator may produce a maximal 10-fold decrease in the affinity of the receptor for a ligand upon saturation of the allosteric sites.

Analysis of variance (ANOVA), a statistical procedure that quantifies differences between means of samples and the extent of variances within and between those means to determine the probability of there being a difference in the samples.

Antagonist, a molecule that interferes with the interaction of an agonist and a receptor protein or a molecule that blocks the constitutive elevated basal response of a physiological system.

Association constant, the ratio of the rate of onset of a molecule to a receptor binding site and the rate of dissociation of the molecule away from that site (reciprocal of K_{eq}; see Affinity).

B_{max}, a term denoting the maximal binding capacity of an experimental binding system, usually a preparation containing receptors (membranes, cells). The magnitude is most often expressed in number of receptors per cell or molar concentration of receptors per milligram protein.

cDNA, complementary DNA copied from a messenger RNA coding for a protein; it is inserted into surrogate host cells to cause them to express the protein.

Cheng–Prusoff correction, published by Cheng and Prusoff (*Biochem. Pharmacol.* **22**, 3099–3108, 1973), this method is used to derive the equilibrium dissociation constant of a ligand–receptor (or enzyme) pair from the experimentally observed IC_{50} (concentration that produces 50% reduction in effect) for that molecule; see Eqn 4.11.

Clone, identical cells (with respect to genetic constitution) derived from a single cell by asexual reproduction. Receptors can be cloned into cells by inserting a gene into the cell line; a colony of cells results that are identical and all have the expressed receptor.

Competitive antagonist, by definition, competitive antagonists compete with the agonist for the same binding domain on the receptor. Therefore, the relative affinities and quantities of the agonist and antagonist dictate which ligand dominates. Under these circumstances, the concentration of agonist can be raised to the point where the concomitant receptor occupancy by the antagonist is insignificant. When this occurs, the maximal response to the agonist is observed, that is, surmountable antagonism results.

Concentration ratio, the ratio of molar concentrations of agonist that produce equal levels of response in a given pharmacological preparation (usually the ratio of EC_{50} concentrations). This term is used most often when discussing antagonism (equiactive concentration of agonist in the absence and presence of an antagonist).

Concentration–response curve, a more specific (and technically correct) term for a dose–response curve done *in vitro*. This curve defines the relationship between the concentrations of a given molecule and the observed pharmacological effect.

Constitutive receptor activity, receptors spontaneously produce conformations that activate G-proteins in the absence of agonists. This activity, referred to as constitutive activity, can be observed in systems in which the receptor expression levels are high and the resulting levels of spontaneously activating receptor species produce a visible physiological response. An inverse agonist reverses this constitutivie activity and thus reduces, in a dose-dependen manner, the spontaneously elevated basal response of a constitutively active receptor system.

Cooperativity, the interaction of molecules on a protein resulting from the mutual binding of those molecules. The cooperativity may be positive (whereby the binding of one of the substances facilitates the interaction of the protein with the other molecule) or negative (binding of one molecule decreases the interaction of the protein with the other molecule).

Cooperativity factor, an allosteric ligand has an effect on a receptor protein mediated through the binding of that ligand to the allosteric binding domain. The intensity of that effect, usually a change in the affinity of the receptor for other ligands or the efficacy of a ligand for the receptor, is quantified by the cooperativity factor. Denoted α, a positive value for α defines a potentiation. Conversely, a fractional value denotes an inhibition. Thus if $\alpha = 0.1$, a 10-fold decrease in the affinity of a tracer ligand for the receptor is produced by the allosteric modulator. Magnitudes of the α factor for a given allosteric molecule are unique for the tracer for receptor function/binding used to measure the interaction; see Chapters 7.4.

Coupling, processes that cause the interaction of molecules with membrane receptors to produce an observable cellular response; see Chapter 2.2.

Cubic ternary complex model, a molecular model (*J. Ther. Biol.* **178**, 151–167, 1996a; **178**, 169–182, 1996b; **181**, 381–397, 1996c) describing the coexistence of two receptor states that can interact with both G-proteins and ligands. The receptor/G-protein complexes may or may not produce a physiological response; see Chapter 3.11.

Degrees of Freedom, statistical term for the number of choices that can be made when fixing values of expected frequency leading to the number of independent comparisons that can be made in a sample of observations.

Desensitization, the reduction in response to an agonist or other physiological stimulation upon repeated instance of stimulation or continued presence of an agonist. Also referred to as tachyphyllaxis.

Dissociation constant, the ratio of the rate of offset of ligand away from a receptor divided by the rate of onset of the ligand approaching the receptor. It has the units of concentration and specifically is the concentration of ligand that occupies 50% of the total number of sites available for ligand binding at equilibrium (see Affinity).

Domain, sequence of amino acids in a protein that can be identified as controlling a specific function, that is, recognition of ligands.

Dose ratio, the concentration of agonist producing the same response in the presence of a given concentration of antagonist divided by the concentration of agonist produc- the same response in the absence of the antagonist. For control EC_{50} for an agonist dose–response EC_{50} in the presence of a given t is 30 nM, then the dose ratio in tration ratio). duction in the number of biologi- urface receptors, enzymes) usually

occurring with repeated stimulation of the system. For example, repeated stimulation of receptors by an agonist can lead to uncoupling of the receptors from stimulus–response mechanisms (due to phosphorylation of the receptors) followed by internalization of the receptor protein into the cell. This latter process is referred to as downregulation of receptors; see Chapter 2.6 and Fig. 5.7.

EC_{50}/ED_{50}, the "effective concentration" of an agonist producing (in this case) 50% maximal response to that particular drug (not necessarily 50% of the maximal response of the system). Other values can be quantified for other levels of response in which case the subscript denotes the response level (i.e., EC_{25} refers to the concentration of agonist producing 25% maximal response to that agonist). ED_{50} is the *in vivo* counterpart of EC_{50} referring to the dose of an agonist that produces 50% maximal effect.

Efficacy, historically, this term was given to agonists to define the property of the molecule that causes the production of a physiological response. However, with the discovery of negative efficacy (inverse agonists) and efficacy related to other properties of receptors that do not involve a physiological response, a more general definition of efficacy is that property of a molecule that causes the receptor to change its behavior toward the host.

E_{max}, Conventional term for the maximal response capable of being produced in a given system.

Equiactive dose ratios, ratios of molar concentrations of drug (usually agonists) that produce the same response in a given system; also referred to as EMR and EPMR; see Chapter 10.2.3.

Equiactive (equieffective) molar concentration (potency) ratios (EMR, EPMR), variants of the term dose ratio or equiactive dose ratios. Usually pertaining to agonists, these are the molar concentrations that produce the same response in a given system. These ratios are dependent on the affinity and efficacy of the agonists and thus are system independent, that is, characterize agonists and receptors in all systems. Care must be taken that the maximal responses of the agonists concerned are equal.

Equilibrium (dissociation) constant, reciprocal of the association constant and affinity; characterizes the binding of a molecule to a receptor. Specifically, it is the ratio of the rate of offset of the molecule away from the receptor divided by the rate of onset toward the receptor. It also is a molar concentration that binds to 50% of the receptor population.

Extended ternary complex model, a modification of the original ternary complex model for GPCRs (*J. Biol. Chem.* **268**, 4625–4636, 1993) in which the receptor is allowed to spontaneously form an active state that can then couple to G-proteins and produce a physiological response due to constitutive activity.

Fade, the time-dependent decrease in response upon prolonged exposure of a biological system to an agonist. Originally, this was defined as the characteristic peak contraction followed by relaxation produced by guinea pig vas deferentia, but the term has also been generalized to include all forms of real time observed loss of responsiveness (often termed tachyphyllaxis). It can be due to desensitization of the receptor or other factors. Fade is generally thought of as a case of decline of response in the continued presence of agonist as opposed to frequent stimulation.

Full agonist, name given to an agonist that produces the full system maximal response (E_{max}). It is a system-dependent phenomenon and should not necessarily be associated with a particular agonist, as an agonist can be a full agonist in some systems and a partial agonist in others.

Functional antagonism, reduction in the responsiveness to a given agonist by activation of cellular mechanisms that produce a counterstimulus to the cell.

Furchgott analysis, a technique [*in* "Advances in Drug Research" (N. J. Harper and A. B. Simmonds, eds.), Vol. 3, pp. 21–55. Academic Press, London, 1996] used to measure the affinity of a full agonist in a functional assay (see Chapters 5.6.2 and 12.2.3).

Gaddum analysis, Gaddum (method of), this method (*Q. J. Exp. Physiol.* **40**, 49–74, 1955) compares equiactive concentrations of an agonist in the absence and presence of a concentration of noncompetitive antagonist that depresses the maximal agonist response. These are compared in a double reciprocal plot (or variant thereof) to yield the equilibrium dissociation constant of the noncompetitive antagonist–receptor complex (see Chapters 6.4 and 12.2.8).

Gaddum equation (competitive antagonism), the pivotal simple equation (see Chapters 6.2 and 6.8.1) describing the competition between two ligands for a single receptor site. It forms the basis for Schild analysis.

Gaddum equation (noncompetitive antagonism), this technique measures the affinity of a noncompetitive antagonist based on a double reciprocal plot of equiactive agonist concentrations in the absence and presence of the noncompetitive antagonist. The antagonist must depress the maximal response to the agonist for the method to be effective; see Chapter 6.4.

Gene, the sequence of DNA that codes for a complete protein.

Genetic polymorphism, due to two or more alleles in a gene leading to more than one phenotype with respect to biological target reactivity to drugs.

Genome, the set of genes for an organism that determines all inherited characteristics. In general, the sequence and location of every gene responsible for coding every protein.

Genotype, the pattern of genes inherited by an individual. The makeup of a biological target due to coding of the gene for that target.

G-proteins, trimeric membrane-bound proteins that have intrinsic GTPase activity and act as intermediaries between 7TM receptors and a host of cellular effectors; see Section 2.2.

Hemiequilibria, a pseudoequilibrium that can occur when a fast-acting agonist equilibrates with a receptor system where a slow acting antagonist is present. The agonist will occupy the nonantagonist bound receptors quickly and then must equilibrate with antagonist bound receptors; this latter process can be extremely slow so as to be essentially irreversible within the time frame of some experiments. Under these conditions, a slow-acting competitive antagonist may appear to be an irreversibly acting antagonist.

Heptahelical receptors, another name for 7 TM receptors or G-protein-coupled receptors. It refers to the motif of the helices of the protein crossing the cell membrane seven times to form intracellular and extracellular domains.

Hyperbola (hyperbolic), a set of functions defining nonlinear relationships between abscissae and ordinates. This term is used loosely to describe nonlinear relationships between the initial interaction of molecules and receptors and the observed response (i.e., stimulus–response cascades of cells).

IC$_{50}$, the concentration (usually molar) of an inhibitor (receptor, enzyme antagonist) that blocks a given predefined stimulus by 50%. It is an empirical value in that its magnitude varies with the strength of the initial stimulus to be blocked.

Insurmountable antagonism, a receptor blockade that results in depression of the maximal response. Under these circumstances, unlike competitive antagonism, no increase in the concentration of agonist will regain the control maximal response in the presence of the antagonist.

Intrinsic activity, a scale of agonist activity devised by Ariens (*Arch. Int. Pharmacodyn. Ther.* **99**, 32–49, 1954) referring to the fractional maximal response to an agonist relative to a standard "full agonist" in the same system (where a full agonist produces the full system maximal response). Thus, a partial agonist that produces a maximal response 50% that of a full agonist has an intrinsic activity (denoted α) of 0.5. Full agonists have $\alpha = 1$ and antagonists $\alpha = 0$.

Intrinsic efficacy, the term efficacy, as defined originally by Stephenson (*Br. Pharmacol.*, **11**, 379–393, 1956), involved agonist and system components. Intrinsic efficacy (as given by Furchgott; "Advances in Drug Research" (N. J. Harper and A. B. Simmonds, eds.), Vol. 3, pp. 21–55. Academic Press, London, 1966) was defined to be a solely agonist-specific quantification of the ability of the agonist to induce a physiological or pharmacological response. Thus efficacy is the product of intrinsic efficacy multiplied by the receptor density (see Chapter 3.5).

Inverse agonist, these ligands reverse constitutive receptor activity. Currently it is thought that this occurs because inverse agonists have a selectively higher affinity for the inactive vs the active conformation of the receptor. It is important to note that while inverse agonist activity requires constitutive activity to be observed, the property of the molecule responsible for this activity does not disappear when there is no constitutive activity. In these cases, inverse agonists function as simple competitive antagonists.

Irreversible antagonists, irreversible ligands have negligible rates of offset (i.e., once the ligand binds to the receptor it essentially stays there). Under these circumstances, receptor occupancy does not achieve a steadystate by rather, increases with increasing exposure time to the ligand. Thus, once a receptor is occupied by the irreversible antagonist, it remains inactivated throughout the course of the experiment.

IUPHAR, an acronym for International Union of Pharmacology, a nongovernment organization of national societies functioning under the International Council of Scientific Unions.

In vitro, Latin *in vitro veritas* (in glass lies the truth) referring to experiments conducted in an artificial environment (i.e., organ bath, cell culture) leading to conditions of fewer and more controllable variables.

In vivo, with reference to *in vitro*, referring to experiments conducted in whole living organisms.

k_1, referring to the rate of onset of a molecule to a receptor with units of $s^{-1} mol^{-1}$.

$k_2 (k_{-1})$, referring to the rate of offset of molecule from a receptor in units of s^{-1}.

K_A, standard pharmacologic convention for the equilibrium dissociation constant of an agonist–receptor complex with units of M. It is the concentration that occupies half the receptor population at equilibrium. It also can be thought of as the reciprocal of affinity.

K_B, convention for the equilibrium dissociation constant of an antagonist–receptor complex usually determined in a functional assay denoting antagonism of a physiological response, although it can be associated with an antagonist when it is used in other types of experiment. It has units of M and is the concentration that occupies half the receptor population at equilibrium. It also can be thought of as the reciprocal of affinity.

K_d, convention for the equilibrium dissociation constant of a radioligand–receptor complex.

K_I, basically the K_B for an antagonist but specifically measured in a biochemical binding study (or enzyme assay).

Ligand, a molecule that binds to a biological receptor.

Ligand binding, a biochemical technique that measures the physical association of a ligand with a biological target (usually a protein); see Chapter 4.2.

Logistic function, generally yields a sigmoidally shaped line similar to that defined by drug dose–response relationships in biological systems. It is defined by $y = (1 + e^{-(a+bx)})^{-1}$. Substituting a as $\log(EC_{50})$ and x as $\log [A]$ leads to the Langmuir adsorption isotherm form of dose–response curves $y = [A]^b/([A]^b + (EC_{50})^b)$.

Log normal distribution, the distribution of a sample that is normal only when plotted on a logarithmic scale. The most prevalent cases in pharmacology refer to drug potencies (agonist and/or antagonist) that are estimated from semilogarithmic dose–response curves. All parametric statistical tests on these must be performed on their ˙ˑhmic counterparts, specifically their expression as a ˉˀale (–log values); see Chapter 1.11.2.

ˉˋates that the rate of a chemical) the concentration (mass) of the

ˌetics, in 1913 L. Michaelis and ˌt the rate of an enzymatic reaction

differed from conventional chemical reactions. They postulated a scheme whereby the reaction of a substrate plus enzyme yields enzyme plus substrate and placed it into the form of the equation: reaction velocity = (maximal velocity of the reaction × substrate concentration)/(concentration of substrate + a fitting constant K_m). The K_m (referred to as the Michaelis-Menten constant) is the concentration of the substrate at which the reaction rate is half the maximal value; it also characterizes the tightness of the binding between substrate and enzyme.

Negative efficacy, by definition, efficacy is that property of a molecule that causes the receptor to change its behavior toward the biological host. Negative efficacy refers to the property of selective affinity of the molecule for the inactive state of the receptor; this results in inverse agonism. Negative efficacy causes the active antagonism of constitutive receptor activity but is only observed in systems that have a measurably elevated basal response due to constitutive activity. It is a property of the molecule and not the system.

Noncompetitive antagonism, if an antagonist binds to the receptor and precludes agonist activation of that receptor by its occupancy, then no amount of agonist present in the receptor compartment can overcome this antagonism and it is termed noncompetitive. This can occur either by binding to the same binding domain as the agonist or another (allosteric) domain. Therefore, this definition is operational in that it does not necessarily imply a molecular mechanism, only a cause and effect relationship. The characteristic of noncompetitive antagonism is eventual depression of the maximal response; however, parallel displacement of agonist dose–response curves, with no diminution of maximal response, can occur in systems with receptor reserve for the agonist; see Chapter 6.4.

Nonlinear regression, a technique that fits a specified function of x and y by the method of least squares (i.e., the sum of the squares of the differences between real data points and calculated data points is minimized).

Nonspecific binding (nsb), binding of a traceable (i.e., radioactive) ligand (in a binding assay designed to measure the specific binding of the ligand) that binds to other components of the experimental system (i.e., other non-related proteins, wall of the vessel). It is defined operationally as the amount of ligand not displaced by an excess (approximately $100 \times K_B$) of a selective antagonist for the biological target; see Chapter 4.2.

Null method, physiological or pharmacological effects are translations of biochemical events by the cell. The null method assumes that equal responses emanate from equal initial stimulation of the receptor; therefore, when comparing equal responses, the complex translation is cancelled and statements about the receptor activity of agonists can be made. Relative potencies of agonists producing equal responses thus are interpreted to be measures of the relative receptor stimuli produced by the agonists at the receptor; see Chapter 5.6.2.

Occupancy, the probability that a molecule will be bound to a receptor at a given concentration. For example, an occupancy of 50% states that, at any one instant, half of the receptors will have a molecule bound and half will not. This is a stochastic process and the actual receptors that are bound change constantly with time. However, at any one instant, the total fraction bound will be the fractional occupancy.

Operational model, devised and published by James Black and Paul Leff (*Proc. R. Soc. Lond. Biol.* **220**,141–162, 1983), this model uses experimental observation to describe the production of a physiological response by an agonist in general terms. It defines affinity and the ability of a drug to induce a response as a value of τ, which is a term describing the system (receptor density and efficiency of the cell to convert an activated receptor stimulus into a response) and the agonist (efficacy). It has provided a major advance in the description of functional effects of drugs; see Chapter 3.6 for further discussion.

Orphan receptor, a gene product that is predicted to be a receptor through structure and spontaneous interaction with G-proteins but for which there is no known endogenous ligand or physiological function.

Outliers, observations that are very inconsistent with the main sample of data, i.e., apparently significantly different from the rest of the data. While there are statistical methods to test whether these values may be aberrant and thus should be removed, caution should be exercised in this practice as these data may also be the most interesting and indicative of a rare but important occurrence.

Partial agonist, whereas a full agonist produces the system maximal response, a partial agonist produces a maximal response that is below that of the system maximum (and that of a full agonist). As well as producing a submaximal response, partial agonists produce antagonism of more efficacious full agonists.

pA_2/pA_X, this negative logarithm of the molar concentration of an antagonist produces a twofold (for a pA_2) shift to the right of an agonist dose–response curve. If the shift is different from 2, then it may be defined as pA_x, where the degree of the shift of the dose–response curve is x (i.e., pA_5 is the –log concentration that produces a fivefold shift to the right of the agonist dose–response curve). The pA_2 is by far the most prevalent value determined, as this also may have meaning on a molecular level (i.e., under certain conditions the pA_2 is also the pK_B for an antagonist).

pD_2, historical term for the negative logarithm of the EC_{50} for an agonist in a functional assay, not often used in present-day pharmacology.

Phenotype, characteristics that result from the expression of a genotype.

pK_B, negative logarithm of the K_B. This is the common currency of antagonist pharmacology, as pK_B values are log normally distributed and thus are used to characterize receptors and antagonist potency.

pK_I, negative logarithm of the K_I, the equilibrium dissociation constant of an antagonist–receptor complex measured in a biochemical binding or enzyme study (also log normally distributed).

Polymorphisms, in pharmacology, these are associated with genetic polymorphisms of biological targets (see Genetic polymorphisms).

Potency, the concentration (usually molar) of a drug that produces a defined effect. Often, potencies of agonists are defined in terms of EC_{50} or pEC_{50} values. The potency usually does not involve measures of maximal effect but rather only in locations along the concentration axis of dose–response curves.

Potentiation, the increase in effect produced by a molecule or procedure in a pharmacological preparation. This can be expressed as an apparent increase in efficacy (i.e., maximal response), potency, both.

Pseudoirreversible antagonism, true irreversible antagonism involves a covalent chemical bond between the antagonist and the receptor (such that the rate of offset of the antagonist from the receptor is zero). However, on the time scale of pharmacological experiments, the rate of offset of an antagonist can be so slow as to essentially be irreversible. Therefore, although no covalent bond is involved, the antagonist is for all intents and purposes bound irreversibly to the receptor.

Receptor reserve, in highly efficiently coupled receptor systems, high-efficacy agonists may produce excess stimulus that saturates cellular stimulus–response mechanisms. Under these conditions, these agonists produce the system maximal response through activation of only a fraction of the existing receptor population. The remaining fraction is thus "spare" or a "reserve" in that irreversible removal of this fraction will cause a shift to the right of the agonist dose–response curve but no diminution of maximum. For example, in a system where the maximal response to an agonist can be attained by activation of 5% of the receptor population, there will be a 95% receptor reserve.

Receptors, in theoretical terms, a receptor is a biological recognition unit that interacts with molecules of other stimuli (i.e., light) to translate information to cells. Receptors technically can be any biological entity such as enzymes, reuptake recognition sites, and genetic material such as DNA; however, the term usually is associated with proteins on the cell surface that transmit information from chemicals to cells. The most therapeutically relevant receptor class is G-protein-coupled receptors, presently comprising 45% of all existing drug therapies.

Recombinant DNA, this is DNA containing new genetic material in an order different from the original. Genetic engineering can be used to do this deliberately to produce new proteins in cells.

Relative intrinsic activity, this actually is redundant, as intrinsic activity itself is defined only in relative terms, i.e., the maximal response of an agonist as a fraction of the maximal response to another agonist.

Relative potency, absolute agonist potency is the product of receptor stimulus (brought about by agonist affinity and efficacy) and the processing of the stimulus by the cell into an observable response. Because this latter process is system (cell type) dependent, absolute potencies are system-dependent measures of agonist activity. However, when comparing two agonists in the system, null procedures cancel these effects; therefore, the relative potency of agonists (provided both are full agonists) are system-independent estimates of agonist activity that can be compared across systems; see Chapter 10.2.3.

Resultant analysis, this procedure, developed by James Black and colleagues (*Br. J. Pharmacol.* **84**,561–571, 1985), allows measurement of the receptor affinity of a competitive antagonist, which has secondary properties that obscure the receptor antagonism; see Chapter 6.6 for further discussion.

Saturation binding, a biochemical procedure that quantifies the amount of traceable ligand (i.e., radioligand) to a receptor protein. It yields the affinity of the ligand and the maximal number of binding sites (B_{max}); see Chapter 4.2.1.

Scatchard analysis, a common linear transformation of saturation binding data used prevalently before the widespread availability of nonlinear fitting software. The Scatchard transformation (see Chapter 4.2.1), while easy to perform, can be misleading and lead to errors.

Schild analysis, this powerful method of quantifying the potency of a competitive antagonist was developed by Heinz Schild (*Br. J. Pharmacol.* **14**,48–58, 1959; see Chapter 6.3). It is based on the principle that the antagonist-induced dextral displacement of a dose–response curve is due to its potency (K_B value) and its concentration in the receptor compartment. Because the antagonism can be observed and the concentration of antagonist is known, the K_B can be calculated.

Schild plot, the relationship between antagonism and concentration is loglinear according to the Schild equation. The tool to determine if this is true experimentally is the Schild plot, namely a regression of log (DR–1) values (where DR is the dose ratio for the agonist in the presence and absence of antagonist) upon the logarithm of the molar concentration of the antagonist. If this regression is linear with unit slope, then the antagonism adheres to the simple competitive model and the intercept of regression is the pK_B. For further discussion, see Chapter 6.3.

Second messenger, these are molecules produced by cellular effectors that go on to activate other biochemical processes in the cell. Some examples of second messengers are cyclic AMP, inositol triphosphate, arachidonic acid, and calcium ion (see Chapter 2.2).

Selectivity, the difference in activity a given biologically active molecule has for two or more processes. Thus, if a molecule has a 10-fold (for example) greater affinity for process A over process B, then it can be said to have selectivity for process A. However, the implication is that the different activity is not absolute, that is, given enough molecule, the activation of the other process(es) will occur.

Sigmoid, the characteristic "S-shaped" curves defined by functions such as the Langmuir isotherm and logistic function (when plotted on a logarithmic abscissal scale).

Spare receptors, another term for receptor reserve (see Receptor reserve).

Specificity, this can be thought of as an extreme form of selectivity (see Selectivity) where, in this case, no increase in the concentration of the molecule will be sufficient to activate the other process(es). This term is often used erroneously in that the extremes of concentration have not been tested (or cannot be tested due to chemical, toxic, or solubility constraints in a particular system) to define what probably is only selectivity.

Stimulus, this is quanta of initial stimulation given to the receptor by the agonist. There are no units to stimulus and it is always utilized as a ratio quantity comparing two or more agonists. Stimulus is not an observable response but is processed by the cell to yield a measurable response.

Stimulus–response coupling, another term for receptor coupling (see Receptor Coupling). It describes the series of biochemical reactions that link the initial activation of the receptor to the observed cellular (or organ) response.

Subtype, often refers to a receptor and denotes a variation in the gene product such that the endogenous ligand is the same (i.e., neurotransmitter, hormone) but the function, distribution, and sensitivity of the receptor subtypes differ. Antagonists often can distinguish receptor subtypes.

Surmountable antagonism, an antagonist-induced shift to the right of an agonist dose–response curve with no diminution of the maximal response to the agonist (observed with simple competitive antagonists and some types of allosteric modulators).

Tachyphyllaxis, the progressive reduction in response due to repeated agonist stimulation (see Desensitization and Fade). The maximal response to the agonist is reduced in tachyphyllaxis (whereas the sensitivity is reduced with tolerance).

Ternary complex (model), this model describes the formation of a complex among a ligand (usually an agonist), a receptor, and a G-protein. Originally described by De Lean and colleagues (*J. Biol. Chem.* **255**, 7108–7117, 1980), it has been modified to include other receptor behaviors (see Chapters 3.8 to 3.11), such as constitutive receptor activity.

Transfection, the transfer of DNA from one cell into another cell. This DNA then replicates in the acceptor cell.

Two-state model, a model of proteins that coexists in two states controlled by an equilibrium constant. Molecules with selective affinity for one of the states will produce a bias in that state upon binding to the system. Two-state theory was conceived to describe the function of ion channels but also has relevance to receptors (see Chapter 3.7).

Uncompetitive antagonism, form of inhibition (originally defined for enzyme kinetics) in which both the maximal asymptotic value of the response and the equilibrium dissociation constant of the activator (i.e., agonist) are reduced by the antagonist. This differs from noncompetitive antagonism where the affinity of the receptor for the activating drug is not altered. Uncompetitive effects can occur due to allosteric modulation of receptor activity by an allosteric modulator (see Chapter 6.4).

Appendices

A.1 Statistical Tables of Use for Assessing Significant Difference

1. t distribution
2. F Distribution $(p, 0.05)$
3. F Distribution $(p < 0.025)$
4. F Distribution $(p < 0.01)$

A.1.1 t Distribution

To determine the 0.05 critical value from t distribution with 5 degrees of freedom, look in the 0.05 column at the fifth row: $t_{(.05,5)} = 2.015048$.

TABLE A.1

t Table with right tail probabilities

df\p	0.40	0.25	0.10	0.05	0.025	0.01	0.005	0.0005
1	0.324920	1.000000	3.077684	6.313752	12.70620	31.82052	63.65674	636.6192
2	0.288675	0.816497	1.885618	2.919986	4.30265	6.96456	9.92484	31.5991
3	0.276671	0.764892	1.637744	2.353363	3.18245	4.54070	5.84091	12.9240
4	0.270722	0.740697	1.533206	2.131847	2.77645	3.74695	4.60409	8.6103
5	0.267181	0.726687	1.475884	2.015048	2.57058	3.36493	4.03214	6.8688
6	0.264835	0.717558	1.439756	1.943180	2.44691	3.14267	3.70743	5.9588
7	0.263167	0.711142	1.414924	1.894579	2.36462	2.99795	3.49948	5.4079
8	0.261921	0.706387	1.396815	1.859548	2.30600	2.89646	3.35539	5.0413
9	0.260955	0.702722	1.383029	1.833113	2.26216	2.82144	3.24984	4.7809
10	0.260185	0.699812	1.372184	1.812461	2.22814	2.76377	3.16927	4.5869
11	0.259556	0.697445	1.363430	1.795885	2.20099	2.71808	3.10581	4.4370
12	0.259033	0.695483	1.356217	1.782288	2.17881	2.68100	3.05454	4.3178
13	0.258591	0.693829	1.350171	1.770933	2.16037	2.65031	3.01228	4.2208
14	0.258213	0.692417	1.345030	1.761310	2.14479	2.62449	2.97684	4.1405
15	0.257885	0.691197	1.340606	1.753050	2.13145	2.60248	2.94671	4.0728
16	0.257599	0.690132	1.336757	1.745884	2.11991	2.58349	2.92078	4.0150
17	0.257347	0.689195	1.333379	1.739607	2.10982	2.56693	2.89823	3.9651
18	0.257123	0.688364	1.330391	1.734064	2.10092	2.55238	2.87844	3.9216
19	0.256923	0.687621	1.327728	1.729133	2.09302	2.53948	2.86093	3.8834
20	0.256743	0.686954	1.325341	1.724718	2.08596	2.52798	2.84534	3.8495
21	0.256580	0.686352	1.323188	1.720743	2.07961	2.51765	2.83136	3.8193
22	0.256432	0.685805	1.321237	1.717144	2.07387	2.50832	2.81876	3.7921
23	0.256297	0.685306	1.319460	1.713872	2.06866	2.49987	2.80734	3.7676
24	0.256173	0.684850	1.317836	1.710882	2.06390	2.49216	2.79694	3.7454
25	0.256060	0.684430	1.316345	1.708141	2.05954	2.48511	2.78744	3.7251
26	0.255955	0.684043	1.314972	1.705618	2.05553	2.47863	2.77871	3.7066
27	0.255858	0.683685	1.313703	1.703288	2.05183	2.47266	2.77068	3.6896
28	0.255768	0.683353	1.312527	1.701131	2.04841	2.46714	2.76326	3.6739
29	0.255684	0.683044	1.311434	1.699127	2.04523	2.46202	2.75639	3.6594
30	0.255605	0.682756	1.310415	1.697261	2.04227	2.45726	2.75000	3.6460
inf	0.253347	0.674490	1.281552	1.644854	1.95996	2.32635	2.57583	3.2905

A.1.1.1 F Distribution

By convention, the numerator degrees of freedom are always given first [switching the order of degrees of freedom changes the distribution, that is, $F_{(10,12)}$ does not equal $F_{(12,10)}$]. For the following F tables, **rows represent denominator degrees of freedom** and **columns represent numerator degrees of freedom**.

A.1.2 F Table for α=0.05

df2/df1	1	2	3	4	5	6	7	8	9	10	12	15	20	24	30	40	60	120	INF
1	161.4476	199.5000	215.7073	224.5832	230.1619	233.9860	236.7684	238.8827	240.5433	241.8817	243.9060	245.9499	248.0131	249.0518	250.0951	251.1432	252.1957	253.2529	254.3144
2	18.5128	19.0000	19.1643	19.2468	19.2964	19.3295	19.3532	19.3710	19.3848	19.3959	19.4125	19.4291	19.4458	19.4541	19.4624	19.4707	19.4791	19.4874	19.4957
3	10.1280	9.5521	9.2766	9.1172	9.0135	8.9406	8.8867	8.8452	8.8123	8.7855	8.7446	8.7029	8.6602	8.6385	8.6166	8.5944	8.5720	8.5494	8.5264
4	7.7086	6.9443	6.5914	6.3882	6.2561	6.1631	6.0942	6.0410	5.9988	5.9644	5.9117	5.8578	5.8025	5.7744	5.7459	5.7170	5.6877	5.6581	5.6281
5	6.6079	5.7861	5.4095	5.1922	5.0503	4.9503	4.8759	4.8183	4.7725	4.7351	4.6777	4.6188	4.5581	4.5272	4.4957	4.4638	4.4314	4.3985	4.3650
6	5.9874	5.1433	4.7571	4.5337	4.3874	4.2839	4.2067	4.1468	4.0990	4.0600	3.9999	3.9381	3.8742	3.8415	3.8082	3.7743	3.7398	3.7047	3.6689
7	5.5914	4.7374	4.3468	4.1203	3.9715	3.8660	3.7870	3.7257	3.6767	3.6365	3.5747	3.5107	3.4445	3.4105	3.3758	3.3404	3.3043	3.2674	3.2298
8	5.3177	4.4590	4.0662	3.8379	3.6875	3.5806	3.5005	3.4381	3.3881	3.3472	3.2839	3.2184	3.1503	3.1152	3.0794	3.0428	3.0053	2.9669	2.9276
9	5.1174	4.2565	3.8625	3.6331	3.4817	3.3738	3.2927	3.2296	3.1789	3.1373	3.0729	3.0061	2.9365	2.9005	2.8637	2.8259	2.7872	2.7475	2.7067
10	4.9646	4.1028	3.7083	3.4780	3.3258	3.2172	3.1355	3.0717	3.0204	2.9782	2.9130	2.8450	2.7740	2.7372	2.6996	2.6609	2.6211	2.5801	2.5379
11	4.8443	3.9823	3.5874	3.3567	3.2039	3.0946	3.0123	2.9480	2.8962	2.8536	2.7876	2.7186	2.6464	2.6090	2.5705	2.5309	2.4901	2.4480	2.4045
12	4.7472	3.8853	3.4903	3.2592	3.1059	2.9961	2.9134	2.8486	2.7964	2.7534	2.6866	2.6169	2.5436	2.5055	2.4663	2.4259	2.3842	2.3410	2.2962
13	4.6672	3.8056	3.4105	3.1791	3.0254	2.9153	2.8321	2.7669	2.7144	2.6710	2.6037	2.5331	2.4589	2.4202	2.3803	2.3392	2.2966	2.2524	2.2064
14	4.6001	3.7389	3.3439	3.1122	2.9582	2.8477	2.7642	2.6987	2.6458	2.6022	2.5342	2.4630	2.3879	2.3487	2.3082	2.2664	2.2229	2.1778	2.1307
15	4.5431	3.6823	3.2874	3.0556	2.9013	2.7905	2.7066	2.6408	2.5876	2.5437	2.4753	2.4034	2.3275	2.2878	2.2468	2.2043	2.1601	2.1141	2.0658
16	4.4940	3.6337	3.2389	3.0069	2.8524	2.7413	2.6572	2.5911	2.5377	2.4935	2.4247	2.3522	2.2756	2.2354	2.1938	2.1507	2.1058	2.0589	2.0096
17	4.4513	3.5915	3.1968	2.9647	2.8100	2.6987	2.6143	2.5480	2.4943	2.4499	2.3807	2.3077	2.2304	2.1898	2.1477	2.1040	2.0584	2.0107	1.9604
18	4.4139	3.5546	3.1599	2.9277	2.7729	2.6613	2.5767	2.5102	2.4563	2.4117	2.3421	2.2686	2.1906	2.1497	2.1071	2.0629	2.0166	1.9681	1.9168
19	4.3807	3.5219	3.1274	2.8951	2.7401	2.6283	2.5435	2.4768	2.4227	2.3779	2.3080	2.2341	2.1555	2.1141	2.0712	2.0264	1.9795	1.9302	1.8780
20	4.3512	3.4928	3.0984	2.8661	2.7109	2.5990	2.5140	2.4471	2.3928	2.3479	2.2776	2.2033	2.1242	2.0825	2.0391	1.9938	1.9464	1.8963	1.8432
21	4.3248	3.4668	3.0725	2.8401	2.6848	2.5727	2.4876	2.4205	2.3660	2.3210	2.2504	2.1757	2.0960	2.0540	2.0102	1.9645	1.9165	1.8657	1.8117
22	4.3009	3.4434	3.0491	2.8167	2.6613	2.5491	2.4638	2.3965	2.3419	2.2967	2.2258	2.1508	2.0707	2.0283	1.9842	1.9380	1.8894	1.8380	1.7831
23	4.2793	3.4221	3.0280	2.7955	2.6400	2.5277	2.4422	2.3748	2.3201	2.2747	2.2036	2.1282	2.0476	2.0050	1.9605	1.9139	1.8648	1.8128	1.7570
24	4.2597	3.4028	3.0088	2.7763	2.6207	2.5082	2.4226	2.3551	2.3002	2.2547	2.1834	2.1077	2.0267	1.9838	1.9390	1.8920	1.8424	1.7896	1.7330
25	4.2417	3.3852	2.9912	2.7587	2.6030	2.4904	2.4047	2.3371	2.2821	2.2365	2.1649	2.0889	2.0075	1.9643	1.9192	1.8718	1.8217	1.7684	1.7110
26	4.2252	3.3690	2.9752	2.7426	2.5868	2.4741	2.3883	2.3205	2.2655	2.2197	2.1479	2.0716	1.9898	1.9464	1.9010	1.8533	1.8027	1.7488	1.6906
27	4.2100	3.3541	2.9604	2.7278	2.5719	2.4591	2.3732	2.3053	2.2501	2.2043	2.1323	2.0558	1.9736	1.9299	1.8842	1.8361	1.7851	1.7306	1.6717
28	4.1960	3.3404	2.9467	2.7141	2.5581	2.4453	2.3593	2.2913	2.2360	2.1900	2.1179	2.0411	1.9586	1.9147	1.8687	1.8203	1.7689	1.7138	1.6541
29	4.1830	3.3277	2.9340	2.7014	2.5454	2.4324	2.3463	2.2783	2.2229	2.1768	2.1045	2.0275	1.9446	1.9005	1.8543	1.8055	1.7537	1.6981	1.6376
30	4.1709	3.3158	2.9223	2.6896	2.5336	2.4205	2.3343	2.2662	2.2107	2.1646	2.0921	2.0148	1.9317	1.8874	1.8409	1.7918	1.7396	1.6835	1.6223
40	4.0847	3.2317	2.8387	2.6060	2.4495	2.3359	2.2490	2.1802	2.1240	2.0772	2.0035	1.9245	1.8389	1.7929	1.7444	1.6928	1.6373	1.5766	1.5089
60	4.0012	3.1504	2.7581	2.5252	2.3683	2.2541	2.1665	2.0970	2.0401	1.9926	1.9174	1.8364	1.7480	1.7001	1.6491	1.5943	1.5343	1.4673	1.3893
120	3.9201	3.0718	2.6802	2.4472	2.2899	2.1750	2.0868	2.0164	1.9588	1.9105	1.8337	1.7505	1.6587	1.6084	1.5543	1.4952	1.4290	1.3519	1.2539
inf	3.8415	2.9957	2.6049	2.3719	2.2141	2.0986	2.0096	1.9384	1.8799	1.8307	1.7522								

A.1.3 Table for α = 0.025

df2/df1	1	2	3	4	5	6	7	8	9	10	12	15	20	24	30	40	60	120	INF
1	647.7890	799.5000	864.1630	899.5833	921.8479	937.1111	948.2169	956.6562	963.2846	968.6274	976.7079	984.8668	993.1028	997.2492	1001.414	1005.598	1009.800	1014.020	1018.258
2	38.5063	39.0000	39.1655	39.2484	39.2982	39.3315	39.3552	39.3730	39.3869	39.3980	39.4146	39.4313	39.4479	39.4562	39.465	39.473	39.481	39.490	39.498
3	17.4434	16.0441	15.4392	15.1010	14.8848	14.7347	14.6244	14.5399	14.4731	14.4189	14.3366	14.2527	14.1674	14.1241	14.081	14.037	13.992	13.947	13.902
4	12.2179	10.6491	9.9792	9.6045	9.3645	9.1973	9.0741	8.9796	8.9047	8.8439	8.7512	8.6565	8.5599	8.5109	8.461	8.411	8.360	8.309	8.257
5	10.0070	8.4336	7.7636	7.3879	7.1464	6.9777	6.8531	6.7572	6.6811	6.6192	6.5245	6.4277	6.3286	6.2780	6.227	6.175	6.123	6.069	6.015
6	8.8131	7.2599	6.5988	6.2272	5.9876	5.8198	5.6955	5.5996	5.5234	5.4613	5.3662	5.2687	5.1684	5.1172	5.065	5.012	4.959	4.904	4.849
7	8.0727	6.5415	5.8898	5.5226	5.2852	5.1186	4.9949	4.8993	4.8232	4.7611	4.6658	4.5678	4.4667	4.4150	4.362	4.309	4.254	4.199	4.142
8	7.5709	6.0595	5.4160	5.0526	4.8173	4.6517	4.5286	4.4333	4.3572	4.2951	4.1997	4.1012	3.9995	3.9472	3.894	3.840	3.784	3.728	3.670
9	7.2093	5.7147	5.0781	4.7181	4.4844	4.3197	4.1970	4.1020	4.0260	3.9639	3.8682	3.7694	3.6669	3.6142	3.560	3.505	3.449	3.392	3.333
10	6.9367	5.4564	4.8256	4.4683	4.2361	4.0721	3.9498	3.8549	3.7790	3.7168	3.6209	3.5217	3.4185	3.3654	3.311	3.255	3.198	3.140	3.080
11	6.7241	5.2559	4.6300	4.2751	4.0440	3.8807	3.7586	3.6638	3.5879	3.5257	3.4296	3.3299	3.2261	3.1725	3.118	3.061	3.004	2.944	2.883
12	6.5538	5.0959	4.4742	4.1212	3.8911	3.7283	3.6065	3.5118	3.4358	3.3736	3.2773	3.1772	3.0728	3.0187	2.963	2.906	2.848	2.787	2.725
13	6.4143	4.9653	4.3472	3.9959	3.7667	3.6043	3.4827	3.3880	3.3120	3.2497	3.1532	3.0527	2.9477	2.8932	2.837	2.780	2.720	2.659	2.595
14	6.2979	4.8567	4.2417	3.8919	3.6634	3.5014	3.3799	3.2853	3.2093	3.1469	3.0502	2.9493	2.8437	2.7888	2.732	2.674	2.614	2.552	2.487
15	6.1995	4.7650	4.1528	3.8043	3.5764	3.4147	3.2934	3.1987	3.1227	3.0602	2.9633	2.8621	2.7559	2.7006	2.644	2.585	2.524	2.461	2.395
16	6.1151	4.6867	4.0768	3.7294	3.5021	3.3406	3.2194	3.1248	3.0488	2.9862	2.8890	2.7875	2.6808	2.6252	2.568	2.509	2.447	2.383	2.316
17	6.0420	4.6189	4.0112	3.6648	3.4379	3.2767	3.1556	3.0610	2.9849	2.9222	2.8249	2.7230	2.6158	2.5598	2.502	2.442	2.380	2.315	2.247
18	5.9781	4.5597	3.9539	3.6083	3.3820	3.2209	3.0999	3.0053	2.9291	2.8664	2.7689	2.6667	2.5590	2.5027	2.445	2.384	2.321	2.256	2.187
19	5.9216	4.5075	3.9034	3.5587	3.3327	3.1718	3.0509	2.9563	2.8801	2.8172	2.7196	2.6171	2.5089	2.4523	2.394	2.333	2.270	2.203	2.133
20	5.8715	4.4613	3.8587	3.5147	3.2891	3.1283	3.0074	2.9128	2.8365	2.7737	2.6758	2.5731	2.4645	2.4076	2.349	2.287	2.223	2.156	2.085
21	5.8266	4.4199	3.8188	3.4754	3.2501	3.0895	2.9686	2.8740	2.7977	2.7348	2.6368	2.5338	2.4247	2.3675	2.308	2.246	2.182	2.114	2.042
22	5.7863	4.3828	3.7829	3.4401	3.2151	3.0546	2.9338	2.8392	2.7628	2.6998	2.6017	2.4984	2.3890	2.3315	2.272	2.210	2.145	2.076	2.003
23	5.7498	4.3492	3.7505	3.4083	3.1835	3.0232	2.9023	2.8077	2.7313	2.6682	2.5699	2.4665	2.3567	2.2989	2.239	2.176	2.111	2.041	1.968
24	5.7166	4.3187	3.7211	3.3794	3.1548	2.9946	2.8738	2.7791	2.7027	2.6396	2.5411	2.4374	2.3273	2.2693	2.209	2.146	2.080	2.010	1.935
25	5.6864	4.2909	3.6943	3.3530	3.1287	2.9685	2.8478	2.7531	2.6766	2.6135	2.5149	2.4110	2.3005	2.2422	2.182	2.118	2.052	1.981	1.906
26	5.6586	4.2655	3.6697	3.3289	3.1048	2.9447	2.8240	2.7293	2.6528	2.5896	2.4908	2.3867	2.2759	2.2174	2.157	2.093	2.026	1.954	1.878
27	5.6331	4.2421	3.6472	3.3067	3.0828	2.9228	2.8021	2.7074	2.6309	2.5676	2.4688	2.3644	2.2533	2.1946	2.133	2.069	2.002	1.930	1.853
28	5.6096	4.2205	3.6264	3.2863	3.0626	2.9027	2.7820	2.6872	2.6106	2.5473	2.4484	2.3438	2.2324	2.1735	2.112	2.048	1.980	1.907	1.829
29	5.5878	4.2006	3.6072	3.2674	3.0438	2.8840	2.7633	2.6686	2.5919	2.5286	2.4295	2.3248	2.2131	2.1540	2.092	2.028	1.959	1.886	1.807
30	5.5675	4.1821	3.5894	3.2499	3.0265	2.8667	2.7460	2.6513	2.5746	2.5112	2.4120	2.3072	2.1952	2.1359	2.074	2.009	1.940	1.866	1.787
40	5.4239	4.0510	3.4633	3.1261	2.9037	2.7444	2.6238	2.5289	2.4519	2.3882	2.2882	2.1819	2.0677	2.0069	1.943	1.875	1.803	1.724	1.637
60	5.2856	3.9253	3.3425	3.0077	2.7863	2.6274	2.5068	2.4117	2.3344	2.2702	2.1692	2.0613	1.9445	1.8817	1.815	1.744	1.667	1.581	1.482
120	5.1523	3.8046	3.2269	2.8943	2.6740	2.5154	2.3948	2.2994	2.2217	2.1570	2.0548	1.9450	1.8249	1.7597	1.690	1.614	1.530	1.433	1.310
inf	5.0239	3.6889	3.1161	2.7858	2.5665	2.4082	2.2875	2.1918	2.1136	2.0483	1.9447								

A.1.4 F Table for α = 0.05

df2/df1	1	2	3	4	5	6	7	8	9	10	12	15	20	24	30	40	60	120	INF
1	4052.181	4999.500	5403.352	5624.583	5763.650	5858.986	5928.356	5981.070	6022.473	6055.847	6106.321	6157.285	6208.730	6234.631	6260.649	6286.782	6313.030	6339.391	6365.864
2	98.503	99.000	99.166	99.249	99.299	99.333	99.356	99.374	99.388	99.399	99.416	99.433	99.449	99.458	99.466	99.474	99.482	99.491	99.499
3	34.116	30.817	29.457	28.710	28.237	27.911	27.672	27.489	27.345	27.229	27.052	26.872	26.690	26.598	26.505	26.411	26.316	26.221	26.125
4	21.198	18.000	16.694	15.977	15.522	15.207	14.976	14.799	14.659	14.546	14.374	14.198	14.020	13.929	13.838	13.745	13.652	13.558	13.463
5	16.258	13.274	12.060	11.392	10.967	10.672	10.456	10.289	10.158	10.051	9.888	9.722	9.553	9.466	9.379	9.291	9.202	9.112	9.020
6	13.745	10.925	9.780	9.148	8.746	8.466	8.260	8.102	7.976	7.874	7.718	7.559	7.396	7.313	7.229	7.143	7.057	6.969	6.880
7	12.246	9.547	8.451	7.847	7.460	7.191	6.993	6.840	6.719	6.620	6.469	6.314	6.155	6.074	5.992	5.908	5.824	5.737	5.650
8	11.259	8.649	7.591	7.006	6.632	6.371	6.178	6.029	5.911	5.814	5.667	5.515	5.359	5.279	5.198	5.116	5.032	4.946	4.859
9	10.561	8.022	6.992	6.422	6.057	5.802	5.613	5.467	5.351	5.257	5.111	4.962	4.808	4.729	4.649	4.567	4.483	4.398	4.311
10	10.044	7.559	6.552	5.994	5.636	5.386	5.200	5.057	4.942	4.849	4.706	4.558	4.405	4.327	4.247	4.165	4.082	3.996	3.909
11	9.646	7.206	6.217	5.668	5.316	5.069	4.886	4.744	4.632	4.539	4.397	4.251	4.099	4.021	3.941	3.860	3.776	3.690	3.602
12	9.330	6.927	5.953	5.412	5.064	4.821	4.640	4.499	4.388	4.296	4.155	4.010	3.858	3.780	3.701	3.619	3.535	3.449	3.361
13	9.074	6.701	5.739	5.205	4.862	4.620	4.441	4.302	4.191	4.100	3.960	3.815	3.665	3.587	3.507	3.425	3.341	3.255	3.165
14	8.862	6.515	5.564	5.035	4.695	4.456	4.278	4.140	4.030	3.939	3.800	3.656	3.505	3.427	3.348	3.266	3.181	3.094	3.004
15	8.683	6.359	5.417	4.893	4.556	4.318	4.142	4.004	3.895	3.805	3.666	3.522	3.372	3.294	3.214	3.132	3.047	2.959	2.868
16	8.531	6.226	5.292	4.773	4.437	4.202	4.026	3.890	3.780	3.691	3.553	3.409	3.259	3.181	3.101	3.018	2.933	2.845	2.753
17	8.400	6.112	5.185	4.669	4.336	4.102	3.927	3.791	3.682	3.593	3.455	3.312	3.162	3.084	3.003	2.920	2.835	2.746	2.653
18	8.285	6.013	5.092	4.579	4.248	4.015	3.841	3.705	3.597	3.508	3.371	3.227	3.077	2.999	2.919	2.835	2.749	2.660	2.566
19	8.185	5.926	5.010	4.500	4.171	3.939	3.765	3.631	3.523	3.434	3.297	3.153	3.003	2.925	2.844	2.761	2.674	2.584	2.489
20	8.096	5.849	4.938	4.431	4.103	3.871	3.699	3.564	3.457	3.368	3.231	3.088	2.938	2.859	2.778	2.695	2.608	2.517	2.421
21	8.017	5.780	4.874	4.369	4.042	3.812	3.640	3.506	3.398	3.310	3.173	3.030	2.880	2.801	2.720	2.636	2.548	2.457	2.360
22	7.945	5.719	4.817	4.313	3.988	3.758	3.587	3.453	3.346	3.258	3.121	2.978	2.827	2.749	2.667	2.583	2.495	2.403	2.305
23	7.881	5.664	4.765	4.264	3.939	3.710	3.539	3.406	3.299	3.211	3.074	2.931	2.781	2.702	2.620	2.535	2.447	2.354	2.256
24	7.823	5.614	4.718	4.218	3.895	3.667	3.496	3.363	3.256	3.168	3.032	2.889	2.738	2.659	2.577	2.492	2.403	2.310	2.211
25	7.770	5.568	4.675	4.177	3.855	3.627	3.457	3.324	3.217	3.129	2.993	2.850	2.699	2.620	2.538	2.453	2.364	2.270	2.169
26	7.721	5.526	4.637	4.140	3.818	3.591	3.421	3.288	3.182	3.094	2.958	2.815	2.664	2.585	2.503	2.417	2.327	2.233	2.131
27	7.677	5.488	4.601	4.106	3.785	3.558	3.388	3.256	3.149	3.062	2.926	2.783	2.632	2.552	2.470	2.384	2.294	2.198	2.097
28	7.636	5.453	4.568	4.074	3.754	3.528	3.358	3.226	3.120	3.032	2.896	2.753	2.602	2.522	2.440	2.354	2.263	2.167	2.064
29	7.598	5.420	4.538	4.045	3.725	3.499	3.330	3.198	3.092	3.005	2.868	2.726	2.574	2.495	2.412	2.325	2.234	2.138	2.034
30	7.562	5.390	4.510	4.018	3.699	3.473	3.304	3.173	3.067	2.979	2.843	2.700	2.549	2.469	2.386	2.299	2.208	2.111	2.006
40	7.314	5.179	4.313	3.828	3.514	3.291	3.124	2.993	2.888	2.801	2.665	2.522	2.369	2.288	2.203	2.114	2.019	1.917	1.805
60	7.077	4.977	4.126	3.649	3.339	3.119	2.953	2.823	2.718	2.632	2.496	2.352	2.198	2.115	2.028	1.936	1.836	1.726	1.601
120	6.851	4.787	3.949	3.480	3.174	2.956	2.792	2.663	2.559	2.472	2.336	2.192	2.035	1.950	1.860	1.763	1.656	1.533	1.381
inf	6.635	4.605	3.782	3.319	3.017	2.802	2.639	2.511	2.407	2.321	2.185	2.039	1.878	1.791	1.696	1.592	1.473	1.325	1.000

A.2 Mathematical Fitting Functions

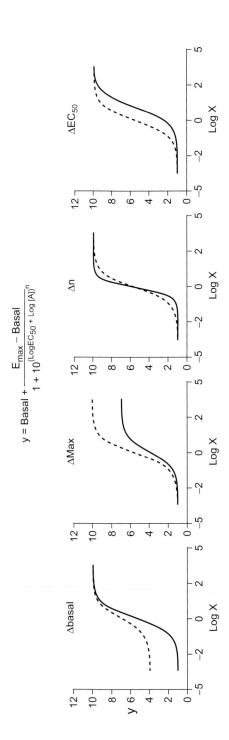

$$y = Basal + \frac{E_{max} - Basal}{1 + 10^{(LogEC_{50} + Log\,[A])^n}}$$

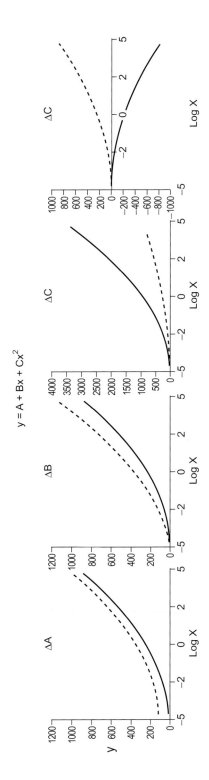

$$y = A + Bx + Cx^2$$

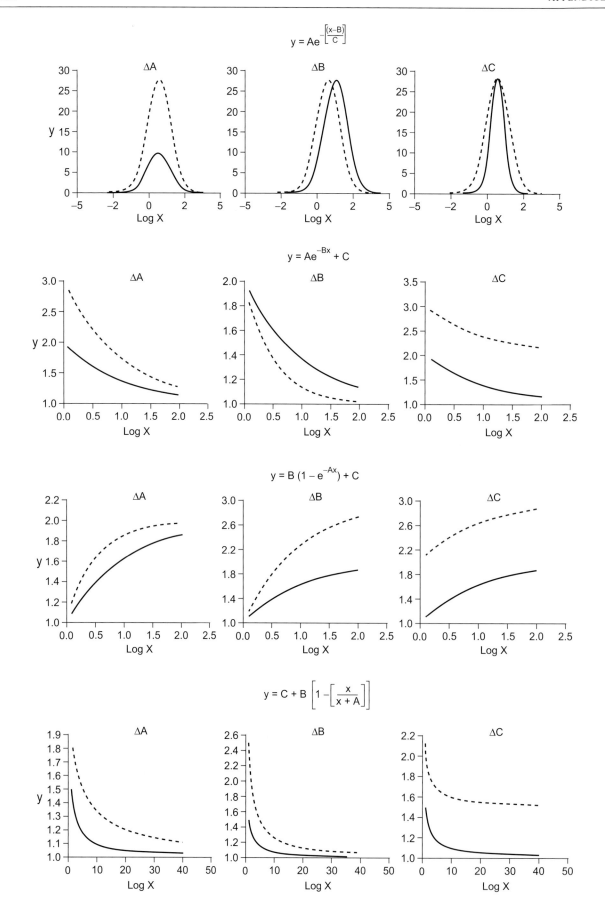

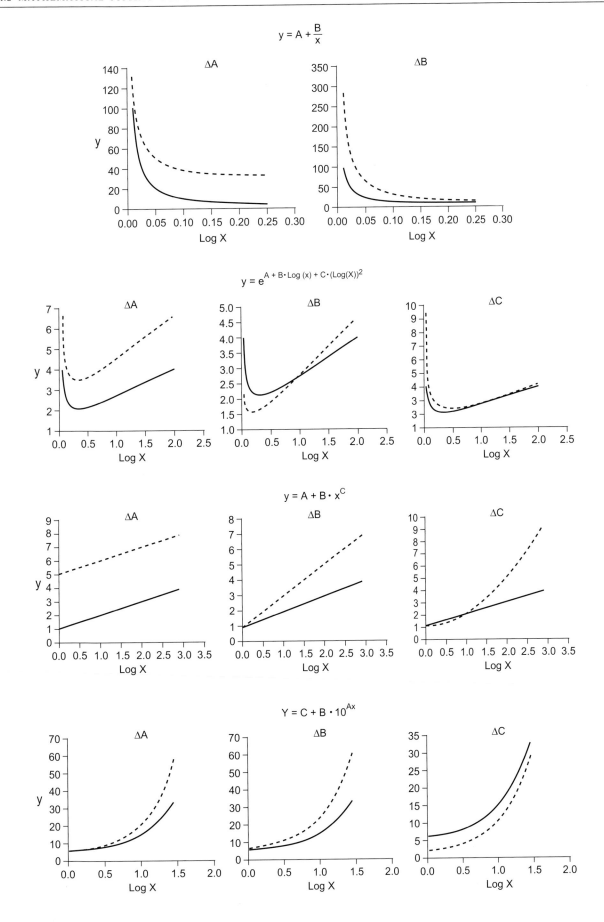

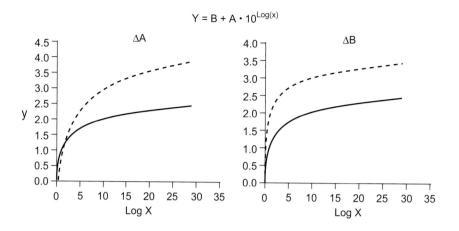

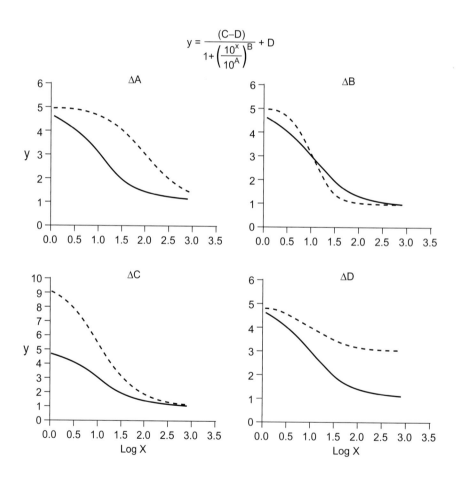

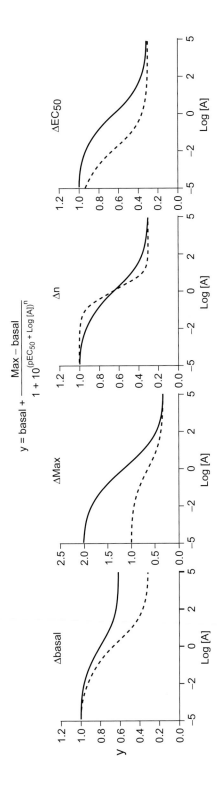

Index

Page numbers with "t" denote tables; those with "f" denote figures